Belt Selection and Application for Engineers

MECHANICAL ENGINEERING

A Series of Textbooks and Reference Books

1. Spring Designer's Handbook, *by Harold Carlson*
2. Computer-Aided Graphics and Design, *by Daniel L. Ryan*
3. Lubrication Fundamentals, *by J. George Wills*
4. Solar Engineering for Domestic Buildings, *by William A. Himmelman*
5. Applied Engineering Mechanics: Statics and Dynamics, *by G. Boothroyd and C. Poli*
6. Centrifugal Pump Clinic, *by Igor J. Karassik*
7. Computer-Aided Kinetics for Machine Design, *by Daniel L. Ryan*
8. Plastics Products Design Handbook, Part A: Materials and Components; Part B: Processes and Design for Processes, *edited by Edward Miller*
9. Turbomachinery: Basic Theory and Applications, *by Earl Logan, Jr.*
10. Vibrations of Shells and Plates, *by Werner Soedel*
11. Flat and Corrugated Diaphragm Design Handbook, *by Mario Di Giovanni*
12. Practical Stress Analysis in Engineering Design, *by Alexander Blake*
13. An Introduction to the Design and Behavior of Bolted Joints, *by John H. Bickford*
14. Optimal Engineering Design: Principles and Applications, *by James N. Siddall*
15. Spring Manufacturing Handbook, *by Harold Carlson*
16. Industrial Noise Control: Fundamentals and Applications, *edited by Lewis H. Bell*
17. Gears and Their Vibration: A Basic Approach to Understanding Gear Noise, *by J. Derek Smith*

18. Chains for Power Transmission and Material Handling: Design and Applications Handbook, *by the American Chain Association*
19. Corrosion and Corrosion Protection Handbook, *edited by Philip A. Schweitzer*
20. Gear Drive Systems: Design and Application, *by Peter Lynwander*
21. Controlling In-Plant Airborne Contaminants: Systems Design and Calculations, *by John D. Constance*
22. CAD/CAM Systems Planning and Implementation, *by Charles S. Knox*
23. Probabilistic Engineering Design: Principles and Applications, *by James N. Siddall*
24. Traction Drives: Selection and Application, *by Frederick W. Heilich III and Eugene E. Shube*
25. Finite Element Methods: An Introduction, *by Ronald L. Huston and Chris E. Passerello*
26. Mechanical Fastening of Plastics: An Engineering Handbook, *by Brayton Lincoln, Kenneth J. Gomes, and James F. Braden*
27. Lubrication in Practice, Second Edition, *edited by W. S. Robertson*
28. Principles of Automated Drafting, *by Daniel L. Ryan*
29. Practical Seal Design, *edited by Leonard J. Martini*
30. Engineering Documentation for CAD/CAM Applications, *by Charles S. Knox*
31. Design Dimensioning with Computer Graphics Applications, *by Jerome C. Lange*
32. Mechanism Analysis: Simplified Graphical and Analytical Techniques, *by Lyndon O. Barton*
33. CAD/CAM Systems: Justification, Implementation, Productivity Measurement, *by Edward J. Preston, George W. Crawford, and Mark E. Coticchia*
34. Steam Plant Calculations Manual, *by V. Ganapathy*
35. Design Assurance for Engineers and Managers, *by John A. Burgess*
36. Heat Transfer Fluids and Systems for Process and Energy Applications, *by Jasbir Singh*
37. Potential Flows: Computer Graphic Solutions, *by Robert H. Kirchhoff*
38. Computer-Aided Graphics and Design, Second Edition, *by Daniel L. Ryan*
39. Electronically Controlled Proportional Valves: Selection and Application, *by Michael J. Tonyan, edited by Tobi Goldoftas*
40. Pressure Gauge Handbook, *by AMETEK, U.S. Gauge Division, edited by Philip W. Harland*
41. Fabric Filtration for Combustion Sources: Fundamentals and Basic Technology, *by R. P. Donovan*
42. Design of Mechanical Joints, *by Alexander Blake*
43. CAD/CAM Dictionary, *by Edward J. Preston, George W. Crawford, and Mark E. Coticchia*

44. Machinery Adhesives for Locking, Retaining, and Sealing, *by Girard S. Haviland*
45. Couplings and Joints: Design, Selection, and Application, *by Jon R. Mancuso*
46. Shaft Alignment Handbook, *by John Piotrowski*
47. BASIC Programs for Steam Plant Engineers: Boilers, Combustion, Fluid Flow, and Heat Transfer, *by V. Ganapathy*
48. Solving Mechanical Design Problems with Computer Graphics, *by Jerome C. Lange*
49. Plastics Gearing: Selection and Application, *by Clifford E. Adams*
50. Clutches and Brakes: Design and Selection, *by William C. Orthwein*
51. Transducers in Mechanical and Electronic Design, *by Harry L. Trietley*
52. Metallurgical Applications of Shock-Wave and High-Strain-Rate Phenomena, *edited by Lawrence E. Murr, Karl P. Staudhammer, and Marc A. Meyers*
53. Magnesium Products Design, *by Robert S. Busk*
54. How To Integrate CAD/CAM Systems: Management and Technology, *by William D. Engelke*
55. Cam Design and Manufacture, Second Edition; with cam design software for the IBM PC and compatibles, disk included, *by Preben W. Jensen*
56. Solid-State AC Motor Controls: Selection and Application, *by Sylvester Campbell*
57. Fundamentals of Robotics, *by David D. Ardayfio*
58. Belt Selection and Application for Engineers, *edited by Wallace D. Erickson*

Mechanical Engineering Software

Spring Design with an IBM PC, *by Al Dietrich*

Mechanical Design Failure Analysis: With Failure Analysis System Software for the IBM PC, *by David G. Ullman*

Belt Selection and Application for Engineers

edited by

Wallace D. Erickson
Gates Rubber Company
Denver, Colorado

MARCEL DEKKER, INC. New York and Basel

Library of Congress Cataloging-in-Publication Data

Belt selection and application for engineers / edited by Wallace D.
Erickson.
p. cm. -- (Mechanical engineering ; 58)
Includes Index.
ISBN 0-8247-7353-5
1. V-Belts. 2. Belts and belting. I. Erickson, Wallace D.,
II. Series: Mechanical engineering (Marcel Dekker, Inc.)
; 58
TJ1105.B45 1987
621.8'52--dc19 87-17349
CIP

MARCEL DEKKER, INC.
270 Madison Avenue, New York, New York 10016

Current printing (last digit):
10 9 8 7 6 5 4 3 2 1

PRINTED IN THE UNITED STATES OF AMERICA

Preface

Many methods of mechanical power transmission are in use today to drive industrial, automotive, agricultural, and domestic equipment and machinery. But, by far the most common means of power transmission is the flexible belt. This is most true when the speeds of the driveR and driveN shafts must be different or where shafts must be separated by some distance.

The trend toward higher speed prime movers and the necessity of a speed ratio to achieve a slower, useful driveN speed is further accentuating the use of belts as the most appropriate method of power transmission.

Belts have numerous advantages over other kinds of power transmission, including: overall economies, cleanliness and no need for lubrication, lower maintenance costs, easy installation, shock loads that can be smoothed out, ability to be used for clutching and variable speed and ability to transmit power between widely spaced shafts.

This is a how-to book written to provide engineers with the most up-to-date information on power transmission using flexible belting. Areas such as fundamental principles, design procedures, application techniques, maintenance and safety, pulleys and sheaves, industry standardization, and belt efficiency are covered in depth.

This book will provide engineers with a step-by-step approach to designing and maintaining drives with flexible belting. The text is supplemented by various photos, charts, graphs, tables, and illustrations that highlight key points.

Perhaps the reader is wondering if it really takes all these pages just to cover the topic of transmitting power with flexible belting. The answer is yes, and there could have been many more. Not all this information will be needed each time a drive is designed or examined, but sooner or later, the engineer will be able to use most of this material.

Wallace D. Erickson

Acknowledgments

Several years and 18 chapters later, I am very pleased and proud to have this textbook in print.

During my nearly 40 years as an engineer, I have never seen a single source book or manual that provides so much information and details on rubber belts for power transmission. It was not a solo effort by any means.

My indebtedness begins with Jim Shepherd, manager of Gates V-Belt Product Application department. Jim enlisted five engineers from his group who wrote chapters on their specialized area of belts.

These individuals are Walt Nawrocki, Dave Roos, Charlie Artus (now retired), Dan Schwartz and Mike Coleman. While managing his engineers' time on this project, Jim also contributed a chapter of his own.

Chapter assignments were spread out over four other engineering departments. Bob Stetson (retired) and Dave Bayne, Automotive Application Engineering; Ham Cox (retired), Belt Test Lab; Ray Elder, Belt Technical; and Bill Westhoff, V-Belt Engineering and Development, contributed valuable chapters.

The creative process of preparing this textbook was assumed by several non-writers who shared their professional talents. Jerry Donovan of Gates Advertising & Communications department, and Jim House of our advertising/public relations agency, Broyles, Allebaugh & Davis, Inc., coordinated our efforts and edited the chapters for the authors.

In addition, Jerry acted as my liaison with the publisher, and gathered all the necessary illustrations.

I also wish to thank my secretary, Lucille Knowles, for her many, many hours of research and typing.

Most of these illustrations were prepared exclusively for the textbook. Special thanks to Bob Schultheis of Advertising & Communications, and Mike Sitz and Larry Bush, of Gates Electrical Utilities Engineering department. Finally, I wish to acknowledge the competent support given by Marcel Dekker's editorial and production staffs. Their patience and assistance are much appreciated.

Wallace D. Erickson

Contributors

CHARLES H. ARTUS*
ROY DAVID BAYNE
MICHAEL S. COLEMAN
H. L. COX
RAY L. ELDER II
WALLACE D. ERICKSON*
WALTER T. NAWROCKI
DAVID E. ROOS
JAMES D. SHEPHERD
DANIEL L. SCHWARTZ
ROBERT E. STETSON*
WILLIAM L. WESTHOFF

The Gates Rubber Company
Denver, Colorado

*Retired

Contents

Belt Selection and Application for Engineers

1

Introduction and Brief History

WALLACE D. ERICKSON *Automotive Products Technical and Development Division, The Gates Rubber Company, Denver, Colorado*

I. BASIC REQUIREMENTS OF POWER TRANSMISSION

A. Definition

Power transmission by belts is described as the transmission of power from a prime mover to one or more driven machines by means of a flexible nonmetallic member. The belt operates in conjunction with a pulley or sheave mounted on each of the rotating shafts.

Power is usually transmitted by frictional forces developed between the belt and the pulley or sheave. The synchronous belt is, however, an exception whereby the belt teeth make positive engagement with the teeth of the pulley.

B. Area of Application

The application scope of power transmission belts is virtually without bounds, ranging from drives which can be held in the palm of the hand to drives several feet in width, and using belts over fifty feet in length. Power transmission capabilities can vary from less than 0.1 horsepower (HP) to several thousand HP as illustrated in Figures 1 and 2.

II. REQUIREMENTS OF POWER TRANSMISSION MEDIA

The normal requirements of power transmission media are listed below along with a brief description of how power transmission belts fulfill these requirements.

A. Dependable Service Life

1. Predictable Performance

The service life of a power transmission belt drive is readily predictable by the knowledgeable application engineer. This insures the machine user of adequate service life without the costly need for unnecessary overdesign.

2. Visual Warning of Failure

A visual inspection of the drive will usually show signs of belt failure well in advance, allowing for an orderly shutdown of the drive for replacement or maintenance.

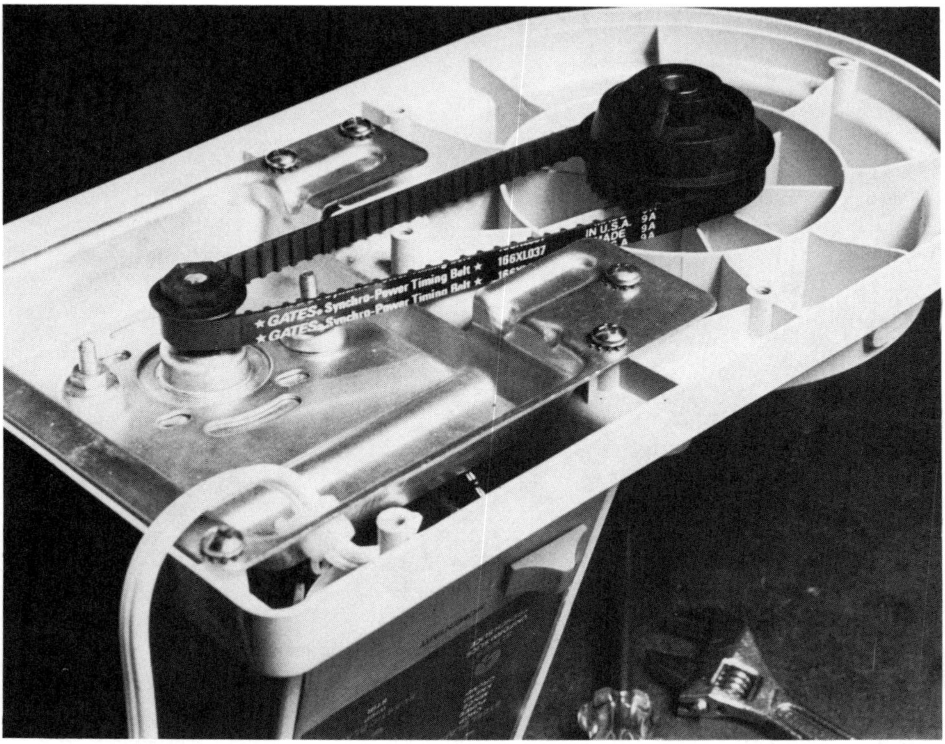

FIGURE 1 A kitchen food processor using an MXL synchronous belt.

B. Speed Ratio Versatility

The rotational speed of the driven machine is usually different from that of the driver unit. Through a wide selection of sheave or pulley diameters, the belt drive allows an easy selection of speed ratios to give the proper driven speed. It also allows the designer to easily multiply the torque on a driver unit to meet the demands of a driven unit in the most efficient manner.

Changes in speed ratio to accommodate different driver requirements can easily be made at any time by simply changing the diameter of one or both of the sheaves or pulleys.

FIGURE 2 A large industrial V-belt drive on a rock crusher is subjected to pulsating loads.

C. Low Maintenance

Power transmission belt drives require no lubrication and run for long periods of time with virtually no maintenance. An occasional check on belt tension and a visual check on the condition of the drive is all that is required.

D. Quiet Operation

The metal-to-metal contact present with gear and chain drives is eliminated entirely with belt drives. The relatively soft belt material mates quietly with the sheave or pulley.

E. Low Initial Cost

The cost of the belt and sheave or pulley components is usually among the lowest for all types of power transmission media. In addition, further savings can often be realized since installation is simple and fast. Housings and cases to provide for lubrication are not needed as with gears and chain.

F. Efficient Operation

The mechanical efficiency of power transmission belt drives is usually 95% or greater for a properly designed drive.

G. Readily Available Replacement Parts

Power transmission belts are stocked by power transmission and rubber parts distributors in a large selection of standard sizes all over the world. This eliminates costly downtime and reduces the replacement parts inventory the user must carry.

III. HISTORICAL HIGHLIGHTS

A. Early Flat Belt and Rope Drives

Power transmission belts have played an important role in the industrial development of the world for more than 200 years. Flat belts of plied-up leather and regular cotton or hemp rope running in V-grooves (sheaves) were predominant in early industrial history. These belts transmitted power from large steam engines or water wheels to various types of production machinery, usually through a series of line shafts.

1. Advantages of Flat Belts

Individual machines were driven from the line shaft by flat belts which could easily be shifted between a driving pulley and an idler pulley to start and stop the machine.

Flat belts have the advantage of being fiexible, bending easily around small pulley diameters. They can be made in a wide variety of widths as required to transmit the necessary power.

2. Limitations of Flat Belts

However, the transmission of power by a flat belt depends on the tangential frictional force developed between the belt and the pulley:

$$F_F + F_N \mu \tag{1}$$

where

F_F = tangential frictional force, lbf or N
μ = coefficient of friction
F_N = normal force between belt and pulley, lbf or N

Since normal force, F_N, is directly proportional to the tension in the belt, a reasonably high coefficient of friction must be maintained to keep belt tensions in an acceptable range.

On flat leather belting it was customary to use belt dressing to increase the coefficient of friction. But in spite of this, belt slippage often occurred, which increased the need for maintenance on the drive. A slipping flat belt could easily "walk off" the pulleys, causing unnecessary machinery downtime and maintenance.

3. Advantages of Wedging

Increasing the normal force, F_N, on a belt drive through the wedging of the belt into a V-shaped pulley groove was a phenomenon first used with cotton or hemp-rope drives. These drives were usually large drives which transmitted power from the prime mover to line shafts. Because of the wedging feature they required less tension than flat belts and would not come off of the sheaves the way flat belts did. Such a drive is illustrated in Figure 3.

At the turn of the twentieth century, the first V-shaped belts were used to replace the round ropes. These belts were made from plied-up leather and did not bend easily around a small sheave diameter. Early literature confirms that these belts achieved a prominent place in power transmission, however, such as the main drive from the engine to the rear wheel on motorcycles.

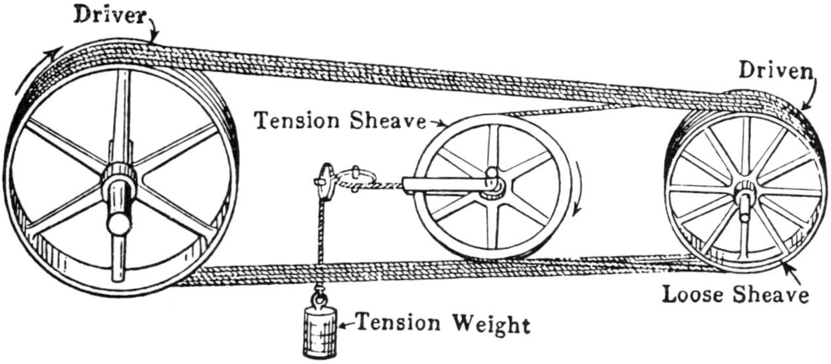

FIGURE 3 Cotton or hemp rope drive running in V-grooved sheaves.

B. Introduction of the Vulcanized Rubber V-Belt

In 1917, Mr. John Gates, co-founder of The Gates Rubber Company, developed the first vulcanized rubber V-belt to replace the round hemp rope used to drive the fan on a 1917 Cole coupe. Within a few years, rubber V-belts with a fabric tensile section for longitudinal strength and stability began to dominate the belt power transmission market. The vulcanized rubber V-belt offered the advantages of operating at reduced tensions due to wedging, reduced stretch because of its longitudinal stability, and elimination of the annoying tendency of the flat belt to track off of the pulleys. The increased normal force due to the wedging of a V-belt is illustrated in Figure 4.

The resultant normal forces (F_N) are much greater than the radial force developed by belt tension. Therefore, the frictional force between the belt and the pulley is increased for the same coefficient of friction.

1. Fast Rise in V-Belt Popularity

With the change in the basic prime mover for industrial machines from the line shaft to the individual electric motor, the shifting capability (by sliding axially) of the flat belt was no longer needed. This, coupled with reasons mentioned above, gave the V-belt a very fast rise in popularity.

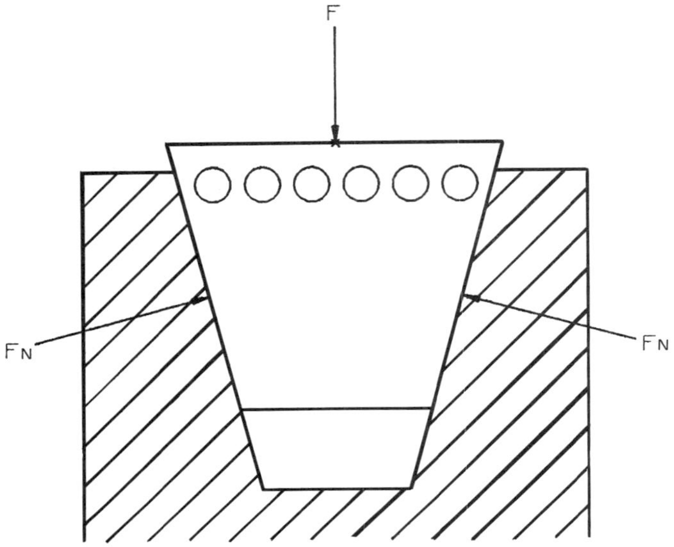

FIGURE 4 The normal force between a V-belt and pulley groove is increased by wedging.

2. Other Variations of the V-Belt

Other variations of the V-belt were developed over the years to handle special applications. The variable speed V-belt is one example. This type of belt operates in pulleys which can be changed in groove width allowing a change in operating diameter. This results in a change in speed ratio between the driver and the driven shafts. Figures 5 and 6 illustrate the basic principle of operation of a variable speed drive. Figure 5 shows the variable driver pulley with the flanges fully open and the belt at the minimum diameter position. This arrangement gives the slowest driven speed.

Figure 6 shows the same variable speed drive with the variable pulley fully closed giving a maximum diameter of the belt on the driver pulley. This arrangement gives the highest driven speed.

The variable pulley can be on either the driver or the driven shafts or, if necessary to obtain a larger speed variation, it can be used on both the driver and the driven.

Another variation of the V-belt is the double V-belt, designed to transmit power from either side of the belt (see Figure 7). This belt is designed for use on serpentine-type drives which consist of several pulleys that rotate in different directions. Such a drive is illustrated in Figure 8.

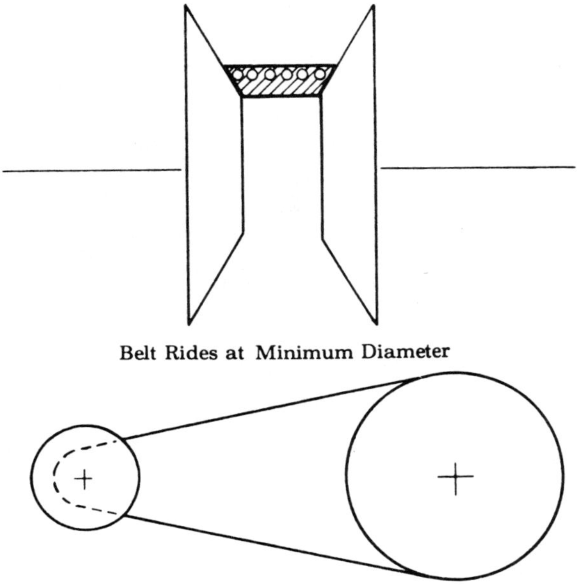

FIGURE 5 Variable speed drive with variable sheave fully open.

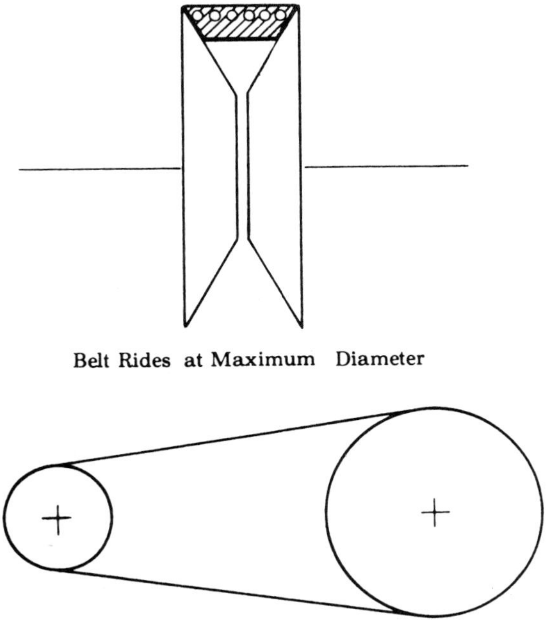

FIGURE 6 Variable speed drive with variable sheave fully closed.

FIGURE 7 Double V-belt for use on serpentine drives.

C. Other Types of Power Transmission Belts

1. V-Ribbed Belt

Another type of belt was introduced in the 1950s which is a cross between a flat belt and a V-belt. This belt is generally called a V-ribbed belt and was introduced under the trade name Poly-V.

This belt is essentially a flat belt with V-shaped ribs projecting from the bottom of the belt which guide the belt and make it more stable than a flat belt (see Figure 9). In the original concept, the V-ribs of the belt completely filled the grooves of the pulley. For this reason, the V-ribbed belt did not have the wedging action of the V-belt and consequently had to operate at higher belt tensions.

Later versions of V-ribbed belts have truncated ribs to more closely emulate the wedging effect of a V-belt. It is still common

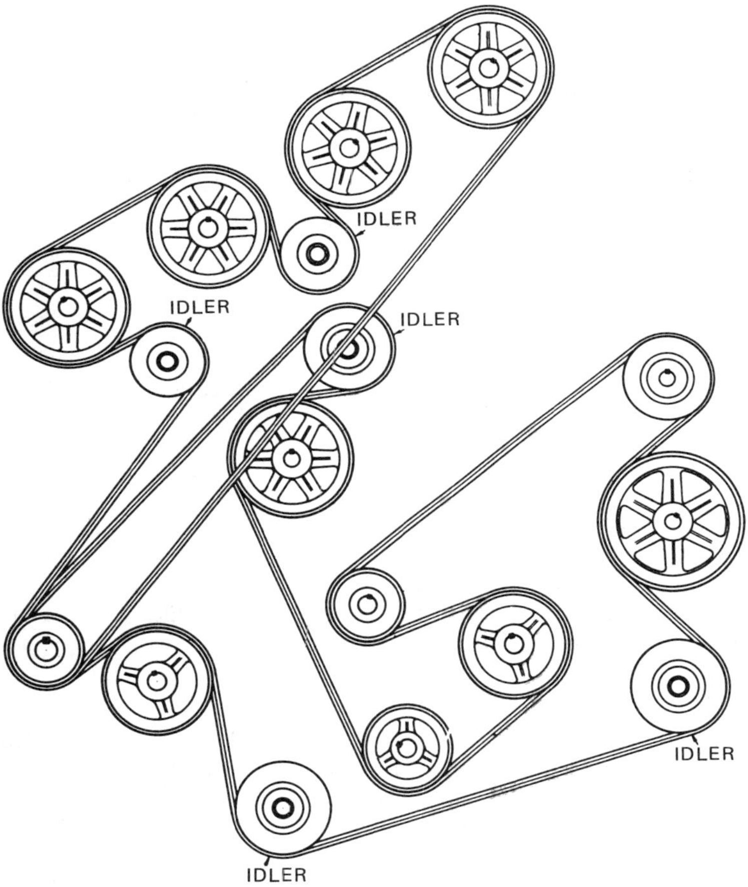

FIGURE 8 Serpentine-type drive using a double V-belt.

practice, however, to use higher tensions with V-ribbed belts than with V-belts because the V-ribbed belt still seats against the rounded land between pulley grooves.

While the V-ribbed belt is not as flexible as a flat belt, the smaller V-ribs still allow for a very flexible belt that performs well on small-diameter pulleys. On certain drives where the shafts and bearings will withstand higher operating tensions, the V-ribbed belt will give good performance on small pulley diameters.

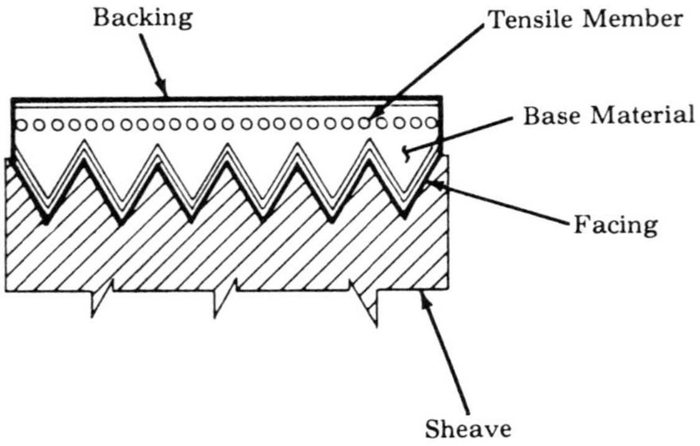

FIGURE 9 A V-ribbed belt with continuous tensile section like a flat belt.

A good example is the domestic clothes dryer. A speed ratio of over 30:1 is required between the motor and the drum. The V-ribbed belt will operate on a small enough motor pulley to achieve this high speed ratio while operating directly on the dryer drum like a flat belt (see Figure 10).

2. Synchronous Belt

A frictional drive like those using V-belts, V-ribbed belts or flat belts does not give exact driven speeds. There is a certain amount of belt "creep" (usually about 0.5% for a V-belt) which makes it impossible to drive machinery such as the indexing heads of machine tools or the camshaft of an internal combustion engine where synchronization is needed. This was a segment of the power transmission market not open to belts until about 1950 when the synchronous belt was developed.

Synchronous belts are capable of transmitting power through the positive engagement of teeth on the belt with corresponding teeth on the pulley. This creates the synchronization between the driver and the driven shafts which is essential on some types of applications. See Figure 11 for a basic description of a synchronous belt mating with the pulley.

Synchronous belt drives have a distinct advantage over gears or chain drives because they can transmit reasonably high loads at a wide variety of speeds with a low noise level and without lubrication. The shock-absorbing characteristics of the low modulus rubber teeth

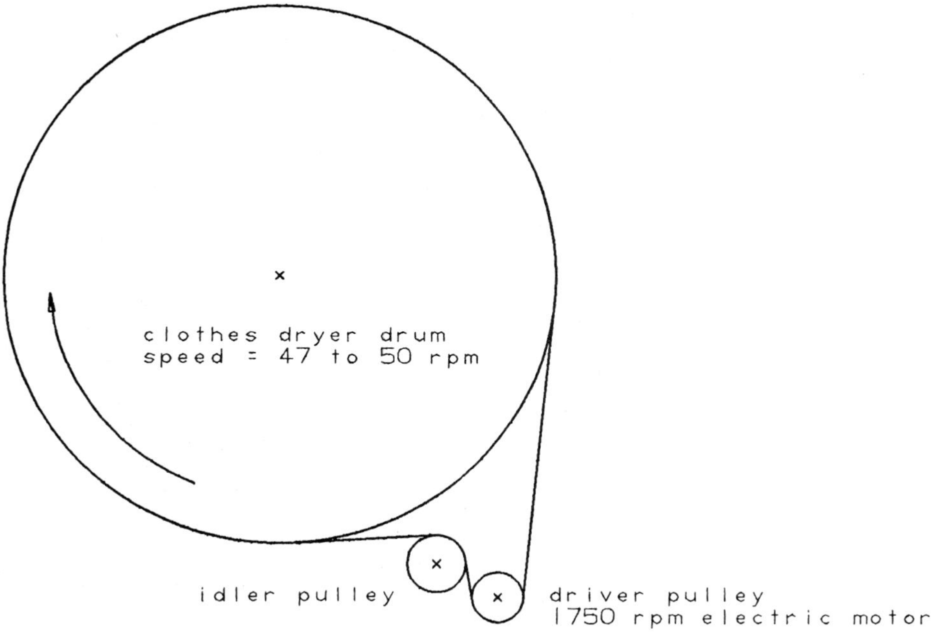

FIGURE 10 A V-ribbed belt driving the drum of a domestic clothes dryer.

against the metal pulley teeth are also an advantage on certain applications.

Synchronous belts can also be used for normal power transmission drives where synchronous speeds are not required. However, on these types of applications, the more popular V-belt generally proves to be a less expensive and more reliable drive.

IV. NOMENCLATURE FOR POWER TRANSMISSION BELTS AND PULLEYS

A general discussion on belt and pulley nomenclature is given in this chapter. More specific detail will be given in the engineering chapters for each specific belt type.

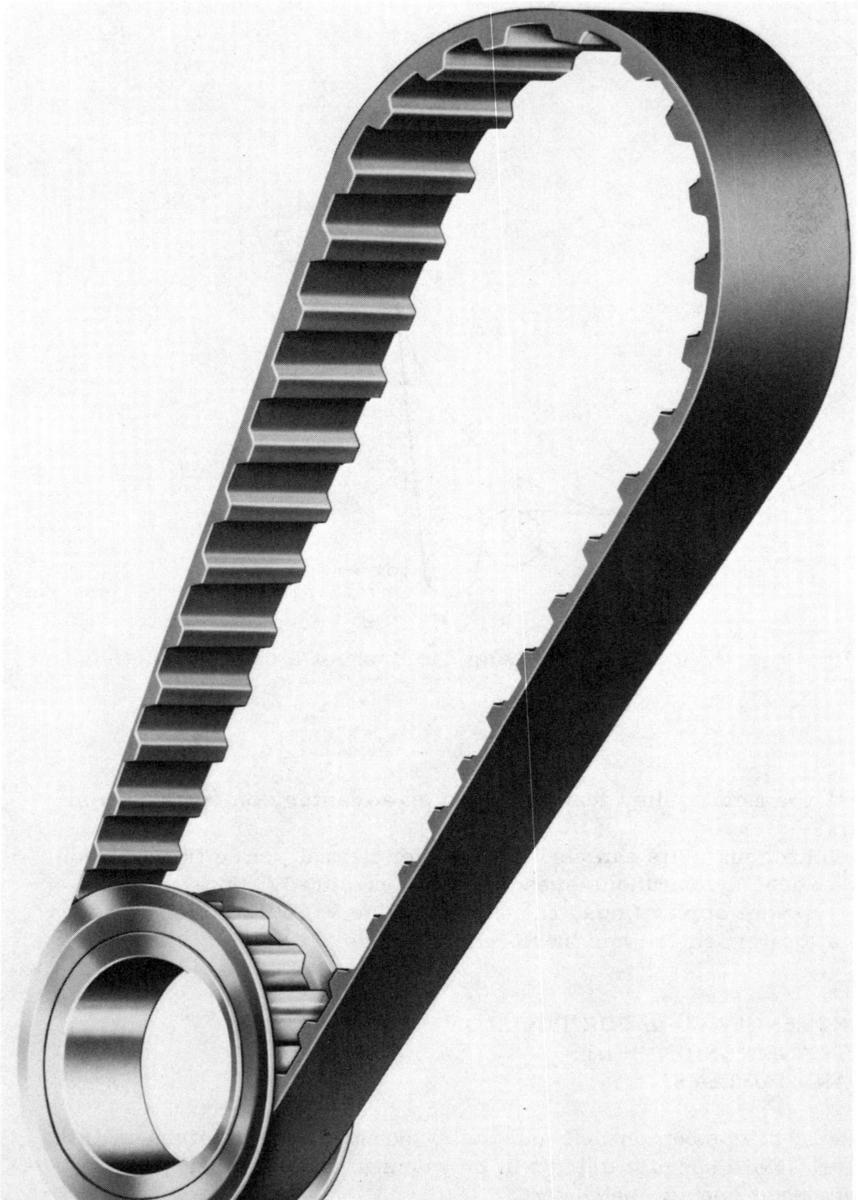

FIGURE 11 The rubber and fabric teeth of a synchronous belt mate positively with the teeth on a synchronous pulley.

In any type of nomenclature system, it is advantageous to be able to physically measure the dimensions of the components within reasonable accuracy, using ordinary tools such as a scale or a tape measure. This will allow the user to determine the required replacement products easily if specification literature is not readily available. This type of system is now used for most types of belt drives.

A. Flat Belts

Flat belts are typically specified by their thickness, their width, and their inside length. The pulleys for a flat belt drive are specified by their outside diameter and their width.

Inside length is specified because it allows for an easy measurement of the length required for a drive (done by wrapping a measuring tape around the outside of the pulleys). While the outside length of the belt can be measured more easily than the inside length, the belt is often broken or stretched so that an accurate measurement is difficult to obtain. It is much easier and more accurate to measure the distance around the pulleys of a drive (the inside length).

The other dimensions of the belt (width and thickness) as well as the outside diameter and width of the pulley can easily be measured with a regular scale or tape. In the case of outside diameter, a Pi tape is accurate and easy to use.

B. V-Belts

1. Inside Length

When V-belts were first introduced into the market place, they were also specified by their inside length — a carry-over from flat belt nomenclature. However, since the inside length of a V-belt does not physically touch a measurable portion of the pulley, there is no advantage to be gained by using the inside length.

2. Pitch Length

As time went by, the length nomenclature for V-belts was gradually changed to the pitch length. The pitch length is the length of the belt at the pitch line or neutral axis of the belt. This length could not readily be measured on either the drive or the belt, however, because it did not correspond to any accessible physical portion of the belt or pulleys.

As with the flat belt, experience dictated that the most practical way to measure the length requirement of a V-belt drive was to wrap a tape measure around the outside diameter of the pulleys. This method gives a length that corresponds closely with the outside length of the belt since the top of the V-belt rides close to the outside diameter of the pulley on most drives.

3. Effective Length

A more precise method of measuring and specifying belt length was needed on some drives where space limitations demanded close control. Slight differences in the position of the top of the belt relative to the outside diameter of the pulley could be critical on certain applications.

To resolve this problem, the effective system of geometrically defining a V-belt drive was established. The effective diameter of a pulley is defined as the diameter at which the groove width is a specified value. Then the effective length of the belt is the length of a plane passing through (and tangent to) the effective diameters of the drive while the belt is at a specified tension.

It was important to introduce the consideration that the length be determined with the belt subjected to a specified tension since this is the way a belt operates on a drive. Some belts elongate more when tensioned on a drive and therefore need to be shorter when measured at no tension. Measuring at a specific tension about the same as that required on the drive eliminates this problem.

By manufacturing pulleys with a top width close to the effective width, the outside diameter (which can be easily measured) will coincide with the effective diameter close enough for most usage. By designing the cross section of the belt so that the top of the belt is close to the top of the pulley, the outside length of the belt as measured with a tape will be close enough to the effective length for most applications.

The effective system allows a determination of belt requirements using simple measuring tools that is close enough for most drives and yet has the capability of accurately defining a drive system on applications where fit is critical.

4. Cross Section Designation

Early V-belts were designated by a letter which was not related quantitatively to the size (width) of the belt, i.e., A, B, C, D, etc. Later nomenclature systems related the cross section designation to the effective width of the groove. Since the top width of the groove is usually close to the effective width, measuring the top width of the groove will tell the user what cross section to order.

For example, if a mechanic measured a groove on a high capacity or narrow industrial drive to be 5/8 in., he would relate this to the 5V standard cross section – the number before the "V" indicating 1/8 in. increments.

In the metric system, the numbering is more direct. The same belt would be called a 15N cross section since the groove width is approximately 15 mm. The "N" designates a narrow or high-capacity type belt. A similar nomenclature system to specify V-belt cross section is used for most types of V-belts as explained in the engineering chapters.

C. V-Ribbed Belts

1. Effective Length

The effective system is also used for V-ribbed belts. The effective width of the groove is at the approximate outside diameter of the pulley. Therefore, the effective length for a V-ribbed drive can be determined reasonably accurately by running a tape measure around the outside of the pulleys on a drive.

2. Cross Section Designation

The cross section of a V-ribbed belt is designated by a letter which is not significant in terms of the size of the belt. However, each cross section has a unique spacing of the ribs on the belt which can easily be measured on the belt. The spacing of the corresponding grooves of the pulley can be measured just as readily.

D. Synchronous Belts

1. Pitch Determination

The size of a synchronous belt section is designated as the pitch of the belt in a manner similar to a power transmission chain. The pitch of a synchronous belt is the distance between the same point on two adjacent teeth when the belt is held in a straight position. The pitch dimension can easily be measured with a scale to determine which of the standard pitches applies to that specific belt.

2. Pitch Length

The most practical belt length and pulley diameter system to use for a synchronous belt is the pitch system. This is because the pitch length of the belt can easily be determined by multiplying the number of teeth by the pitch.

The same is true for determining the pitch diameter of a pulley. The number of teech multiplied by the pitch gives the pitch circumference which can be divided by pi to get the pitch diameter.

2

Basic Principles Involved in Transmitting Power with Flexible Belting

WALLACE D. ERICKSON *Automotive Products Technical and Development Division, The Gates Rubber Company, Denver, Colorado*

I. BASIC DEFINITIONS

In transmitting power by any medium, it is important to have a thorough knowledge of the fundamental mechanics involved.

A. Newton's Laws of Motion

As with most mechanical phenomena, the transmission of power is based on the three laws of motion developed by Isaac Newton about three hundred years ago. These laws are:

A body at rest remains at rest and a body in motion continues to move at a constant speed and in a straight line unless compelled to change by external forces.

The time rate of change (acceleration) of a body in motion is proportional to a force acting on the body and is in the direction of the force.

To every action there is always an equal and opposite reaction.

1. Application to Rotating Objects

These basic laws of motion can be readily extended to rotating objects. This is necessary in the fundamental mechanics of power transmission.

A rotating body with a constant angular velocity about an axis will continue to rotate with the same angular velocity until acted upon by an unbalanced torque.

When an unbalanced torque is applied to a rotating body, the angular acceleration produced is proportional to the magnitude of the unbalanced torque and inversely proportional to the moment of inertia of the body.

When an unbalanced torque is applied to a body A along a given axis by a body B, there will be an equal and opposite reactionary torque on body A along the same axis.

How these laws of motion apply to power transmission will become evident as the basic fundamentals are defined.

B. Application of a Force

1. Basic Considerations

Each of the three laws of motion developed by Newton center around the force. A force can be a push or a pull and can be applied in such a way as to give a body linear motion or rotational motion as is necessary in power transmission by belts. The classical formula for a force which applies to Newton's second law is:

$$\text{force} = \text{mass} \times \text{acceleration}$$

2. Work

Work is performed when a force acting on a body causes it to move – either in a linear direction or rotationally around an axis. The basic formula for work is:

$$W = FL \tag{1}$$

where

W = work performed
F = force exerted
L = distance moved

Note that work is independent of time and also independent of the mass of the body being moved through the distance L.

The units of work are usually foot pounds in the English system and Newton meters in the SI metric system. When a force of 1 lb is exerted to move an object 1 ft then 1 ft lb of work has been done.

It is important to note that even though forces may be exerted on a body, no work is being done unless the body moves. A good example might be a man hoisting water from a well. If he is unable to lift the heavy bucket of water by pulling with all his might, no work has been done. Although it might be hard to convince the man of this, no work is done until the bucket of water is moved through a distance.

3. Power

Whenever work is performed, we are usually interested in how long it takes to perform the work. The rate of doing work is defined as power. The faster a given amount of work is performed, the greater the power requirement.

If the work described in Figure 1 is performed in 1 min, then

$$\text{power required} = \frac{1 \text{ ft lb}}{\text{min}} \tag{2}$$

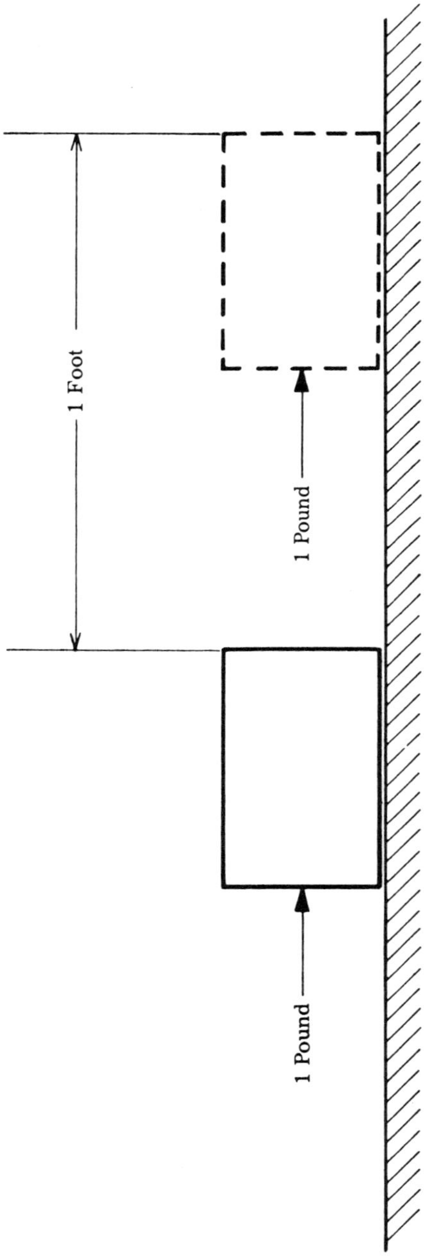

FIGURE 1 Sketch of a body being moved by force.

Horsepower: In the United States, power is usually specified in terms of horsepower (HP), a unit of measurement developed by James Watt. Watt needed a unit of measure to describe the power capabilities of his steam engine. Since most work was performed by horses in those days, Watt determined by experimentation that a powerful draft horse could perform 33,000 ft lb of work per min on a continuous basis.

Therefore,

$$1\ \text{HP} = \frac{33{,}000\ \text{ft lb}}{\text{min}} \tag{3}$$

This definition gives the following formula for calculating HP:

$$\text{HP} = \frac{FL}{33{,}000\ t} \tag{4}$$

where

HP = horsepower
F = force exerted (lbf)
L = distance moved (ft)
t = time required (min)

This definition of power is still being used in the United States today.

Watt: The unit of power specified by the Organization for International Standardization (ISO) is, however, the watt:

$$1\ \text{W} = \frac{1\ \text{Nm}}{\text{s}} \tag{5}$$

See Chapter 16 for a discussion of the ISO SI units of measure pertaining to power transmission.

Metric Horsepower: A metric horsepower is often used in European countries. This unit of power is defined as:

$$1\ \text{metric HP} = \frac{75\ \text{kg m}}{\text{s}} \tag{6}$$

When converted to conventional English system units, metric HP is defined as:

$$1\ \text{metric HP} = \frac{32{,}556\ \text{ft lb}}{\text{min}} \tag{7}$$

This is very close to the English HP unit of power defined before.

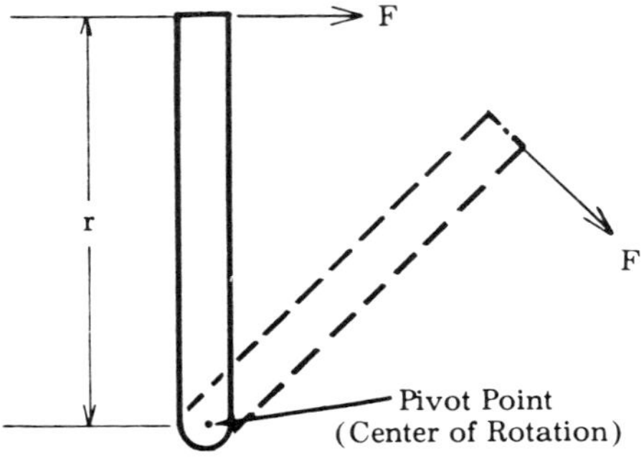

FIGURE 2 Torque being produced by a force acting on a lever arm.

4. Torque

Basic definition: Torque is defined as the moment of a force system which tends to produce rotation. If a force, F, is applied to a lever arm, r, a torque, Q, is produced about an axis of rotation (see Figure 2).

The formula for torque is:

$$Q = Fr \tag{8}$$

Power and torque relationship: Since power transmission by belts is involved with rotating bodies, it is convenient to express power in terms of torque. The same basic definitions apply. The force-distance relationship, however, is expressed in terms of torque acting through an angle of rotation.

The basic definition of power applies to bodies rotated about an axis by a torque in the same manner as it does to bodies being moved by a force in a linear direction. For example, if a weight, W, is being lifted by a rope and pulley system a distance, L, by a force, F, in a given time, t, the formula in Equation (4) would apply (see Figure 3).

The power that is transmitted by moving the weight through the distance, L, in a given time can also be transmitted by applying a torque to the pulley shaft.

We can easily derive the HP formula in terms of torque by considering Figure 4 in a somewhat different way.

If the force, F, is applied through a distance equal to the radius, r, of the pulley, the distance the weight, W, is moved, L, equals r and the basic formula for HP (Equation 4) becomes:

$$HP = \frac{Fr}{33{,}000\ t} \tag{9}$$

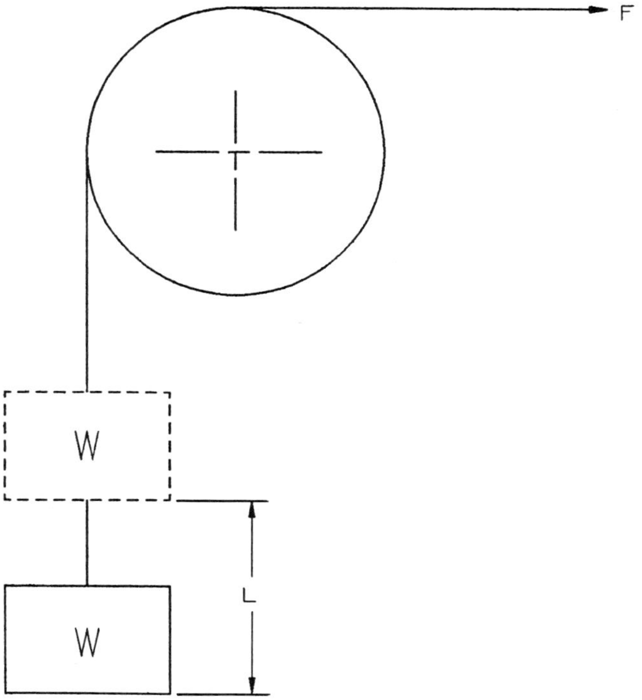

FIGURE 3 Lifting a weight by a pulley system.

From Equation (8):

$$F = \frac{Q}{r}$$

Substituting this definition for F into Equation (9) we have:

$$HP = \frac{(Q/r)\ r}{33{,}000\ t}$$

or

$$HP = \frac{Q}{33{,}000\ t} \tag{10}$$

where

Q = torque (lb ft)
t = time required to rotate a shaft a distance r (min)

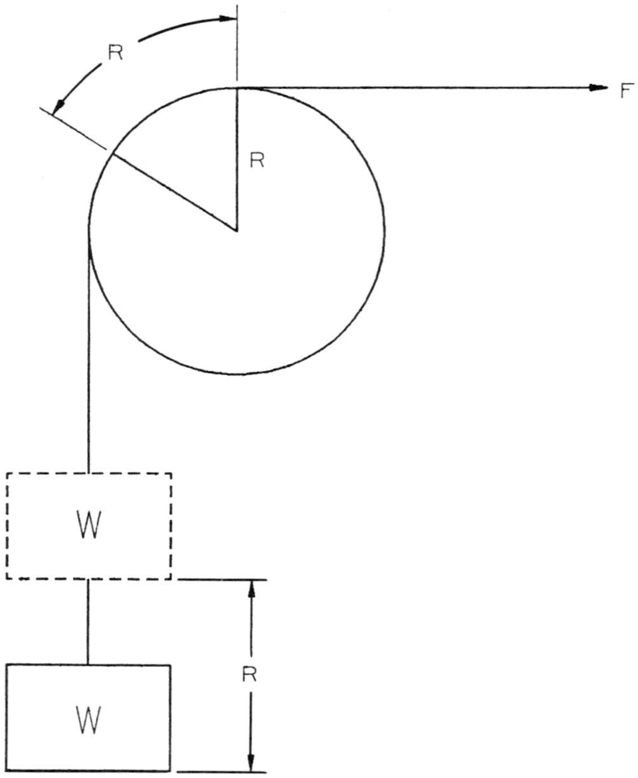

FIGURE 4 Torque being applied through a distance equal to the pulley radius.

The angle through which the pulley rotates to move the weight a distance equal to the pulley radius is called a radian (rad) (see Figure 5).

The HP required to rotate the pulley at any given speed is obtained by simply multiplying Equation (10) by the speed in rad/min:

$$HP = \frac{Q \text{ (rad/min)}}{33{,}000} \tag{11}$$

In power transmission engineering, the speed of a rotating pulley is usually expressed in revolutions per minute (rpm) rather than rad/min. Since there are 2π radians per revolution, the HP formula for

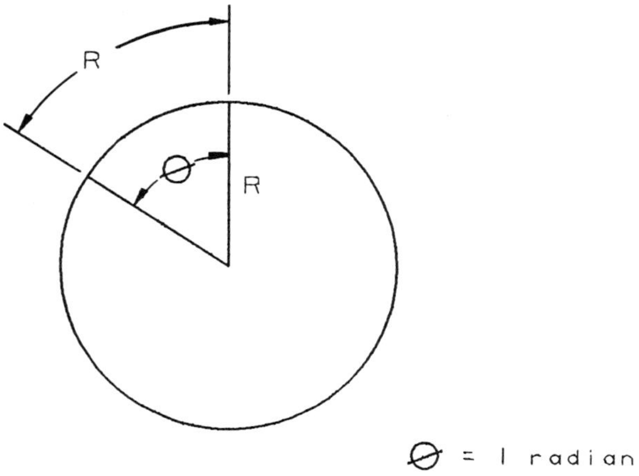

FIGURE 5 Definition of a radian.

speed in rpm can be obtained by dividing both the numerator and denominator of Equation (11) by 2π:

$$HP = \frac{Q\left(\frac{\text{rad/min}}{2\pi}\right)}{33{,}000/2\pi} = \frac{Q \text{ rpm}}{5252} \tag{12}$$

where

Q = torque (lb ft)
rpm = revolutions per minute

Note the unit of measure for torque is pound feet (lb ft) whereas the unit of measure for work is foot pounds (ft lb). This is to emphasize that there is a fundamental difference between the two: work is a scalar quantity involving the product of force and distance acting in the same direction whereas torque is a vector quantity since the force might be at right angle to the distance or radius arm.

II. TRANSMITTING TORQUE AND POWER WITH BELTS

A. Speed Ratio and Torque Multiplication

The torque-producing capability of a prime mover can be multiplied as necessary to obtain the torque required for a particular machine. This is done through the speed ratio of a belt drive.

TABLE 1 Comparison of Different Speed Motors

Motor horse-power	Motor rpm	Torque capability (lb ft)	NEMA frame number	Motor dimensions (in.)	
				Length	Width
1 1/2	870	9.06	184T	13.8	7.5
1 1/2	1160	6.79	182T	12.8	6.5
1 1/2	1750	4.50	145T	11.8	6.0
1 1/2	3450	2.28	143T	10.8	6.0

For example, an air compressor requires 8 lb ft of torque for operation at the optimum speed of 1000 rpm. So from Equation (12), the HP requirement is:

$$\text{HP} = \frac{(8)(1000)}{5252} = 1.52$$

Because belt drives are easily adaptable to various speed ratios by using different diameter pulleys, a machine designer can select from a wide variety of 1 1/2 HP motors. Table 1 gives a comparison of four 1 1/2 HP electric motors based on the National Electric Manufacturers Association (NEMA) Standard MG-1.

The speed ratio of a belt drive is defined as follows:

$$\text{speed ratio} = \frac{\text{driveR rpm}}{\text{driveN rpm}} \tag{13}$$

Therefore, the four 1 1/2 HP motors being considered would require a speed ratio as shown in Table 2. This table also lists the torque the motor is capable of delivering to the driveN machine which is calculated by this formula:

$$\text{torque at driveN} = (\text{torque of driveR})(\text{speed ratio}) \tag{14}$$

As shown in Table 1, the 3450 rpm motor is considerably smaller than the 870 rpm and is designed to develop only 2.28 lb ft of torque rather than the 9.06 lb ft of torque for the 870 rpm motor. Because a wide variety of speed ratios is easily attainable for a belt drive, the lower torque can easily be multiplied so that it is capable of satisfying the requirements of the driveN unit.

TABLE 2 Adapting Different Speed Motors to a DriveN Machine

Motor horsepower	Motor rpm	Torque capability (lb ft)	Speed ratio for driveN rpm of 1000	Torque capability at driveN unit (lb ft)
1 1/2	870	9.06	0.87	7.88
1 1/2	1160	6.79	1.16	7.88
1 1/2	1750	4.50	1.75	7.88
1 1/2	3450	2.28	3.45	7.87

By having a wide variety of speed ratios available to him, the machine designer has the option to use the electric motor of his choosing considering availability, size, and cost. The higher rpm motors are not only smaller but are usually lower in initial cost and operate at a higher efficiency.

The relationship between speed ratio and pulley diameter is given by this formula:

$$\text{speed ratio} = \frac{\text{driveN pitch diameter}}{\text{driveR pitch diameter}} \tag{15}$$

Note that the pitch diameter of the pulley must be used in calculating the speed ratio. This is not always the same as the nomenclature diameter as explained in later chapters.

B. Tension Requirements in Transmitting Torque

Torque is transmitted with a belt drive by a force on the rim of the pulley. This force is applied at the pitch diameter of the pulley and is commonly called the effective pull (see Figure 6).

Since most belt drives are friction devices, a tension must be maintained on both spans of the drive to keep the belt in firm contact with the pulley. The effective pull is defined as the difference between the tension in the tight side of the drive and the slack side of the drive as given in the following formula:

$$T_e = T_T - T_S \tag{16}$$

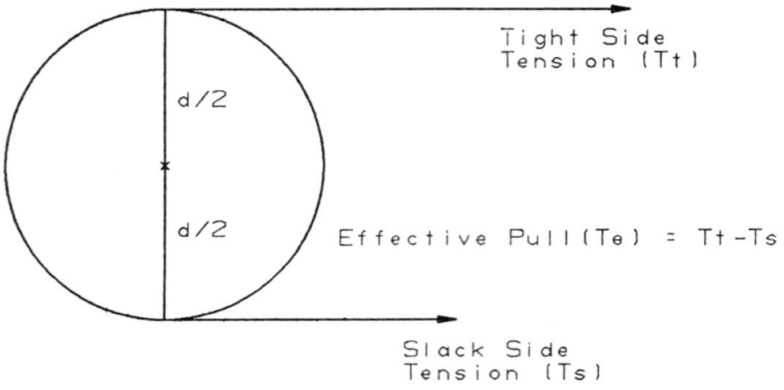

FIGURE 6 Torque being transmitted by effective pull.

where

T_e = effective pull (lbf)
T_T = tight side tension (lbf)
T_S = slack side tension (lbf)

The effective pull is the same whether applied to the driveR pulley or the driveN pulley since it is based on the tensions in the spans of belt between the pulleys. So since torque is (by our basic definitions) the product of the length of the lever arm and the force applied to the lever arm, it can be calculated by this formula:

$$\text{torque} = \frac{d}{2}(T_e) \qquad (17)$$

where

T_e = effective pull (lbf)
d = pitch diameter of pulley (in.)

As can be seen in Figure 6, the torque output of the driveR will impart a given effective pull in the drive depending on the pitch diameter of the driveR (see Equation 17). As this effective pull is then applied to the driveN pulley, it will result in a torque at the driveN pulley based on the diameter of the driveN pulley. This results in a torque multiplication which is a function of the ratio of the two pulleys or the speed ratio.

As evident from Figure 6 and Equation (17), the larger the pulley diameters, the lower the effective pull will be to transmit a given amount of torque. By using pulley diameters as large as possible, the tensions required in the belt drive will be lower which will be reflected favorably in the design of all machine elements: the belt drive, shafts, bearings, and frame work.

III. BASIC PRINCIPLES OF A BELT AS A FRICTIONAL FORCE DEVICE

A. Tension Ratio and Frictional Force

All power transmission belt drives except for synchronous belt drives transmit power through a frictional force that is developed between the belt and the pulley. The classic formula for frictional force is:

$$F_f = F_n \mu \qquad (18)$$

where

F_f = frictional force
F_n = normal force
μ = coefficient of friction between the two materials

In the case of power transmission belt drives, the normal force between the belt and the pulley must be provided by the tension in the belt. For this reason, there must always be tension in the slack side of the drive (T_S) so that the belt will be held firmly against the pulley.

1. Definition of Tension Ratio

The previously defined formula (Equation 16) for effective pull (T_e) takes this into consideration when the slack side tension (T_S) is subtracted from the tight side tension (T_T) to get the pull on the rim of the pulley through which torque is transmitted.

In order to provide the proper tensions in the drive to transmit torque or power it is important that the ratio between the tight side and slack side tensions be within certain limits. The tension ratio on most drives is controlled by the amount of tension that is put into the belt when the drive is not running. This is commonly referred to as static tension since the drive is in a static state at the time the tension is applied and measured. (See Chapter 11 for more detail on tensioning drives.)

The looser the belt is installed on a drive, the higher the tension ratio will be when the drive is running and transmitting a given amount of torque. As previously shown, the effective pull (T_e) is directly

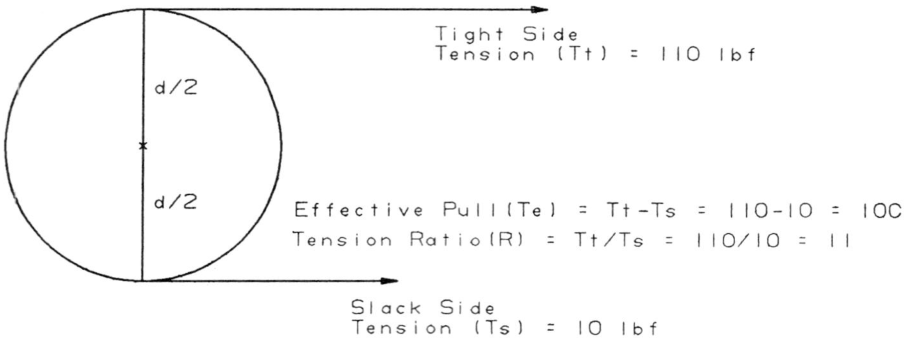

FIGURE 7 Belt drive with low tensions and a high tension ratio.

proportional to the amount of torque being transmitted on a given drive. Since the effective pull is equal to the tight side tension (T_T) minus the slack side tension (T_S), the lower these individual tensions are, the higher the tension ratio (R) will be.

For example, if the required effective pull in a belt drive is 100 lbf, Figures 7 and 8 will illustrate what happens to the tension ratio.

2. Derivation of the Classical Tension Ratio Formula for Flat Belts

The classical formula for the tension ratio (R) at which a flat belt must operate to keep from slipping is:

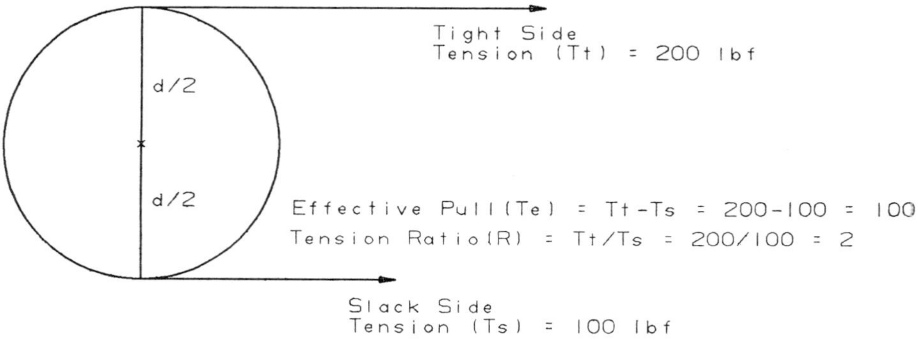

FIGURE 8 Belt drive with high tensions and a low tension ratio.

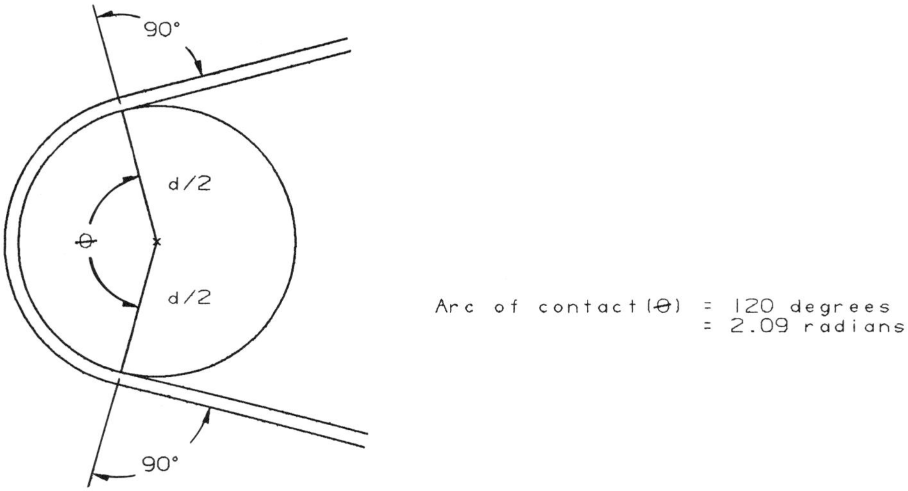

FIGURE 9 Arc of contact between flat belt and pulley.

$$R = \frac{T_T}{T_S} = e^{\mu \theta} \tag{19}$$

where

T_T = tight side tension (lbf)
T_S = slack side tension (lbf)
μ = coefficient of friction between belt and pulley as shown in Figure 9 (rad)

This formula is derived by considering the forces acting on a segment of belt and an increment of the angle θ (see Figure 10).

T represents the tension force acting on the segment from the slack side of the drive whereas the tension acting on the segment from the tight side is $T + \Delta T$. These forces can be broken down into two components acting in the X and Y directions. The Y components which act in a radial direction are counteracted by a force ΔF_N which is the normal force between the belt and the pulley. The frictional force between the belt and the pulley ($\mu \, \Delta F_N$) acts opposite the direction the belt would move relative to the pulley by the difference between the forces in the X direction.

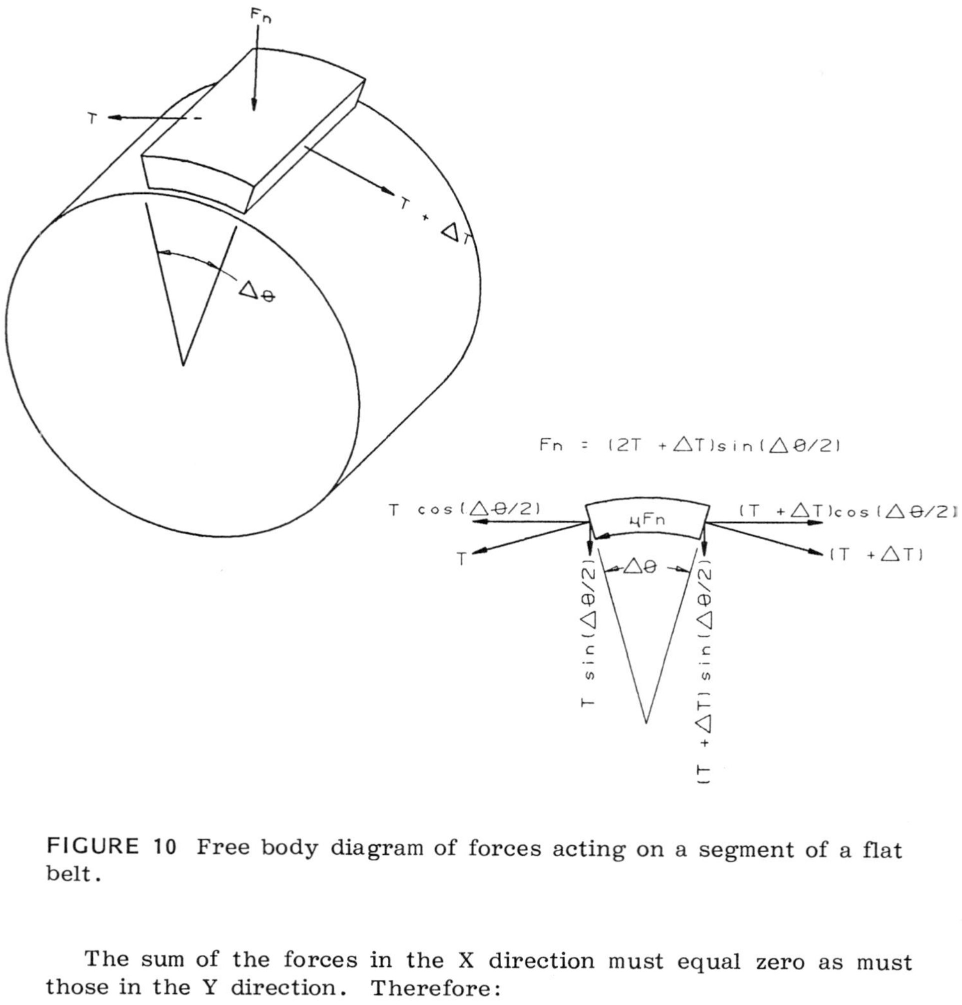

FIGURE 10 Free body diagram of forces acting on a segment of a flat belt.

The sum of the forces in the X direction must equal zero as must those in the Y direction. Therefore:

$$F_X = 0 = (T + \Delta T)\cos\left(\frac{\Delta\theta}{2}\right) - T\cos\left(\frac{\Delta\theta}{2}\right) - \mu\ \Delta F_N \tag{20}$$

$$F_Y = 0 = (T + \Delta T)\sin\left(\frac{\Delta\theta}{2}\right) + T\sin\left(\frac{\Delta\theta}{2}\right) - \Delta F_N \tag{21}$$

Simplifying these formulas:

$$\Delta T\cos\left(\frac{\Delta\theta}{2}\right) - \mu\ \Delta F_N = 0 \tag{22}$$

$$(2T + \Delta T)\ \sin\left(\frac{\Delta\theta}{2}\right) - \Delta F_N = 0 \tag{23}$$

Divide Equations (22) and (23) by $\frac{\Delta\theta}{2}$:

$$\frac{\Delta T}{\Delta\theta/2}\cos\left(\frac{\Delta\theta}{2}\right) - \frac{\Delta F_N}{\Delta\theta/2} = 0 \tag{24}$$

$$(2T + \Delta T)\ \frac{\sin(\Delta\theta/2)}{\Delta\theta/2} - \frac{\Delta F_N}{\Delta\theta/2} = 0 \tag{25}$$

Take the limit as $\Delta\theta$ approaches 0:

$$\cos\left(\frac{\Delta\theta}{2}\right) \to 1 \qquad \frac{\sin(\Delta\theta/2)}{\Delta\theta/2} \to 1$$

$$\frac{\Delta F_N}{\Delta\theta/2} \to \frac{dF_N}{d\theta/2} \qquad \frac{\Delta T}{\Delta\theta/2} \to \frac{dT}{d\theta/2}$$

and $\quad \Delta T \to 0$

Then in the limit equations (24) and (25) become:

$$dT - \mu\ dF_N = 0 \tag{26}$$

$$Td\theta - dF_N = 0 \tag{27}$$

Combining Equations (26) and (27):

$$\frac{dT}{T} = \mu d\theta \tag{28}$$

Integrating between limits:

$$\int_{T_S}^{T_T} = \frac{dT}{T} = \mu\int_0^{\theta} d\theta \tag{29}$$

$$\frac{T_T}{T_S} = e^{\mu\theta} \tag{19}$$

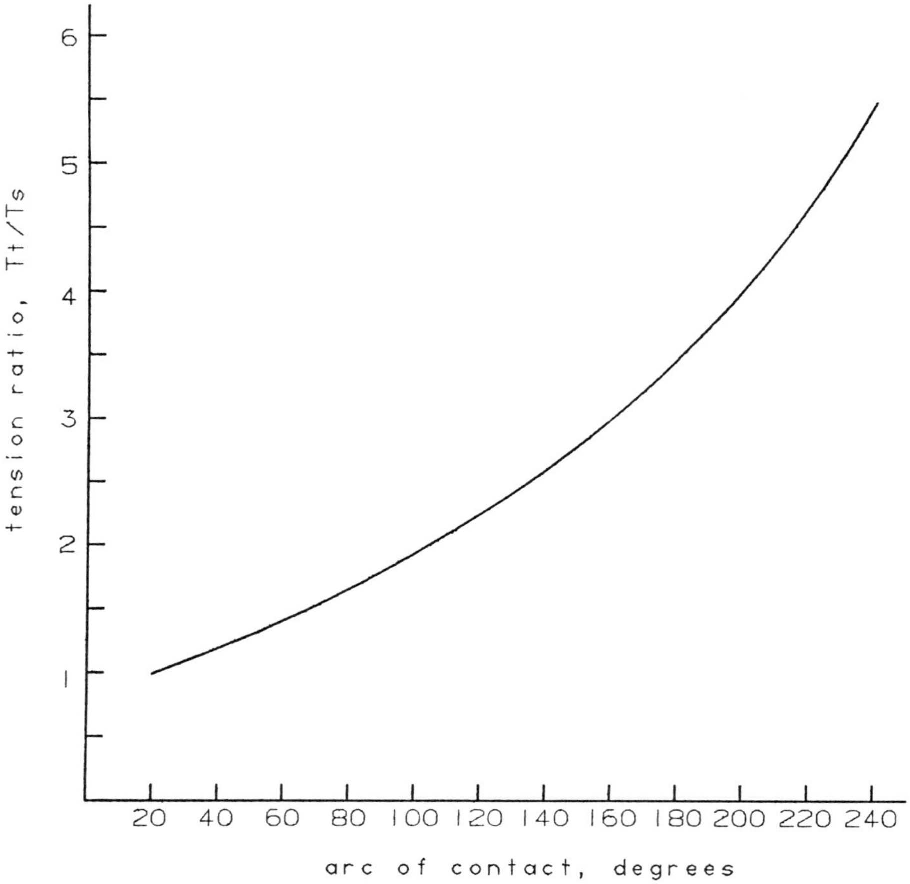

FIGURE 11 Tension ratio as a function of arc of contact for a flat belt.

While coefficient of friction values as high as 1.0 can often be obtained between rubber and steel or cast iron, under laboratory conditions, it is not practical to depend on this high a value in the operation of a power transmission drive. Environmental conditions such as moisture, dust, and other factors will reduce the coefficient of friction on an operating drive. Therefore, a value of 0.4 is generally accepted as being a practical working value.

A coefficient of friction, μ, of 0.4 will give tension ratio (R) values for a flat belt drive as shown in Figure 11.

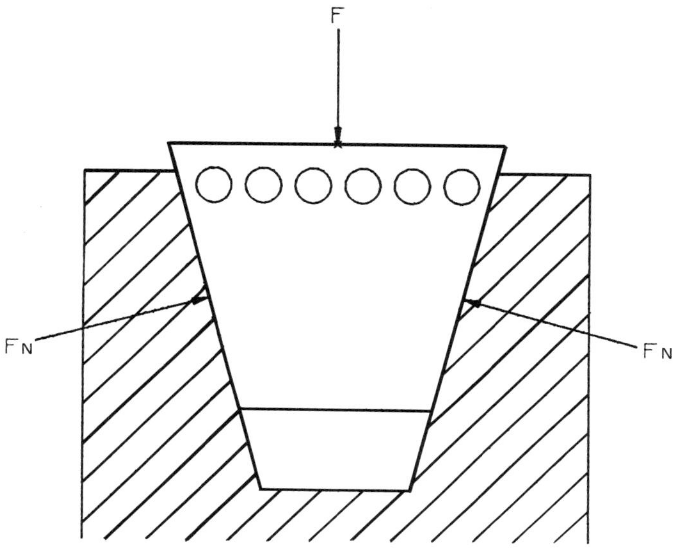

FIGURE 12 Multiplication of normal force, F_N, due to the wedging of a V-belt.

Flat belt drives are usually designed to operate at a tension ratio of 2.5 for an arc of contact of 180°. This will give ample margin for the normal reduction in tension that occurs as the belt operates.

B. Wedging Effect of a V-Belt

1. Classical Formula for Tension Ratio of a V-Belt

The frictional force between a V-belt and a pulley is increased over that of a flat belt because of the increase in the normal force (F_N) due to the wedging. Figure 12 illustrates how the radial force, F, on a segment of a V-belt as the result of the belt tensions is multiplied by the wedging to give a higher normal force, F_N.

The classical tension ratio formula for a V-belt drive is derived in the same manner as the flat belt drive formula except that the wedging is taken into consideration. This formula is:

$$R = \frac{T_T}{T_S} = e^{k\mu\theta} \tag{30}$$

where

- R = tension ratio
- T_T = tight side tension (lbf)
- T_S = slack side tension (lbf)
- k = wedging factor
- μ = coefficient of friction between V-belt and pulley
- θ = arc of contact between V-belt and pulley (rad)

This formula is in popular use throughout the world to describe the relationship between the tension ratio required for a V-belt drive and the arc of contact of the belt with the pulley. It has proven to be reasonably accurate through years of actual use in designing and maintaining V-belt drives.

2. Shortcomings of the Classical Formula

The theory described earlier for a flat belt does not, however, accurately represent the mechanics of a V-belt in a groove because, unlike the flat belt, the V-belt has radial movement as the tension in the belt changes from slack side tension (T_S) to tight side tension (T_T) through the arc of contact on the pulley. The magnitude of this radial movement depends on many factors, including the coefficient of friction between the belt and the pulley, the transverse and longitudinal rigidity of the belt, and the diameter of the pulley.

The mechanics of a V-belt operating in a groove are very complex and have been the subject of much work by many mathematical analysts within the past several years. There have been several interesting articles written on the subject including the following by Dr. B. G. Gerbert of Lund Technical University, Lund, Sweden.

Pressure Distribution and Belt Deformation in V-belt Drives
Tensile Stress Distribution in the Cord of V-belts
Force Conditions, Slip and Power Loss of V-belt Drives
Adjustable Speed V-belt Drives - Mechanical Properties and Design

C. Design Tension Ratio

1. Definition and Purpose of Design Tension Ratio

At the time a belt drive is designed, it is important to be able to predict within reasonable accuracy the belt tensions at which the drive will be required to operate. This is necessary so that the rotating shafts on the driveR and driveN unis can be properly sized and fitted with adequate bearings. It also enables the mechanic installing the belt to determine the correct installation tension.

The dynamic tensions that are in a drive as it operates are the tight side tension (T_T) and the slack side tension (T_S). These tensions are related to the torque or power being transmitted by the following formulas which have been derived from those given earlier in this chapter.

$$Q = \frac{d}{2}(T_T - T_S) \tag{31}$$

$$HP = \frac{d\ (T_T - T_S)\ rpm}{126051} \tag{32}$$

where

Q = torque lb ft)
HP = horsepower
T_T = tight side tension (lbf)
T_S = slack side tension (lbf)

If either one of these formulas is solved simultaneously with the proper tension ratio formula (depending on whether the drive is a flat belt or a V-belt), the dynamic tensions (T_T and T_S) can be calculated. The formulas for tension ratio are:

For a flat belt (Equation 19):

$$R = \frac{T_T}{T_S} = e^{\mu\theta}$$

Formula in terms of torque (Q):

$$T_T = \frac{2Q}{d\left(1 - 1/e^{\mu\theta}\right)} \tag{33}$$

Formula in terms of horsepower (HP):

$$T_T = \frac{126051\ HP}{(d)(rpm)\left(1 - 1/e^{\mu\theta}\right)} \tag{34}$$

(Definition of symbols as previously given.)

For a V-belt (Equation 30):

$$R = \frac{T_T}{T_S} = e^{k\mu\theta}$$

Formula in terms of torque (Q):

$$T_T = \frac{2Q}{d\left(1 - 1/e^{k\mu\theta}\right)} \tag{35}$$

Formula in terms of horsepower (HP):

$$T_T = \frac{126051 \text{ HP}}{(d)(rpm)(1 - 1/e^{k\mu\theta})} \tag{36}$$

(Definition of symbols as previously given.)

While the arc of contact, θ, can easily be determined for any given drive, it is much more difficult to determine a good working value of coefficient of friction, μ. While values of coefficient of friction as high as 1.0 can often be obtained between rubber and polished steel under ideal laboratory conditions, it is impractical to depend on values anywhere near this high for drives operating in a normal environment which may contain some moisture, dust and other adverse conditions.

Furthermore, most drives operate with the driveR and driveN units locked in place. This means that the initial tension with which the belt is installed will decay as the drive is operated. This is due to a gradual elongation of the tensile cords of the belt and also, in the case of a V-belt, the seating of the belt in the groove due to transverse compression of the V-belt cross section. The V-belt and corresponding groove will also wear, causing the belt to seat deeper in the groove resulting in a reduction of tension in the belt.

To compensate for tension decay, the belt must be installed at a lower tension ratio (higher tensions) so that it will operate for a reasonable length of time before the tension decays to the point where slippage may occur.

It has been determined through extensive laboratory testing and experience on actual applications that the average tension ratio at which a flat belt drive will operate the best is 2.5. Because of the wedging effect of a V-belt, the average tension ratio can be considerably higher and has been determined to average 5.0.

Both of these values are based on a drive with an arc of contact of 180° or Pi rad.

By putting the value of 2.5 for the average tension ratio, R, of a flat belt drive with 180° arc of contact, into Equation (19), an average working coefficient of friction can be calculated.

$$R = 2.5 = e^{\mu\theta}$$

$$\mu = \frac{\ln 2.5}{\pi} = 0.2917$$

By using this working value of coefficient of friction, the *design tension ratio* can be calculated for any drive based on the specific arc of contact, θ.

$$R_D = e^{0.2917\theta} \tag{37}$$

where

R_D = design tension ratio (flat belt drive)
e = base of natural logarithm
θ = arc of contact between flat belt and pulley (rad)

By using the design tension ratio, the average dynamic tension at which the drive is expected to operate over the normal period of time between retensions can be calculated. The design tension ratio will also form the basis of calculating the static tension which should be put into a drive to insure proper operation. (See Chapter 11 for a detailed discussion of static tension.)

By using the average tension ratio of 5.0 at which a V-belt will operate with an arc of contact of 180°, we can calculate a working value of coefficient of friction. But in the case of a V-belt, we also have the wedging factor to consider.

$$R = 5.0 = e^{k\mu\theta}$$

where

$$k = \frac{\ln 5.0}{\;} = 0.5123$$

The *design tension ratio* for a V-belt drive can be calculated by:

$$R_D = e^{0.512300} \tag{38}$$

where

R_D = design tension ratio (V-belt drive)
e = base of natural logarithm
θ = arc of contact between V-belt and pulley (rad)

2. Effect of Design Tension Ratio on Dynamic Tensions

The tight side tension (T_T) required for a drive can be calculated using the design tension ratio by the following formulas.

For a flat belt drive with the arc of contact given in degrees, the formula would be:

$$T_T = \frac{126051 \text{ HP}}{(d)(rpm)(1 - 1/e^{0.0050910\theta})} \quad (39)$$

For a V-belt drive:

$$T_T = \frac{126051 \text{ HP}}{(d)(rpm)(1 - 1/e^{0.0089410\theta})} \quad (40)$$

Figure 13 illustrates the effect the tension ratio has on the tight side tension of a drive.

It can be shown by Figure 13 that the tight side tension in a flat belt drive operating at a design tension ratio of 2.5 is much higher than the tight side tension in a V-belt drive operating at a design tension ratio of 5.0. For a flat belt with an arc of contact of 180°, the tight side tension is 167% of the effective pull. For a V-belt with the same arc of contact it is only 125%.

It is interesting to note that as the tension ratio approaches 1, the tight side tension as a percent of the effective pull approaches infinity. On the other end of the curve, as the tension ratio approaches infinity, the tight side tension approaches 100% of the effective pull which means the slack side tension would be zero.

Of course, with the lower tension ratios, the slack side tension is also higher, which means much greater shaft loads for the drive that has the lower tension ratio. For example, a flat belt drive transmitting 100 HP on a driver pitch diameter of 5 in. at 1750 rpm with a speed ratio of 1.0 would have the following shaft loading from the dynamic tensions.

For a flat belt drive:

$$T_T = \frac{(126051)(100)}{(5.0)(1750)(1 - 1/2.5)} = 2401 \text{ lbf}$$

$$T_S = \frac{T_T}{2.5} = \frac{2401}{2.5} = 960 \text{ lbf}$$

Total shaft load = $T_T + T_S$ = 3361 lbf

For a V-belt drive:

$$T_T = \frac{(126051)(100)}{(5.0)(1750)(1 - 1/5)} = 1801 \text{ lbf}$$

$$T_S = \frac{T_T}{5} = \frac{1801}{5} = 360 \text{ lbf}$$

Total shaft load = $T_T + T_S$ = 2161 lbf

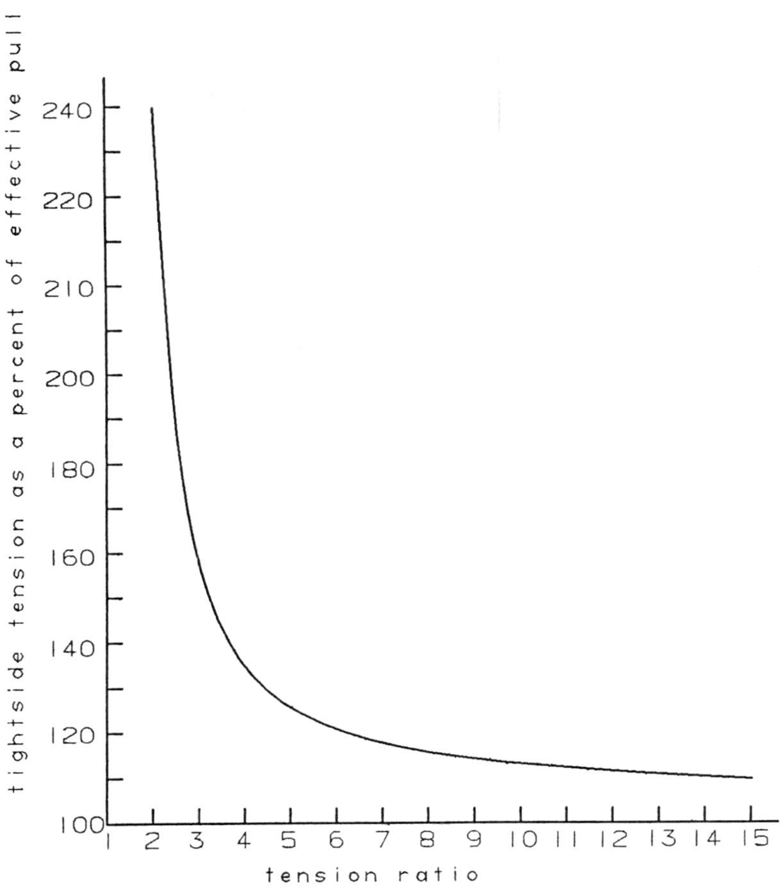

FIGURE 13 Tight side tension as a function of tension ratio.

IV. RELATIONSHIP BETWEEN TENSION RATIO AND SPEED LOSS

Because a flat belt or a V-belt drive is not a positive engagement device but instead depends on frictional force in the transmission of power, there is a small amount of speed loss between the driveR and driveN machines. If a belt is properly tensioned, this speed loss is very small and is generally referred to as creep.

A. Definition of Belt Creep

A power transmission belt is made from relatively low modulus materials. There is a significant change in the length of an increment of belt as it progresses around the pulley from the slack side to the tight side. This change in length results in the slight change in the rotational speed of the driveN pulley relative to the driveR pulley which is referred to as creep.

1. Depending on Tension Ratio

Since creep is dependent on the change in belt length between the slack side and tight side of the drive, it is influenced by the tension ratio. The higher the tension ratio, the greater the difference between the tight side tension (T_T) and the slack side tension (T_S) and, therefore, the greater the amount of creep.

If, for example, a drive is operating at a tension ration of 1, the slack side and tight side tensions would be the same so there would be no creep. Of course, there would also be no transmission of power.

Figure 14 shows a typical relationship between tension ratio and percent of rpm change between the driveR and driveN machines.

B. Definition of Belt Slip

If the tension ratio in a drive becomes too high (tensions too low) the belt will slip on the pulleys. This is a different phenomenon than belt creep as there is significant movement between the belt and the pulley because the normal force on which the frictional force is dependent is not sufficient to handle the torque being transmitted.

Belt slip is often associated with a screeching noise and a significant heat buildup in the drive.

1. Dependency on Arc of Contact

The tension ratio at which a drive will slip depends on the arc of contact (θ) between the belt and the pulley. This can be seen from Equations (19) and (30).

Figure 15 is an extension of the curve in Figure 14 which relates percent rpm change to tension ratio. When tension ratio increases to a sufficiently high value, slip will occur. This causes an abrupt change in the slope of the curve with the percent change in rpm increasing rapidly.

C. Controlling Tension Ratio

1. Fixed Center Drives

Most belt drives operate with a fixed center distance between the driveR and driveN shafts (locked centers) and with no automatic

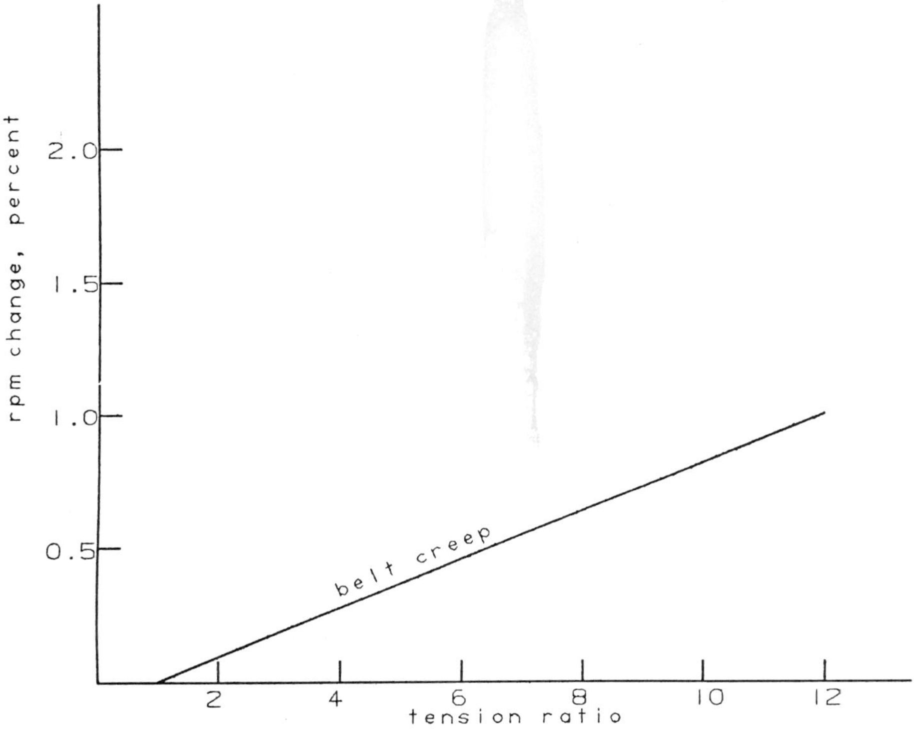

FIGURE 14 Percent change in rpm between driveR and driveN machines as a function of tension ratio (creep).

way by which to maintain tension. The proper value of static tension is put into the belt drive when it is standing still so that when operating at the required load and speed, the operating or dynamic tensions will provide a tension ratio low enough to prevent belt slip. Chapter 11 gives the details on applying and measuring static tension.

It is important to clarify the definition of the tight side (T_T) and slack side (T_S) tensions. These dynamic tensions are the forces by which the belt is pulling at the rim of the pulley to transmit the required torque or power. While these tensions influence the internal tensions within the belt itself, they are only a part of the internal tension. Centrifugal tension and bending tension are also part of the internal tension within the tensile cords of a belt.

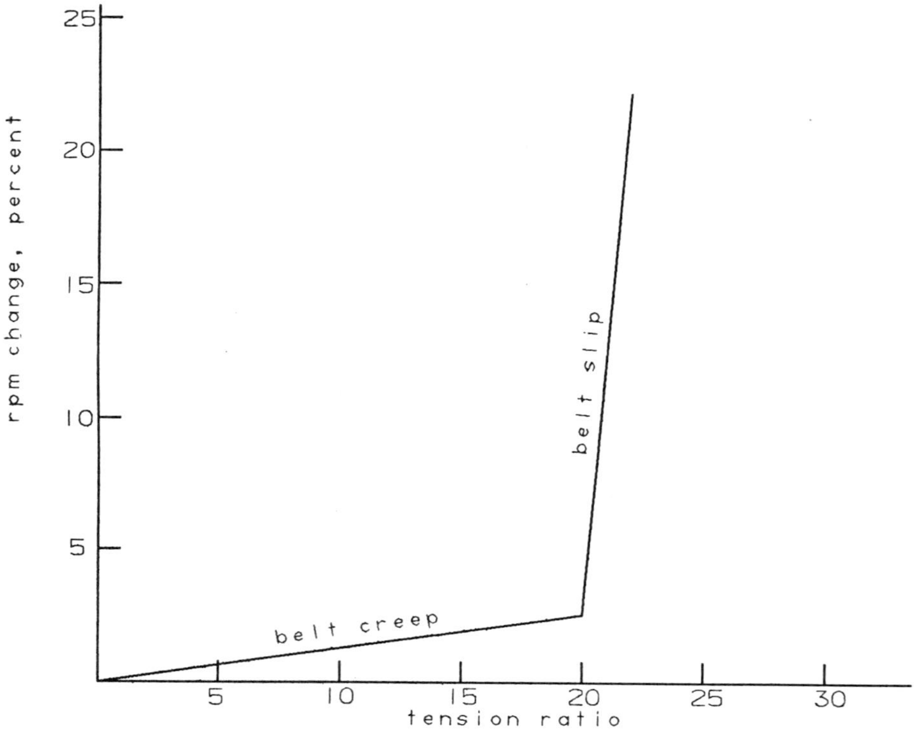

FIGURE 15 Percent change in rpm between driveR and driveN machines as a function of tension ratio (creep and slip).

2. Automatic Tensioning Devices

The dynamic tensions in a drive can be controlled by a variety of automatic tensioning devices. The most popular of these devices is an idler pulley which is held against the belt with a constant force usually generated by a spring or a weight.

The idler is usually applied to the span of belt that requires the lowest tensions (slack side tension). The amount of tension required in the span when the drive is operating at the required load will determine the force that needs to be applied to the idler. Figure 16 shows the vector force diagram by which the idler force (F) is calculated.

$$F = 2T_S \sin \frac{\theta}{2} \qquad (40)$$

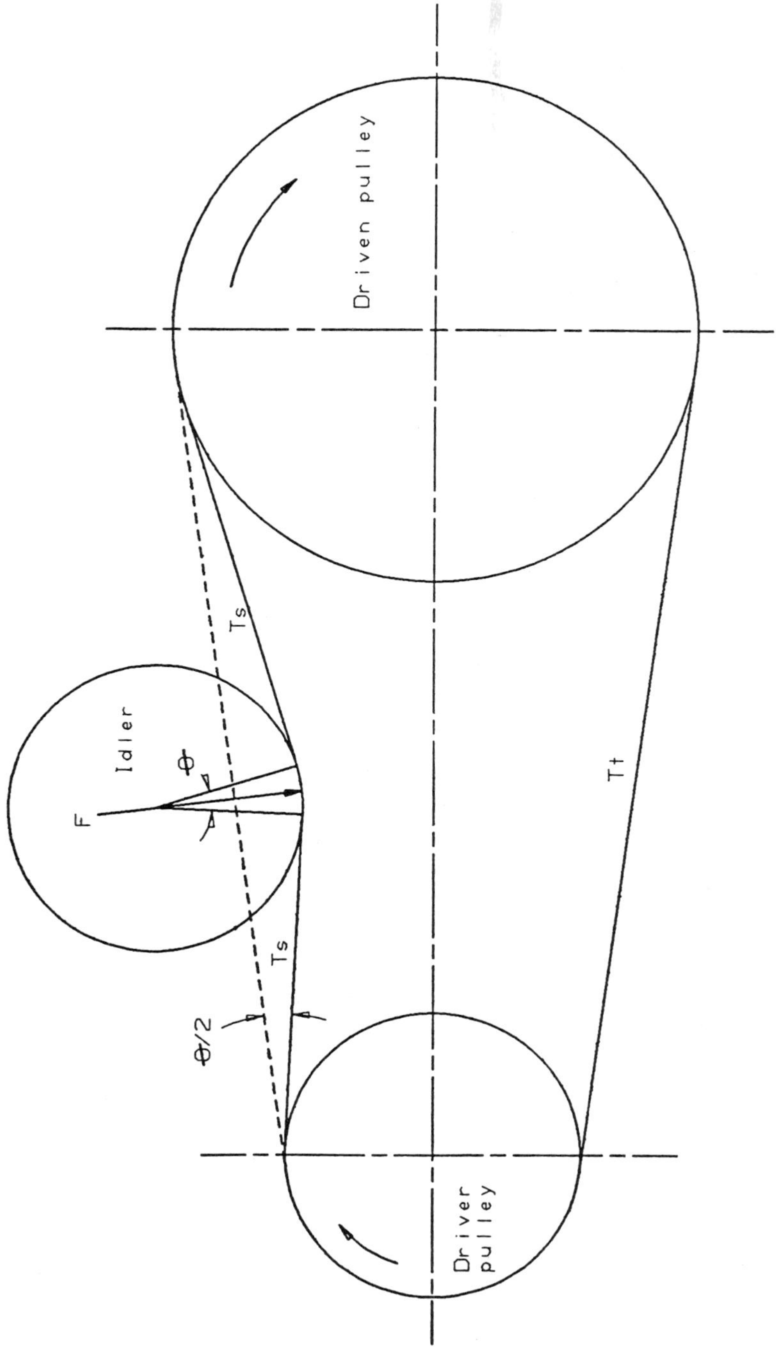

FIGURE 16 Method of calculating idler force required to provide a given belt tension.

where

F = force on idler (lbf)
T_S = desired tension in belt span (lbf)
θ = arc of contact of belt on idler (deg)

The advantage of devices which automatically maintain dynamic tensions in a drive is that they provide a more constant control of tensions over a long period of time and require very little maintenance. Tensions can be set to the required value initially and will remain reasonably constant over the life of the belts. Tensions in a fixed center drive, on the other hand, must be initially set somewhat higher than required so that as tension decays from belt seating and stretch, the tension will be adequate. A periodic resetting of the tension is usually required.

D. Influence of Belt Size and Number of Belts on Required Tension Ratio

In the preceding discussion on how tension ratio effects the creep and slip characteristics of a drive, the only consideration given to the design of the drive was to distinguish between a flat belt and a V-belt and to adjust for arc of contact between the belt and the pulley. It is evident from Equations (19) and (30) that neither the size of the V-belt or flat belt nor the number of belts on the drive enter into the picture.

This is because the frictional force developed between the belt and the pulley is a function of only the normal force and the coefficient of friction. Areas of contact, which would be influenced by the size of the belts or the number of belts on the drive are not a factor.

This is a very important consideration that is often overlooked when designing with multiple V-belt drives. It is the total drive tension that is the controlling factor. It must be set high enough so that the tension ratio is within the reasonable operating range for the torque being transmitted.

If fewer belts are used on a drive, each belt must be tensioned to a higher value so that the total drive tension is in an acceptable range. This is a factor often overlooked when designing with modern-day high capacity belts.

V. RELATIONSHIP BETWEEN TENSION RATIO AND TORQUE LOAD FOR A SYNCHRONOUS BELT DRIVE

A synchronous belt does not transmit torque through a frictional force as does a V-belt or flat belt. Torque is transmitted through

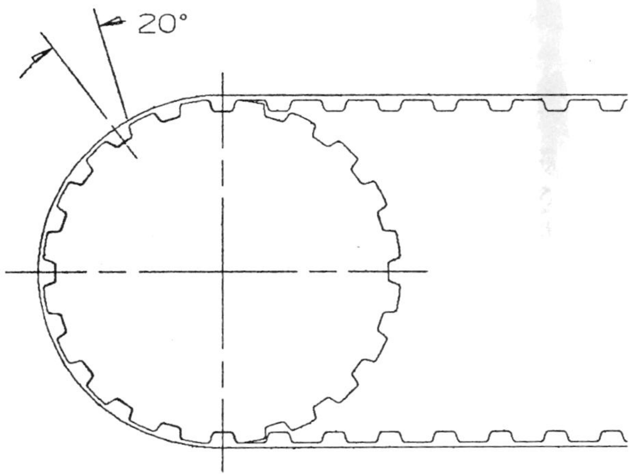

FIGURE 17 Synchronous belt in toothed pulley.

the positive locking of the belt teeth with the teeth on the pulley -- much the same way that a chain drive operates.

This does not mean, however, that a synchronous belt drive operates with a slack side tension (T_S) equal to zero which would correspond to a tension ratio of infinity.

A. How Tension Ratio Varies with Load

A synchronous belt profile is illustrated in Figure 17. Note that the mating flanks of the belt and the pulley tooth through which torque is transmitted have an angle of 20°. As torque is transmitted by a tangential pull of the belt on the rim of the pulley, a radial force vector is developed which tends to lift the belt tooth out of the space between the pulley teeth. This radial force is shown graphically in Figure 18.

This radial force is counteracted by an increase in the total tension in the belt. The higher the torque being transmitted the higher the total tension in a synchronous belt becomes. While a synchronous belt can be installed at low tensions, as torque is applied the belt will tighten in direct relationship to the amount of torque. As the belt tightens, the tension ratio goes down so it is unusual to observe a synchronous belt drive operating with a tension ratio larger than 8.0.

If there is too much slack in the drive when the belt is installed, the belt teeth will move radially as the necessary tension is built up in the belt to counteract the radial force from the torque being

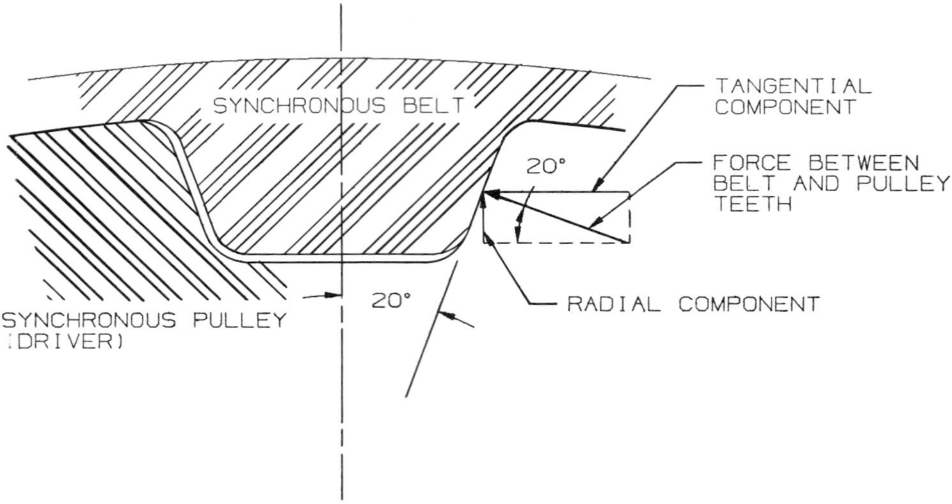

FIGURE 18 Radial force developed on a synchronous belt tooth as torque is transmitted.

transmitted. The belt tooth will not have full contact with the pulley tooth which will result in premature wear of the teeth as well as abnormal stress on the tooth. Obviously if the belt tooth is allowed to move too far radially because of low installation tension, the drive will jump teeth or ratchet.

B. Total Tension in a Synchronous Belt Drive

It is a common misconception that a synchronous drive operates at a low total tension or with a low bearing or shaft load. This is because it is thought of as a positive engagement drive rather than a frictional force drive and because the tension at which a synchronous belt is installed (static tension) is often much less than for a V-belt or flat belt drive.

However, as described in Section V.A. of this chapter, the synchronous belt develops an increased tension as torque is applied so that the operating (dynamic) tensions in the tight and slack sides of the drive are approximately the same as those for a V-belt drive.

This needs to be taken into consideration when calculating shaft pull for sizing shafts and bearings as described in Chapter 11.

3

Industrial Drive Design Using V-Belts

WALLACE D. ERICKSON *Automotive Products Technical and Development Division, The Gates Rubber Company, Denver, Colorado*

I. HEAVY-DUTY INDUSTRIAL DRIVES

A. Definition

Heavy-duty industrial V-belt drives are defined as those drives designed for integral power transmission (prime movers of 1 or more HP at 1750 rpm). These drives operate on a continual basis and under such conditions should give at least 3 years of service with very little maintenance.

Heavy-duty industrial V-belt drives usually use V-belts in multiple which allow for the transmission of high HP (sometimes over 1000 HP) on reasonably compact pulley or sheave diameters. It is quite common for these types of drives to use 10 belts in multiple and some large high-HP drives can use 20 or more belts operating in unison on the same drive.

Heavy-duty industrial V-belt drives are used on a wide range of industrial equipment used in factories, mining, construction, lumber, oil field, and other types of industries. They are often used on mass-produced assemblies such as air-compressor units or they may be designed for individual use for power transmission as industrial equipment is combined into a specific plant or industrial operation. Figures 1 and 2 give typical examples of industrial type drives.

1. Performance Requirements

A typical industrial V-belt drive is expected to operate for 3 years if it runs continuously – 24 hours/day for seven days/week. Maintenance should consist of an occasional inspection of the moving parts and a check of belt tension.

If the machine operates on an 8 hour/day basis, the drive is expected to last for approximately 10 years with, of course, periodic inspections and the checking and resetting of the drive tension as necessary.

For more detail on the installation, maintenance and replacement of power transmission belts, see Chapter 12.

2. Basic Types of Industrial V-Belts

There are two basic types of industrial V-belts that are designed to operate in multiple: classical V-belts and narrow V-belts.

Classical V-belts: Classical multiple V-belts are so named because they have been in existence for many years and have become the stalwart belt of the industrial power transmission industry. They are commonly made in the five cross sections shown in Figure 3.

Larger classical cross sections have been made on a limited basis over the years. These consist of the F, G, and H cross sections with top widths of 2, 2-1/2, and 3-1/4 in., respectively. They have to be made to order, however, so they are not used for new drives today; only for replacement on existing drives.

Classical industrial belts were first standardized by the Rubber Manufacturers Association and the Mechanical Power Transmission Association in 1955 in the five cross sections A through E. As the

FIGURE 1 Narrow industrial V-belt drive on a suction pump used for dredging sand and gravel.

power capacity of all belts has increased over the years, the E section has become less popular and has been discontinued in the metric-SI unit section of the latest standard. For more details on the industry standards that apply to power transmission belts, see Chapter 17.

FIGURE 2 Classical industrial V-belt drive on a paper sheet forming machine.

Narrow V-belts: The industrial narrow V-belts were designed to serve the same application area as the classical V-belts and to do so with a lighter, more compact drive. They were introduced in 1959 by the Gates Rubber Company and were standardized by Rubber Manufacturers Association and the Mechanical Power Transmission Association a few years later.

The narrow industrial belts, like the classical belts, are designed to operate in multiple. The line of belts consists of only three cross sections, however, because they are of a higher HP capacity and through the proper selection of sheave diameter, the three cross sections will serve the same power transmission needs as the five classical cross sections. Figure 4 shows the three narrow belt cross sections.

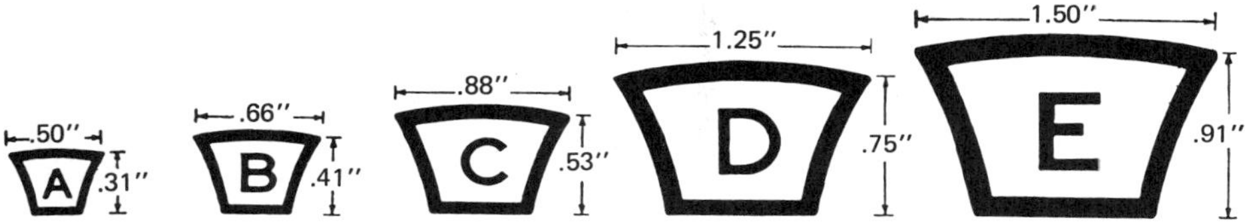

FIGURE 3 Classical industrial V-belt cross sections.

The operating ambient temperature range for both classical and narrow industrial V-belts is between -30°F to 140°F. Both lines of belts also will withstand a reasonable exposure to oil as well as dust.

B. Nomenclature

1. V-Belts

Nomenclature for industrial V-belts consists of one or more digits which represent the belt cross section followed by two or more numbers which designate the belt length. This nomenclature system is described in the industry standards IP-20 and IP-22 published jointly by the Rubber Manufacturers Association, the Mechanical Power Transmission Association, and the Rubber Association of Canada.

Conventional USA units: For classical industrial V-belts, the cross section designation is a letter from A through E as shown in Figure 3. The letter cross section designation is followed by two or three numerals which designate the belt length. For example:

B 60

Cross section designation — Belt length designation

FIGURE 4 Narrow industrial V-belt cross sections.

The exact definition of the belt length designation is rather complex because of the way in which the classical belt line has developed over the span of many years. As explained in Chapter 1, the early-day classical V-belts had a length designation that was based on the inside belt length – the same as a flat belt. These early-day belts were made in lengths up through 215 in. inside circumference.

When it was decided later to specify belt length as a pitch length, the same length designation was maintained to avoid the obsolescence of inventory and the confusion that would result from a change in nomenclature. Because of this, there is a large difference between the pitch length and the length designation for belts of 215 in. or less. For example, the B60 belt has a pitch length of 61.8 in.

The sizes over 215 in. were added after the change to pitch length was made. Therefore, the belt length designation numbers were set up as pitch lengths so they agree very closely with the pitch lengths. For example, a B270 belt has a pitch length of 270.3 in.

Even on these longer length belts, the pitch length and the length designation do not agree exactly. This is because a change was made several years ago in the pitch line location of the belt. The pitch line in the belt, as determined by the placement of the tensile section, was raised, which resulted in a pitch length slightly longer than the original pitch length that formed the basis for the length designation. Again it was decided it would be impractical to change the length designation.

The narrow industrial V-belt line was designed on the effective system which is described in Chapter 1. Therefore, the belt nomenclature shows the cross section and the effective length in one-tenth of an inch. For example:

5V 600

Cross section designation — — Belt length designation

The 5V cross section designation means the nominal width is 5/8 in. The length designation of 600 means the effective length is 60.0 in.

Metric-SI units: When metric-SI nomenclature was established for the classical industrial belts, the geometric identification was changed to the effective system. This means, then, that both the classical and narrow belt types are on the same system and the advantages of the effective system as explained in Chapter 1 can be fully utilized.

The following examples show how the length designation in metric-SI units apply. For classical belts:

16 C 1600

groove width (mm) — effective length (mm)

code letter for classical

For narrow belts:

15 N 1600

groove width (mm) — effective length (mm)

code letter for narrow

2. Sheaves

Standard nomenclature for sheaves is not included in the industry standards. When ordering sheaves, the following information needs to be included:

Cross section designation
Sheave diameter (pitch diameter for classical sheaves in conventional USA units; outside diameter for narrow sheaves and all metric sheaves)
Number of grooves
Bore of hub or bushing
Keyseat dimensions if not standard as specified in Table 1. Keyseat dimensions are illustrated in Figure 5.

C. Stock Components

1. Belts

Industrial V-belt drive components are readily available from industrial distributors' stocks all over the United States or from the various V-belt and sheave manufacturers. Most of the standard belt sizes and sheave diameters specified in the joint industry standards published by the Rubber Manufacturers Association and the Mechanical Power Transmission Association are available as stock off-the-shelf items for quick replacement or to enable a user to easily design his own power transmission drive.

Table 2 lists the standard lengths for each of the cross sections in both the classical and narrow belt lines. In addition to these sizes, some manufacturers and distributors often stock other non-standard sizes which exhibited a high sales frequency. In designing a new drive, however, it is wise to use standard, readily available components to be assured of easily available replacement parts.

TABLE 1 Keyseat Dimensions (In.). (See Figure 5.)

Shaft Diameter	Width, w_k*	Depth, h_k +0.015 −0.000
Up Through 7/16 (0.44)	3/32 (0.094)	3/64 (0.047)
Over 7/16 (0.44) To and Incl 9/16 (0.56)	1/8 (0.125)	1/16 (0.062)
Over 9/16 (0.56) To and Incl 7/8 (0.88)	3/16 (0.188)	3/32 (0.094)
Over 7/8 (0.88) To and Incl 1-1/4 (1.25)	1/4 (0.250)	1/8 (0.125)
Over 1-1/4 (1.25) To and Incl 1-3/8 (1.38)	5/16 (0.312)	5/32 (0.156)
Over 1-3/8 (1.38) To and Incl 1-3/4 (1.75)	3/8 (0.375)	3/16 (0.188)
Over 1-3/4 (1.75) To and Incl 2-1/4 (2.25)	1/2 (0.500)	1/4 (0.250)
Over 2-1/4 (2.25) To and Incl 2-3/4 (2.75)	5/8 (0.625)	5/16 (0.312)
Over 2-3/4 (2.75) To and Incl 3-1/4 (3.25)	3/4 (0.750)	3/8 (0.375)
Over 3-1/4 (3.25) To and Incl 3-3/4 (3.75)	7/8 (0.875)	7/16 (0.438)
Over 3-3/4 (3.75) To and Incl 4-1/2 (4.50)	1 (1.000)	1/2 (0.500)
Over 4-1/2 (4.50) To and Incl 5-1/2 (5.50)	1-1/4 (1.250)	5/8 (0.625)
Over 5-1/2 (5.50) To and Incl 6-1/2 (6.50)	1-1/2 (1.500)	3/4 (0.750)
Over 6-1/2 (6.50) To and Incl 7-1/2 (7.50)	1-3/4 (1.750)	3/4 (0.750)
Over 7-1/2 (7.50) To and Incl 9 (9.00)	2 (2.000)	3/4 (0.750)
Over 9 (9.00) To and Incl 11 (11.00)	2-1/2 (2.500)	7/8 (0.875)
Over 11 (11.00) To and Incl 13 (13.00)	3 (3.000)	1 (1.000)

*Tolerance on widths, w_k (inches). For widths up through 1/2 (0.500)...+0.002, -0.000. For widths over 1/2 (0.500) to and incl 1 (1.000)...+0.003, -0.000. For widths over 1 (1.000)...+0.004, -0.000.
Source: The industry standard IP-22 published jointly by the Rubber Manufacturers Association, the Mechanical Power Transmission Association, and the Rubber Association of Canada.

2. Sheaves

Table 3 lists the standard sheave diameters for the narrow cross sections. These standard diameters are shown in the standard which is published jointly by the Rubber Manufacturers Association and the Mechanical Power Transmission Association.

In addition to the standard diameters, Table 3 also shows the number of grooves which are commonly stocked by most sheave manufacturers in each of the cross sections.

There are no published industry standard diameters for the classical cross sections although most of the sheave manufacturers make the same common diameters. Table 4 lists the classical sheave diameters and number of grooves that are stocked by most of the sheave manufacturers.

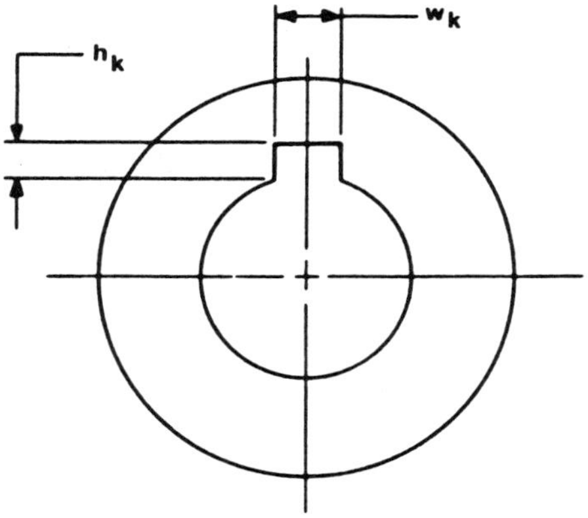

FIGURE 5 Keyseat dimensions.

D. Specific Types of V-Belts

1. Banded

A banded industrial V-belt is constructed with a rubber-impregnated fabric envelope or band wrapped completely around the belt. This is the most popular type of industrial belt and is manufactured in most of the industrial nations of the world.

Figure 6 shows a cross-sectional view of two types of banded industrial belts. Figure 6a shows a single-unit tensile cord which consists of individual cabled cords laid close together in a single plane. This type of construction is becoming increasingly popular since a high tensile strength can be obtained with all of the cords acting in a common pitch plane.

The other type construction (Figure 6b) is called a multiple-cord-type belt since the tensile cords are smaller cords woven together in a sheet of fabric. Two or more sheets of these smaller fabric tensile cords form the tensile or load-carrying component of the belt. This type of tensile cord is losing popularity to the single-unit cord because it does not have the capability of forming a single cord plane about which the belt can flex over a small sheave diameter.

While it is possible to attain a high ultimate tensile strength by using several plies of multiple-cord tensile, the functional advantages of this high tensile strength cannot always be utilized because all cords do not act in a common pitch plane.

TABLE 2 Standard Industrial V-Belt Lengths

Classical V-Belts						Narrow V-Belts			
Standard length designation	Standard pitch lengths for cross sections					Standard length designation	Standard effective lengths for cross sections		
	A	B	C	D	E		3V	5V	8V
26	27.3					250	25.0		
31	32.3					265	26.5		
35	36.3	36.8				280	28.0		
38	39.3	39.8				300	30.0		
42	43.3	43.8				315	31.5		
46	47.3	47.8				335	33.5		
51	52.3	52.8	53.9			355	35.5		
55	56.3	56.8				375	37.5		
60	61.3	61.8	62.9			400	40.0		
68	69.3	69.8	70.9			425	42.5		
75	76.3	76.8	77.9			450	45.0		
80	81.3					475	47.5		
81		82.8	83.9			500	50.0	50.0	
85	86.3	86.8	87.9			530	53.0	53.0	
90	91.3	91.8	92.9			560	56.0	56.0	
96	97.3		98.9			600	60.0	60.0	
97		98.8				630	63.0	63.0	
105	106.3	106.8	107.9			670	67.0	67.0	
112	113.3	113.8	114.9			710	71.0	71.0	

120	121.3	121.8	122.9	123.3		750	75.0	75.0	
128	129.3	129.8	130.9	131.3		800	80.0	80.0	
144		145.8	146.9	147.3		850	85.0	85.0	
158		159.8	160.9	161.3		900	90.0	90.0	
173		174.8	175.9	176.3		950	95.0	95.0	
180		181.8	182.9	183.3	184.5	1000	100.0	100.0	100.0
195		196.8	197.9	198.3	199.5	1060	106.0	106.0	106.0
210		211.8	212.9	213.3	214.5	1120	112.0	112.0	112.0
240		240.3	240.9	240.8	241.0	1180	118.0	118.0	118.0
270		270.3	270.9	270.8	271.0	1250	125.0	125.0	125.0
300		300.3	300.9	300.8	301.0	1320	132.0	132.0	132.0
330			330.9	330.8	331.0	1400	140.0	140.0	140.0
360			360.9	360.8	361.0	1500		150.0	150.0
390			390.9	390.8	391.0	1600		160.0	160.0
420			420.9	420.8	421.0	1700		170.0	170.0
480				480.8	481.0	1800		180.0	180.0
540				540.8	541.0	1900		190.0	190.0
600				600.8	601.0	2000		200.0	200.0
660				660.8	661.0	2120		212.0	212.0
						2240		224.0	224.0
						2360		236.0	236.0
						2500		250.0	250.0
						2650		265.0	265.0
						2800		280.0	280.0
						3000		300.0	300.0
						3150		315.0	315.0

TABLE 3 Standard Narrow Sheaves

3V		5V		8V	
Outside diameter	Number of grooves[a]	Outside diameter	Number of grooves[a]	Outside diameter	Number of grooves[a]
2.65	1, 2, 3,	7.10	2, 3, 4, 5,	12.50	All
2.80	and 4	7.50	6, and 8	13.20	diameters
3.00				14.00	available
3.15		8.00	2, 3, 4, 5,	15.00	in
3.35		8.50	6, 8, and	16.00	4, 5, 6, 8,
3.55[b]		9.00	10	17.00	10, and 12
3.65		9.25		18.00	grooves
4.00[b]		9.75		19.00	
4.12		10.00		20.00	
4.50		10.30		21.20	
		10.60		22.40	
4.75	1, 2, 3, 4,	10.90		23.60[b]	
5.00	5, 6, 8,	11.20		24.80	
5.30	and 10	11.80		30.00	
5.60		12.50		31.50[b]	
6.00		13.20		35.50	
6.30[b]		14.00		40.00	
6.50		15.00		44.50	
6.90		16.00		50.00[b]	
8.00		18.70		52.00[b]	
10.00		20.00		63.00[b]	
10.60		21.20		71.00[b]	
12.50[b]		23.60		79.00[b]	
14.00		25.00		99.00[b]	
16.00[b]		28.00			
19.00					
20.00[b]		31.50	3, 4, 5, 6,		
		37.50	8, and 10		
25.00	2, 3, 4, 5,	40.00			
	6, 8, and	44.50			
	10	50.00			
		63.00			
31.50[b]	3, 4, 5, 6,	71.00			
33.50	8, and 10				

[a]Number of grooves are not specified in industry standards.
[b]This diameter may not be commonly stocked.

TABLE 4 Commonly Stocked Classical Sheaves

Multi-duty A or B[a]			B section		C section		D section	
Pitch diameter A	Pitch diameter B	Number of grooves	Pitch diameter	Number of grooves	Pitch diameter	Number of grooves	Pitch diameter	Number of grooves
3.0	3.4[b]	All	4.6	1, 2, 3, 4,	7.0	2, 3, 4, 5,	12.0	All
3.2	3.6[b]	diameters	5.0	5, and 6	7.5	and 6	13.0	diameters
3.4	3.8[b]	available	5.1		8.0	2, 3, 4, 5,	13.5	available
3.6	4.0[b]	in	5.2		8.5	6, 8, and	14.0	in
3.8	4.2[b]	1, 2, 3, 4,	5.6			10	14.5	4, 5, 6, 8,
4.0	4.4	5, and 6					15.0	10, and 12
4.2	4.6	grooves	6.0	1, 2, 3, 4,	9.0	2, 3, 4, 5,	15.5	grooves
4.8	5.2		6.4	5, 6, 8,	9.5	6, 8, 10,	16.0	
5.0	5.4		6.8	and 10	10.0	and 12	18.0	
5.2	5.6		7.4		10.5		22.0	
5.6	6.0		8.6		11.0		27.0	
6.0	6.4		9.4		12.0		33.0	
6.4	6.8		11.0		13.0		40.0	
7.0	7.4		12.4		14.0		48.0	
8.2	8.6		15.4		16.0			
9.0	9.4		18.4		18.0			
10.6	11.0				20.0			
12.0	12.4		20.0	2, 3, 4, 5,	24.0			
15.0	15.4		25.0	6, 8, and	30.0	3, 4, 5, 6,		
18.0	18.4		30.0	10	36.0	8, 10, and		
			38.0		44.0	12		
					50.0			

[a]Multi-duty grooves are designed to accept either A or B section classical V-belts.
[b]Recommended for use with notched or cogged belts only.

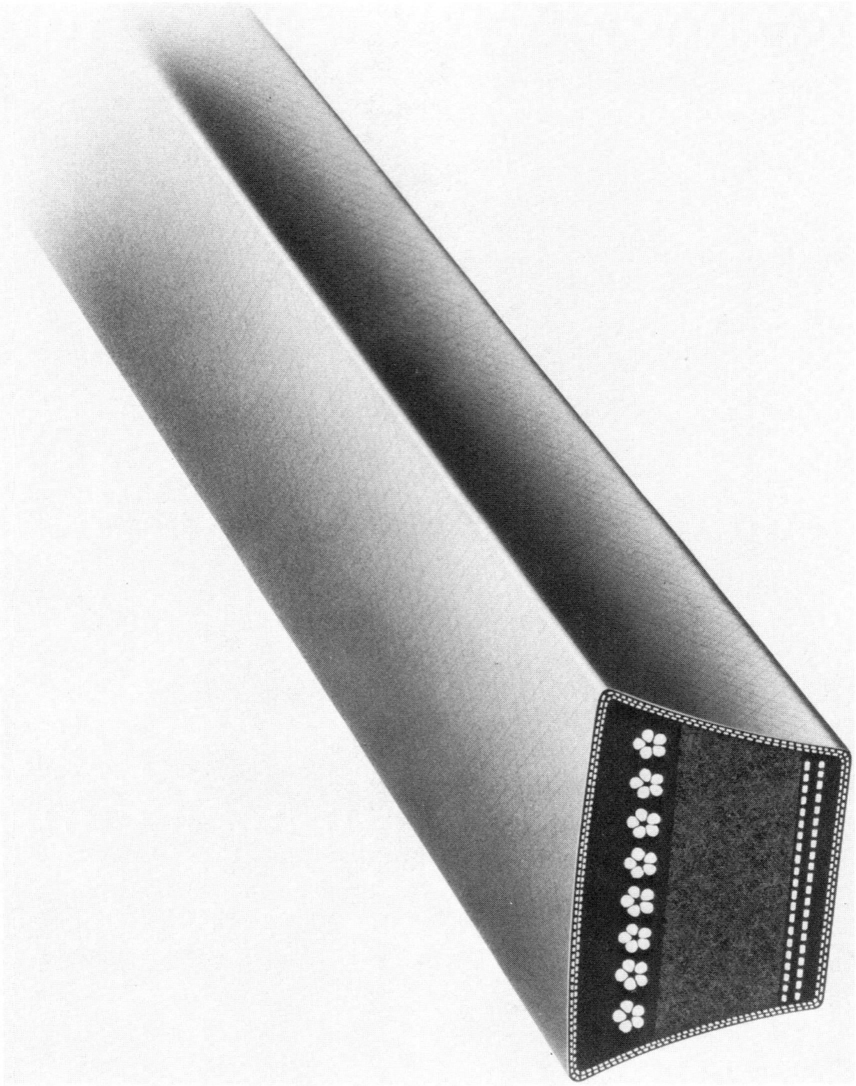

(a)

FIGURE 6 (a) Banded industrial V-belts in single unit cord and (b) multiple cord construction.

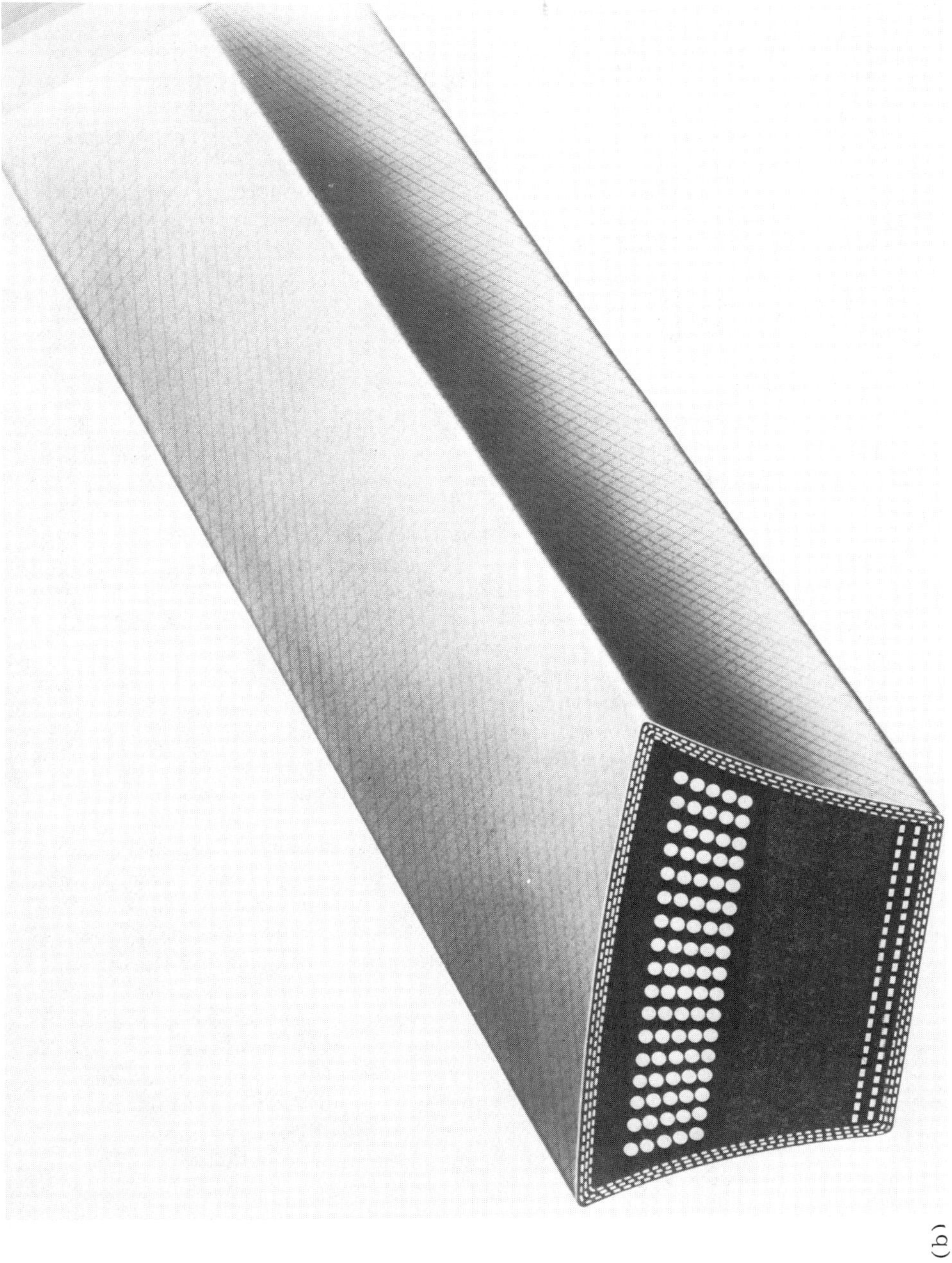

(b)

2. Molded-Notch or Cog

The molded-notch or cog-type industrial belt is gaining popularity because of its higher power transmission capability – particularly on small sheave diameters. Since this belt is a high-capacity belt, it is generally made with a single-unit tensile member. As with the banded belt, the most popular tensile cord material is polyester.

Figure 7 shows the construction details of the molded-notch-type belt. As the name implies, the notches are molded into the undercord of the belt during the curing process. Note that there is no band or cover around the belt. The tensile cord and other components of the belt are held in place through the effective use of high quality adhesion gums.

3. Joined Type

Rather than operating individual belts in multiple on a drive, it is often advantageous to apply a joined belt which consists of the individual belts banded together with a tieband across the top. This type of belt is shown in Figure 8.

As shown in Figure 8, the joined belt can be either a banded (a) or molded notch (b) construction. The basic "V" portion of the belt is essentially the same as the individual belts of the banded or molded notch constructions discussed earlier in this chapter. A tieband across the belts holds them together.

Joined belts are used on specific drives where individual belts whip or vibrate due to impulse load caused by torque or velocity fluctuations of the driver or driven machines. On drives with this phenomenon, individual belts operating in multiple can interfere with each other as they whip or vibrate. This can cause the belts to turn over in the groove or to come off the drive.

When two or more V-belts are joined together with a tieband across the top of the belts, the lateral rigidity of the joined belt becomes much greater than that of an individual belt. The lateral movement of the belt is significantly reduced, thereby eliminating the twisting or misalignment which can cause the belt to turn over or come off the drive.

Although joined belts cost a little more than individual belts, they are used on a regular basis on drives with impulse or shock loading such as piston pumps, air compressors, and rock crushers. The joined belt has solved belt instability problems on these types of drives and has contributed significantly to trouble-free power transmission since it was introduced by The Gates Rubber Company in 1966.

4. High Modulus Type

Industrial type V-belts have been produced on a limited basis with tensile cords of a high modulus material for many years. These belts

FIGURE 7 Bandless molded notch or cog industrial V-belt.

(a)

(b)

FIGURE 8 Joined type industrial V-belt in (a) banded and (b) bandless molded notch construction.

have been used primarily on special applications where the amount of movement of the driver or driven machine is limited and the low elongation of the high modulus tensile cord is advantageous.

While a drive with belts having a high modulus tensile cord will require less total movement of the driver or driven machine to maintain proper belt tension, they will usually require more frequent adjustment to compensate for belt and sheave wear. Since there is very little elongation of the belt as tension is applied, a slight amount of seating of the belt in the groove due to wear will relax the tension very quickly.

For many years, steel cable tensile cords were used in this type of belt. Steel cable cords have a high modulus as well as a high breaking strength. They do not have a good resistance to flex-fatiguing, however.

Fiberglass has also been used for this type of belt because it is somewhat better than steel cable from a flex-fatigue standpoint.

Modern day high modulus belts use aramid fibers for the tensile cords, combining the properties of high modulus with reasonably good flex-fatigue properties. Through the proper use of aramid fiber tensile cords, it is possible to increase the power transmission capabilities of a V-belt. They are being used quite extensively on drives on slush pumps for oil drilling rigs and other applications where high loads must be transmitted in limited space.

E. Power-Rating Formulas

1. Basic Formulas

The power formulas in use today throughout most of the industrial world are developed from the basic load-life analysis work begun by The Gates Rubber Company in the late 1940s and early 1950s. Power ratings based on this work were first published in an industry standard by the Rubber Manufacturers Association and the Mechanical Power Transmission Association in 1955.

The basis of this early work related the flex-fatigue life of a V-belt to the cyclic repetition of peak stresses in the belt caused by the tension needed to turn the sheaves against the torque load, the bending tension imposed as the belt goes around the sheave, and the centrifugal tension which is developed as the belt travels in a circular path around the sheave. The peak tension is the sum of these three tensions and the higher the peak tension, the fewer cycles the belt will complete before failure.

Since the peak stress or tension in any segment of belt increases to a maximum as the segment enters or leaves a sheave on the tight side of the drive (bending + tight side + centrifugal tensions), a stress cycle occurs each time the segment passes over a sheave. Knowing the belt speed of a drive and the length of the belt, it is possible to relate the load-life characteristics to some desired service life which is suitable for a drive design for industrial-type service.

A service life requirement of from 3–5 years has been established through experience as being practical for industrial drives. With this particular service life requirement in mind, and considering that the tight side tension component of the peak tension can be related to the power transmitted, a power-rating formula can be developed.

The first formulas were in HP units and were of the following basic form:

$$P_r = \left[C_1 \left(\frac{S}{L}\right)^{0.09} - \frac{C_2}{d_a} - C_3 S^2 \right] S \tag{1}$$

where

P_r = HP rating of one belt
S = belt speed (ft/min/1000)
L = belt length (in.)
d_a = equivalent or average diameter of both sheaves (in.)
C_1, C_2, C_3 = parameters based on specific characteristics of the belt

The equivalent diameter (d_a) is defined as the diameter that will have the same flex-fatiguing effect on the belt when considered for both sheaves as the actual two sheave diameters will have. The smaller the sheave diameters, the higher the bending stresses in the belt and the lower the capacity left in the belt to transmit power.

The third term is the centrifugal tension which is a function of the mass of the belt per unit length and the square of the belt speed. Again this is a parasitic stress or tension and subtracts from the basic capacity of the belt.

Formulas of this type were accepted as the standard of the industry and published for the classical industrial belts of cross sections A, B, C, D, and E in the 1955 edition of the joint standard by the Rubber Manufacturers Association and the Mechanical Power Transmission Association. The values for the parameters C_1, C_2, and C_3 were established by imperial methods based on laboratory testing under actual loaded conditions. (See Chapter 13 for a more detailed description of laboratory testing of V-belts.) Table 5 shows an example

TABLE 5 Horsepower Ratings for "A" Section Classical V-Belt From 1955 Edition of RMA/MPTA Standard

Belt Speed (Feet per Min.)	Equivalent Diameter (d_e)											
	2.6	2.8	3.0	3.2	3.4	3.6	3.8	4.0	4.2	4.4	4.6	4.8	5.0 and over
200	.16	.18	.20	.21	.23	.24	.25	.26	.27	.28	.28	.29	.30
400	.26	.30	.34	.37	.40	.42	.44	.46	.48	.50	.51	.53	.54
600	.34	.40	.46	.51	.55	.59	.62	.65	.68	.70	.72	.74	.76
800	.41	.50	.57	.63	.69	.74	.78	.82	.86	.89	.92	.95	.97
1000	.47	.57	.66	.74	.81	.88	.93	.98	1.03	1.07	1.11	1.14	1.17
1200	.52	.64	.75	.85	.93	1.01	1.07	1.13	1.19	1.24	1.28	1.32	1.36
1400	.56	.70	.83	.94	1.04	1.13	1.20	1.27	1.34	1.40	1.45	1.50	1.54
1600	.59	.76	.90	1.03	1.14	1.24	1.33	1.41	1.48	1.55	1.61	1.66	1.71
1800	.61	.80	.95	1.10	1.23	1.34	1.44	1.53	1.61	1.69	1.75	1.82	1.87
2000	.62	.83	1.01	1.17	1.31	1.43	1.55	1.65	1.74	1.82	1.89	1.95	2.03
2200	.63	.86	1.06	1.23	1.38	1.52	1.64	1.75	1.85	1.94	2.02	2.10	2.17
2400	.62	.87	1.09	1.28	1.44	1.59	1.73	1.85	1.95	2.05	2.14	2.23	2.30
2600	.60	.87	1.11	1.31	1.50	1.66	1.80	1.93	2.05	2.16	2.25	2.34	2.43
2800	.57	.87	1.12	1.34	1.54	1.71	1.87	2.01	2.13	2.25	2.35	2.45	2.54
3000	.53	.85	1.12	1.36	1.57	1.75	1.92	2.07	2.20	2.33	2.44	2.54	2.64
3200	.48	.82	1.11	1.36	1.58	1.78	1.96	2.12	2.26	2.40	2.52	2.63	2.73
3400	.42	.77	1.08	1.35	1.59	1.80	1.99	2.16	2.31	2.45	2.58	2.70	2.80
3600	.34	.72	1.04	1.33	1.58	1.80	2.00	2.18	2.35	2.50	2.63	2.75	2.87
3800	.25	.65	.99	1.29	1.56	1.80	2.01	2.20	2.37	2.53	2.67	2.80	2.92
4000	.15	.57	.93	1.25	1.53	1.77	2.00	2.20	2.38	2.54	2.69	2.83	2.96
4200	.03	.47	.85	1.18	1.48	1.74	1.97	2.18	2.37	2.54	2.70	2.85	2.98
4400	—	.36	.76	1.11	1.41	1.69	1.93	2.15	2.35	2.53	2.70	2.85	2.99
4600	—	.23	.65	1.01	1.33	1.62	1.87	2.10	2.31	2.50	2.67	2.83	2.98
4800	—	.09	.52	.90	1.24	1.54	1.80	2.04	2.26	2.46	2.64	2.80	2.95
5000	—	—	.38	.77	1.12	1.43	1.71	1.96	2.19	2.39	2.58	2.75	2.91
5200	—	—	.22	.63	.99	1.32	1.61	1.87	2.10	2.32	2.51	2.69	2.85
5400	—	—	.04	.47	.85	1.18	1.48	1.75	2.00	2.22	2.42	2.61	2.78
5600	—	—	—	.29	.68	1.03	1.34	1.62	1.87	2.10	2.31	2.51	2.68
5800	—	—	—	.09	.49	.85	1.18	1.47	1.73	1.97	2.18	2.38	2.57
6000	—	—	—	—	.29	.66	.99	1.29	1.56	1.81	2.04	2.24	2.43

Notes:

1. Some of the smaller d_e values shown above are obtainable only with sheave diameters smaller than the minimum recommended in Table 1. Ratings for such small equivalent diameters are given in case it is necessary to use diameters smaller than the recommended minimum.
2. Belt speed is figured using the pitch diameter and R.P.M. of one of the sheaves on the drive.
3. If belt speeds are over 5000 feet per minute the sheaves may require special construction or special material. They may also require dynamic balancing. Such applications should be referred to the manufacturer.

of these early HP ratings as published in the standard. Note that the ratings are a function of belt speed and equivalent diameter.

Continued testing and analysis of test results confirmed that these early power rating formulas were basically correct but that a slightly different form more nearly represented the capability of a V-belt to transmit power over sheaves of a given size and at a given belt speed. This new form was developed by The Rubber Company in the early 1950s and was published as an industry standard in 1962. This basic power rating equation is:

$$P_r = \left[K_1 - \frac{K_2}{d_a} - K_3S^2 - \log\left(\frac{S}{L}\right)\right] S \qquad (2)$$

where the variables are the same as those shown for Equation (1).

2. Speed Ratio Correction

In Equations (1) and (2) the equivalent diameter takes into consideration that the drive may have two unequal sheave diameters or, in other words, that it has a speed ratio different than one. Since the bending tension induced in a belt as it conforms to the diameter of the pulley is greater for a smaller pulley diameter than it is for a larger, the HP transmission capability of a belt is reduced as the sheave diameter becomes smaller.

On a drive with two sheaves, a segment of V-belt can be thought of as operating one half of the time on the smaller diameter and one half of the time on the larger diameter. Since the belt has more HP transmission capacity on the larger diameter than on the smaller, a balance has to be determined between the two diameters which represents the HP transmission capability of the drive as a whole. This balance can be achieved through the use of an equivalent diameter (d_a) in the bending tension term of the equation.

Equation (3) gives the relationship between the speed ratio of the drive and the smaller of the two sheave diameters.

$$d_a = \frac{d}{1 + \frac{d}{A}\log\left[\frac{1 + 10^{-A/d(R-1/R)}}{2}\right]} \qquad (3)$$

where

d_a = equivalent diameter (in.)
d = pitch diameter of smaller sheave (in.)
R = speed ratio of drive
A = parameter based on specific characteristics of belt

To make the HP equation easier to use, Equation (3) was transformed into a form using an add-on HP rather than an equivalent diameter. At the same time, the variables of small pitch diameter (d) and the rpm of the faster shaft/1000 (r) were used instead of the belt speed/1000 (S). The relationship among these variables is:

$$S = \frac{\pi}{12} dr \tag{4}$$

where

S = belt speed (ft/min/1000)
d = pitch diameter of smaller sheave
r = rpm/1000 of faster shaft

Using the equation for belt speed (4) along with the add-on HP for speed ratio, Equation (2) can be written as:

$$P_r = \left[K_1' - \frac{K_2'}{d} - K_3'(dr)^2 - K_4' \log\left(\frac{dr}{L}\right)\right] dr + \Delta P_r \tag{5}$$

where

P_r = HP rating of one belt
d = pitch diameter of smaller sheave (in.)
r = rpm/1000 of faster shaft
L = belt length (in.)
ΔP_r = add-on HP rating for speed ratio
K_1', K_2', K_3', K_4' = parameters based on specific characteristics of the belt

Equation (5) will yield the same results as Equation (2) but it is somewhat more convenient to use since the values of (d) and (r) are basic known quantities for a specific drive. They can be used directly in the HP-rating formula without having to calculate the belt speed with Equation (4).

Equation (5) uses the pitch diameter of the smaller sheave (d) in the term for the bending tension (K_2'/d). The additional HP rating that results from one sheave being larger than the other is represented by the add-on HP rating (ΔP_r) which is calculated from Equation (6):

$$\Delta P_r = (K_4')(dr) \log\left[\frac{1 + 10^{-K_2'/dK_4'(1/SR - 1)}}{2}\right] \tag{6}$$

where

ΔP_r = add-on HP rating for speed ratio
d = pitch diameter of smaller sheave (in.)
r = rpm/1000 of faster shaft
SR = speed ratio of drive
K_2', K_4' = parameters based on specific characteristics of the belt

Note that in this equation (as well as Equation (3) for equivalent diameter), the speed ratio must always be equal to or greater than 1. Therefore, the speed ratio is defined as the rpm of the faster shaft divided by the rpm of the slower shaft, regardless of which one is the driver or the driven.

So that the add-on HP rating equation is easier to use and to tabulate, the small sheave diameter (d) can be replaced by an average small sheave diameter for the specific belt cross section being considered without significantly affecting the accuracy of the equation. When this is done, Equation (6) becomes:

$$\Delta P_r = K_2' r \left[1 - \frac{1}{K_{SR}}\right] \tag{7}$$

where the variables and parameters are the same as those listed in Equation (6), except

K_{SR} = speed ratio factor based on an average pitch diameter for a specific belt

Replacing ΔP_r in Equation (5) with the right hand side of Equation (7) yields the following HP-rating equation:

$$P_r = \left[K_1' - \frac{K_2'}{d} - K_3'(dr)^2 - K_4' \log\left(\frac{dr}{L}\right)\right] dr + K_2' r \left[1 - \frac{1}{K_{SR}}\right] \tag{8}$$

where

P_r = HP rating of one belt
d = pitch diameter of smaller sheave (in.)
r = rpm/1000 of faster shaft
L = belt length (in.)
K_{SR} = speed ratio factor
K_1', K_2', K_3', K_4' = parameters based on specific characteristics of the belt

3. Belt Length Correction

For the convenience of the designer using industrial HP ratings, it is convenient to tabulate the main portion of the ratings in terms of the smaller sheave diameter and the rpm of the faster shaft. The add-on HP for speed ratio can then be tabulated in terms of speed ratio and the rpm of the faster shaft.

In order to make these convenient tables, it is necessary to develop a separate method of calculating the effect of belt length on the HP rating.

Belt length is important in determining the power transmission capabilities of a belt because it determines the number of flex cycles a given segment of belt is subjected to over the 3–5-year industrial life span when the belt is traveling at a given speed. The longer the belt length, the fewer the flex cycles and the more heavily the belt can be loaded.

The belt length can be taken out of Equation (8) and considered as a separate add-on rating as we have done with the add-on HP rating for speed ratio. However, practice has found it more convenient to use a HP-rating multiplier to adjust the rating for the length of the belt on the drive.

Equation (9) shows the accepted method by which the length correction factor is calculated. This factor is used as a multiplier to adjust for belt length and results in a corrected HP rating which is very close to the one calculated from Equation (8). A mathematical approximation is necessary in order to use a multiplier rather than an add-on correction but the difference is not significant in the overall design of a drive.

$$C_L = \frac{P + \log L_1}{P + \log L_o} \tag{9}$$

where

C_L = belt length correction factor
L_o = base belt length on which the basic ratings are based (in.)
L_1 = belt length of specific drive (in.)
P = parameter based on average small pitch diameter, average rpm and specific characteristics of the belt

Note that the basic HP rating as calculated from Equation (8) is based on a base length of belt (L_o). If the belt length of the drive is equal to the base length, the length correction factor (multiplier) is 1.00. If the specific belt length is shorter than the base length, the correction factor is less than 1.00. If the specific length is longer, the correction factor is larger than 1.00.

4. Arc of Contact Correction

As discussed in Chapter 2, a V-belt needs to operate at a higher tension (lower tension ratio) if the arc of contact is less than 180°. This is to assure that the drive will transmit the required HP without slipping.

The equation for determining the required tension ratio for any specific arc of contact from Chapter 2 is:

$$R = e^{0.51230\,\theta} \tag{10}$$

where

R = tension ratio, tight side tension/slack side tension
θ = arc of contact on the small pulley (rad)

The basic HP-rating equation is based on an arc of contact of 180° (π rad) since this is the way in which belts are tested in the laboratory to determine their load-life characteristics. When V-belts are applied to an application where the arc of contact is less than 180°, the increased tension required in the drive will reduce the amount of tension capacity left in the belt to transmit HP. Therefore, the HP rating of the belt must also be corrected for arc of contact.

From the equations in Chapter 2, we can see that the tight side tension required to transmit HP with a V-belt drive is:

$$T_T = \frac{126051\ \mathrm{HP}}{(d)(\mathrm{rpm})(1 - (1/e^{0.5123\,\theta}))} \tag{11}$$

where

T_T = tight side tension (lbf)
HP = horsepower being transmitted
d = pitch diameter (in.)
rpm = revolutions per minute of shaft
θ = arc of contact (rad)

Since $e^{0.5123\,\theta} = R$, we can write the equation:

$$T_T = \frac{126\ HP}{(d)(r)(1 - (1/R))} \tag{12}$$

where

$r = rpm/1000$

If we arrange Equation (12) in terms of a multiplier to be applied to the HP, we would have:

$$T_T = \frac{126\ (HP)(R/(R-1))}{(d)(r)} \tag{13}$$

From Equation (10), the multiplier (R/R-1) can be calculated for any arc of contact as shown in Table 6.

But since the base HP was established for an arc of contact of 180° or π rad, it would be more convenient to adjust the multiplier so that it would be 1.00 when the arc of contact is 180°. This can be easily done if we equate the multiplier R(R-1) to a new arc correction factor (commonly called Factor G) so that when R/(R-1) = 1.25, Factor G = 1.00.

$$G = 1.00 = (0.8)(1.25) = (0.8)(R/(R-1)) \tag{14}$$

The values for the arc correction factor (G) are given in Table 7 for various arcs of contact.

For industrial-type drives, which almost always use only two sheaves, the arc of contact can be calculated for the small sheave diameter by this equation:

$$\theta = 2 \cos^{-1} \left(\frac{D-d}{2\ C}\right) \tag{15}$$

where

θ = arc of contact on the small sheave (deg)
D = diameter of large sheave (in.)
d = diameter of small sheave (in.)
C = center distance between shafts (in.)

Table 7 also includes values for (D-d)/C so that factor G can be read directly from these values which geometrically describe the drive.

TABLE 6 Arc Correction Factor, R/(R-1)

Arc of contact (deg)	R/(R-1)
50	2.77
60	2.41
70	2.15
80	1.96
90	1.81
100	1.69
110	1.60
120	1.52
130	1.46
140	1.40
150	1.35
160	1.31
170	1.28
180	1.25
190	1.22
200	1.20
210	1.18
220	1.16
230	1.15
240	1.13
250	1.12
260	1.11
270	1.10
280	1.09
290	1.08
300	1.07

5. Summary of Power-Rating Formulas

In summary, the power-rating equations in use today for all types of industrial multiple V-belts (both classical and narrow cross sections) are the following:

$$P_r = \left[K_1' - \frac{K_2'}{d} - K_3'(dr)^2 - K_4' \log dr\right] dr + K_2' r \left(1 - \frac{1}{K_{SR}}\right) \qquad (16)$$

TABLE 7 Arc of Contact Factors (Factor G)

(D-d)/C	Arc of contact (θ) on small sheave degrees	Factor G
0.00	180	1.00
0.10	174	0.99
0.20	169	0.97
0.30	163	0.96
0.40	157	0.94
0.50	151	0.93
0.60	145	0.91
0.70	139	0.89
0.80	133	0.87
0.90	127	0.85
1.00	120	0.82
1.10	113	0.80
1.20	106	0.77
1.30	99	0.73
1.40	91	0.70
1.50	83	0.65

where

P_r = HP rating for one belt of average length operating with a 180° arc of contact
d = pitch diameter of smaller sheave (in.)
r = rpm of faster shaft/1000
K_{SR} = speed ratio factor
K_1', K_2', K_3', K_4' = parameters depending on the physical properties of a specific V-belt type and cross section

The HP rating calculated from the above Equation (16) will be multiplied by the following correction factors:

$$C_L = \frac{P + \log L_1}{P + \log L_o} \tag{17}$$

where

C_L = belt length correction factor
L_o = base belt length on which the basic ratings are based (in.)
L_1 = specific belt length for the drive
P = parameter based on average drive conditions and belt characteristics

$$\text{Factor G} = 0.8\left(\frac{R}{R-1}\right) \tag{18}$$

where

Factor G = arc of contact correction factor
R = tension ratio = $e^{0.5123\,\theta}$ where θ = arc of contact on the small sheave diameter (rad)

F. Drive Design Procedure

1. Selection of Belt Type and Cross Section

Either the classical industrial-type belt or the narrow industrial-type belt can be used for most industrial multiple V-belt drives. These belts are described earlier in this chapter.

The narrow belt type is usually considered for new types of drives where there is not the incentive of maintaining consistency in drive components to simplify inventories, maintenance, and operating procedures.

This is because the narrow belts have a higher power transmission capacity per inch of width and will usually operate on smaller sheave diameters. This will result in a saving of space which is often very important to the machine designer.

The narrow industrial-type belt will usually result in a drive that is lower in cost and will often allow the use of higher speed, lower cost electric motors. This is because the narrow belts will operate on smaller diameters and therefore permit higher speed ratios between the motor and the driven machine.

The easiest way to determine the required cross section to use in transmitting the required HP is to use the cross section selection charts shown in the industrial drive design manuals published by the various V-belt and sheave manufacturers. Examples of these charts are shown in Figure 9 (for narrow belts) and Figure 10 (for classical belts).

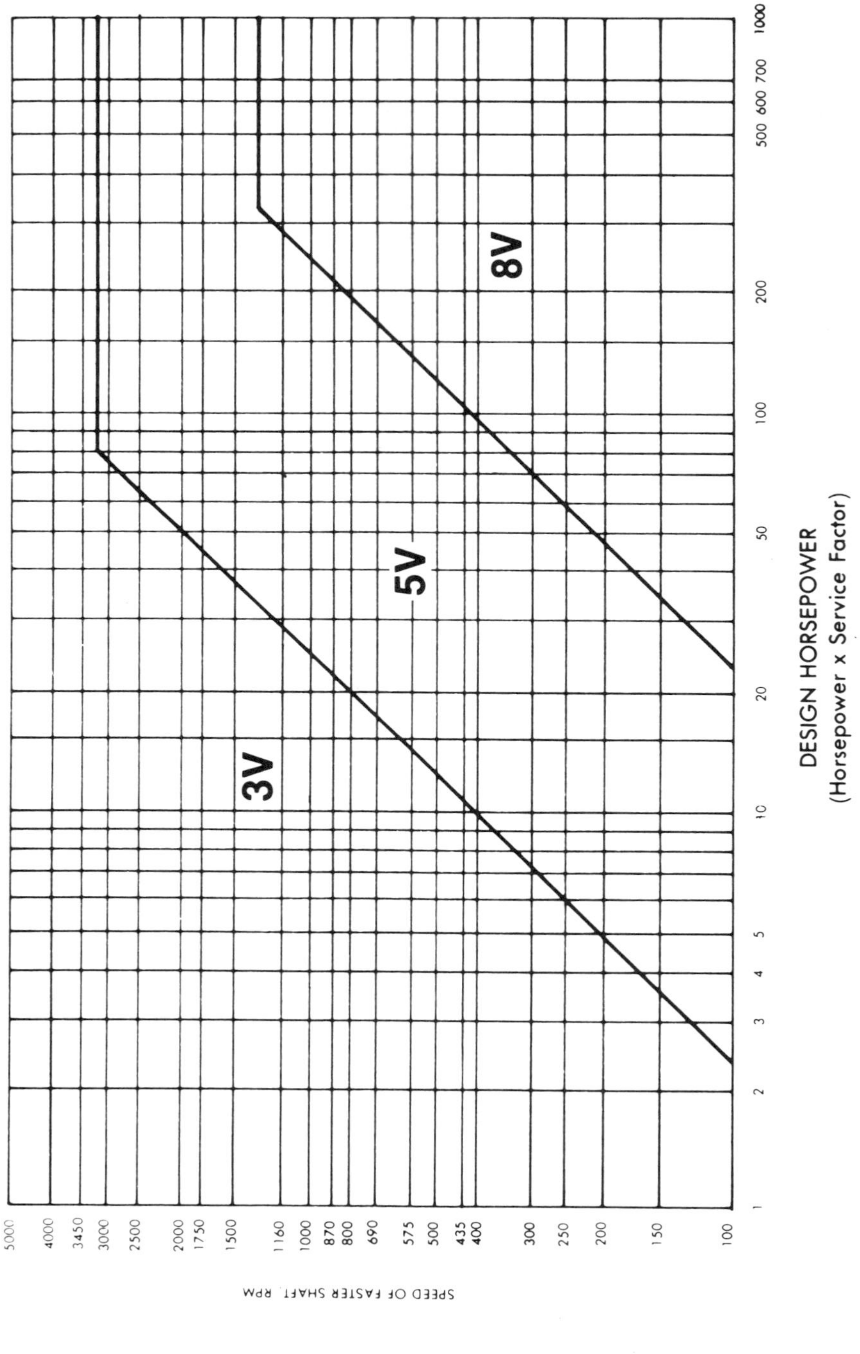

FIGURE 9 Cross section selection chart for narrow industrial belts.

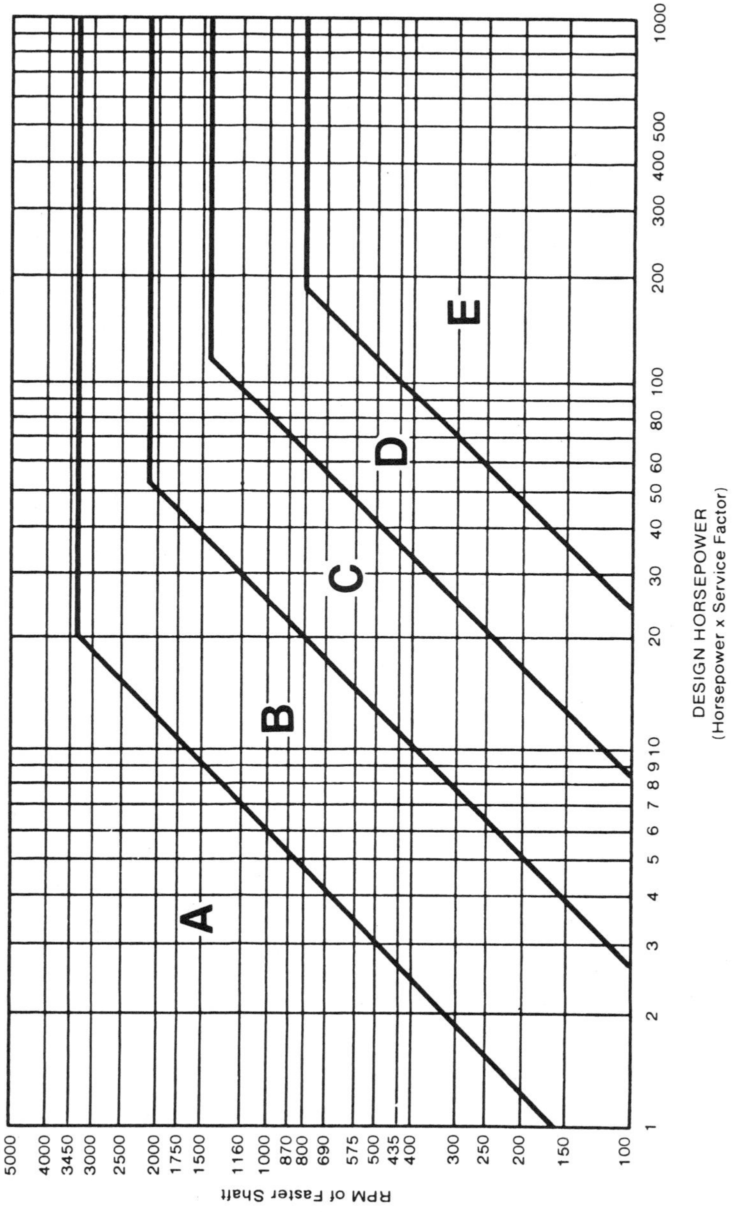

FIGURE 10 Cross section selection chart for classical industrial belts.

It is recommended that the designer use the drive design manuals that are available without cost from the V-belt and sheave manufacturers. It is not the intent of this section to duplicate the commercially available design manuals but rather to explain in more detail the procedures used in industrial drive design so that the designer will have a better understanding of the engineering apsects of his task.

It will be noted that the divisional lines between the various belt cross sections of these charts are constant torque lines. This is because the scale of both the X and Y axes are logarithmic and the divisional lines are at a 45° angle to either axis.

A V-belt drive is essentially a constant torque device so that this assumption is reasonably valid. To illustrate this more clearly, the actual HP/rpm of faster shaft relationship is shown in Figure 11 for the individual diameters of the 3V, 5V, and 8V narrow industrial cross sections.

Note that the specific diameter curves are essentially straight lines at the lower rpm values but that they have a slightly steeper slope than the constant torque line. This is because of the term "K_4' log dr" in the power rating formula which compensates for the speed of the belt in relation to the number of flex-fatigue cycles the belt will be subjected to in the 3–5-year service life.

At the higher rpm values, the curves begin to bend to the left and eventually an rpm is reached where the HP the belt can transmit is decreasing instead of increasing. This is because of the centrifugal tension that builds in the belt as the rpm increases (K_3' $(dr)^2$ term in the formula). Since the centrifugal tension is a function of the square of the diameter times rpm, the centrifugal tension can become very high as rpm increases and there will be little or no capacity left in the belt to transmit HP.

These individual diameter lines are based on the HP rating per belt so the point where one section ends and the larger section begins will not reflect the divisional line between the two cross sections on the finding chart. This is because the cross section finding charts reflect the stock number of grooves that are available in each section. For example, the larger 3V sheave diameters are available in up to 10 grooves for which the HP rating would be higher than the smallest 5V sheave diameter using the minimum number of grooves which is two. This means there is a significant area in which any two cross sections overlap. It is advisable to explore both cross sections to determine which is best suited for the drive.

2. Design Horsepower

The cross section selection charts (Figures 9 and 10) are based on design HP. The design HP is rarely the actual HP to be transmitted

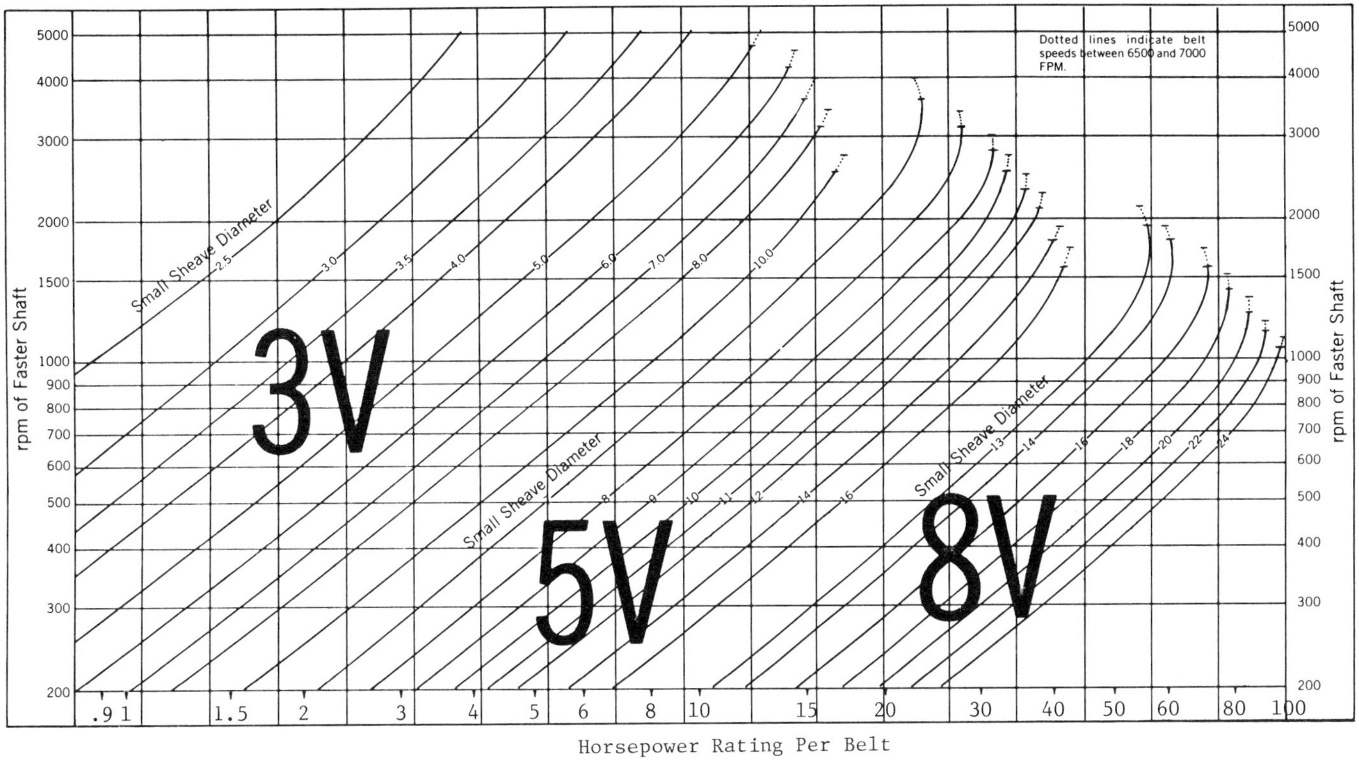

FIGURE 11 Cross section selection chart showing lines for individual sheave diameters.

on a steady state basis because the power surges or peaks during the operating cycle of the machine must be taken into consideration very carefully.

Since the relationship between the fatigue life of a V-belt and the peak tensions the belt is subjected to is an exponential function, the detrimental effect of peak or shock loads on a drive can be quite significant. Therefore the goal is to select a design HP that would give the same belt life if the drive is operated only at the design HP that the actual drive will give when operating under normal conditions.

Most industrial drives are used with electric motors as the prime mover. Most electric motors develop starting torques which are several times the normal running torque. Therefore if a drive starts frequently, the design HP would need to be significantly higher than the rated HP of the motor to compensate for the high starting torque.

3. Service Factors

The design HP of a drive is established by selecting the proper service factor for the prime mover and driven machine combination as well as the operating conditions of the drive. The normal steady state load on the drive (usually the nameplate HP rating of an electric motor) is multiplied by the service factor to obtain the design HP.

Table 8 gives a listing of various types of prime movers and drive machines and a suggested service factor relating to combinations of these units as well as operating conditions. This table is taken from the joint standard on narrow belts (IP-22) published by the Rubber Manufacturers Association, the Mechanical Power Transmission Association, and the Rubber Association of Canada.

It will be noted in Table 8 that prime movers that are capable of developing high starting torques require a higher service factor. The same is true concerning driven machines such as hammer mills that impose a shock load on the V-belt drive. While it is impossible to list all types of driven machines in a table such as this, the designer can usually relate the operation of his machine to one of those in the table and determine an acceptable service factor.

Besides those operating conditions specifically mentioned in the table, the designer should consider:

Frequency of peak loads such as starting torques
Adverse environmental conditions such as heat or dust
Pulsating or shock loads which must be transmitted by the belt

4. Geometric Drive Design

The geometric design of the drive consists of selecting the most suitable sheave diameters and the corresponding belt length that will

TABLE 8 Service Factors for Industrial-Type Drives

TYPES OF DRIVEN MACHINES	TYPES OF DRIVING UNITS: AC Motors; Normal Torque, Squirrel Cage, Synchronous and Split Phase. DC Motors; Shunt Wound. Multiple Cylinder Internal Combustion Engines.			TYPES OF DRIVING UNITS: AC Motors; High Torque, High Slip, Repulsion-Induction, Single Phase Series Wound and Slip Ring. DC Motors; Series Wound and Compound Wound. Single Cylinder Internal Combustion Engines. Line Shafts. Clutches.		
	Intermittent Service (3-5 Hours Daily or Seasonal)	Normal Service (8-10 Hours Daily)	Continuous Service (16-24 Hours Daily)	Intermittent Service (3-5 Hours Daily or Seasonal)	Normal Service (8-10 Hours Daily)	Continuous Service (16-24 Hours Daily)
Agitators for Liquids Blowers and Exhausters Centrifugal Pumps and Compressors Fans up to 10 HP Light Duty Conveyors	1.0	1.1	1.2	1.1	1.2	1.3
Belt Conveyors For Sand, Grain, etc. Dough Mixers Fans Over 10 HP Generators Line Shafts Laundry Machinery Machine Tools Punches-Presses-Shears Printing Machinery Positive Displacement Rotary Pumps Revolving and Vibrating Screens	1.1	1.2	1.3	1.2	1.3	1.4
Brick Machinery Bucket Elevators Exciters Piston Compressors Conveyors (Drag-Pan-Screw) Hammer Mills Paper Mill Beaters Piston Pumps Positive Displacement Blowers Pulverizers Saw Mill and Woodworking Machinery Textile Machinery	1.2	1.3	1.4	1.4	1.5	1.6
Crushers (Gyratory-Jaw-Roll) Mills (Ball-Rod-Tube) Hoists Rubber Calenders-Extruders-Mills	1.3	1.4	1.5	1.5	1.6	1.8

Note: The use of a service factor of 2.0 is recommended for equipment subject to choking. For devices used in grain and milling properties, refer to Mill Mutual Fire Prevention Bureau Bulletin VB-601.
Source: Standard IP-20 published jointly by the Rubber Manufacturers Association, the Mechanical Power Transmission Association, and the Rubber Association of Canada.

satisfy the space requirements and give the required speed of the driven unit.

Using standard sheave diameters: V-belt and sheave manufacturers publish drive selection tables to help the designer select the proper sheave diameters and belt length. The tables show the driven rpm that can be obtained from the popular standard electric motor speeds using various combinations of standard sheave diameters. These tables also show a speed ratio column which can be used when the prime mover is not a standard electric motor.

Table 9 is an example of a drive selection table. Note that there are often several choices of sheave diameters that will give driven speeds within an acceptable range of the desired speed. The driven speeds shown in selection tables are the calculated values using the ratio of the pitch diameters and do not take into consideration belt creep as discussed in Chapter 2. Belt creep is usually about 0.5%.

Most applications do not require that the driven speed be at an exact value so the designer should select a pair of sheave diameters that give a driven rpm close to his required value but also a pair that is compatible to his other drive requirements.

For example, a designer desires to drive a machine at 375 rpm with a 1750 rpm induction motor. As shown in Table 9, he could select a 4.12 in. driver with a 19.00 in. driven for a calculated rpm of 376. If he wants to make his machine as compact as possible, he may select a 2.35 in. motor sheave with a 10.60 in. driven sheave for a calculated driven speed of 381 rpm. Although the smaller diameter drive will require a greater number of belts, it will probably be lower in cost. This aspect will be discussed more thoroughly in Chapter 15.

Once a pair of sheave diameters have been selected, the drive selection tables shown in Table 9 can be used to find the belt length that will give a center distance between driver and driven shafts that is the closest to the desired value. The standard belt number is shown as a column heading above the center distance value corresponding to each diameter pair. For the diameters of 2.35 and 10.60 in. used in the example, a 3VX500 (50 in. effective length) would be a good choice if a center distance of 14 in. is desired.

Non-standard drives: While drive selection tables as illustrated in Table 9 will be used for all geometric design involving standard components, sometimes it is necessary to use nonstandard sheave diameters or belt lengths. This occurs when one of the sheaves is already on hand or when a special sheave needs to be used to be compatible with other mechanical requirements of the driven machine such as flywheel inertia.

It is necessary to calculate the desired speed ratio for a nonstandard drive by dividing the rpm of the faster shaft by the rpm of the slower. This ratio is then applied to the sheave pitch diameters. As previously stated, the nomenclature of the classical industrial belts is the pitch diameter so these can be used directly. The nomenclature diameter for the narrow industrial belts is outside diameter so it is necessary to subtract 0.05 in. for 3V, 0.10 in. for 5V, and 0.20 in. for 8V to obtain the pitch diameter.

TABLE 9 Drive Selection Table For Geometric Design of Narrow Industrial V-Belt Drives

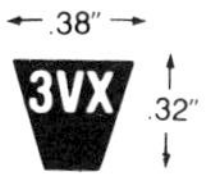

DriveN Speed For Motor Speed of 870 RPM	1160 RPM	1750 RPM	3450 RPM	Sheave Outside Diameters Small Sheave	Large Sheave	Speed Ratio	V-Belt Number and Center Distance 3VX 250	3VX 265	3VX 280	3VX 300	3VX 315	3VX 335	3VX 355	3VX 375	3VX 400	3VX 425	3VX 450	3VX 475	3VX 500	3VX 530	3VX 560
214	286	431	850	2.65	10.60	4.06								7.3	8.7	10.1	11.4	12.7	14.0	15.6	17.1
208	277	418	823	6.00	25.00	4.19															
207	276	416	819	8.00	33.50	4.21															
206	274	414	816	3.35	14.00	4.23													10.0	11.7	13.3
204	272	411	810	4.50	19.00	4.26															
202	269	406	800	2.50	10.60	4.31								7.4	8.8	10.2	11.5	12.8	14.1	15.7	17.2
193	258	389	767	3.15	14.00	4.50													10.1	11.8	13.4
193	258	389	767	5.60	25.00	4.50															
190	253	381	752	2.35	10.60	4.59								7.4	8.9	10.3	11.6	12	14.2	15.8	17.3
187	249	376	740	4.12	19.00	4.66															
184	245	370	729	3.00	14.00	4.73													10.2	11.9	13.5
183	244	368	726	5.30	25.00	4.75															
178	238	359	707	6.90	33.50	4.88															
177	236	356	703	2.20	10.60	4.91								7.5	9.0	10.3	11.7	13.0	14.3	15.9	17.4
173	230	347	685	5.00	25.00	5.04															
172	229	345	680	2.80	14.00	5.07													10.3	12.0	13.7
168	224	337	665	6.50	33.50	5.19															
165	221	333	656	3.65	19.00	5.26															
164	218	330	650	4.75	25.00	5.31															
162	216	326	642	2.65	14.00	5.37												8.9	10.4	12.1	1
155	207	312	615	4.50	25.00	5.61															
155	206	311	614	6.00	33.50	5.62															
153	204	308	606	2.50	14.00	5.69												9.0	10.		
152	202	305	601	3.35	19.00	5.74															
144	192	290	572	5.60	33.50	6.03															
143	191	288	568	2.35	14.00	6.07															
142	190	286	565	3.15	19.00	6.11															
142	189	285	563	4.12	25.00	6.13															

Source: Courtesy of The Gates Rubber Company.

When a pair of diameters that will give the proper driven speed and fit within the required space has been selected, the belt length can be calculated using either the exact belt length formula or the more commonly used approximate formula.

Figure 12 shows a sketch of a V-belt drive and the important dimensions used to derive the exact belt length formula. From this figure we have:

$$\frac{L_e}{2} = h + \left(\pi - \frac{\theta}{2}\right)\frac{D_e}{2} + \left(\frac{\theta}{2}\right)\left(\frac{d_e}{2}\right) \tag{19}$$

where

L_e = effective length of belt (in.)
h = span length (in.) = $(D_e - d_e/2) \tan \theta/2$
D_e = effective diameter of larger sheave (in.)
d_e = effective diameter of smaller sheave (in.)
θ = arc of contact of belt on smaller sheave (rad) (see Equation (15))

Substituting the equation for "h" and simplifying yields:

$$L_e = D_e - d_e \left(\tan \frac{\theta}{2} - \frac{\theta}{2}\right) + \pi D_e \tag{20}$$

where variables are as stated for Equation (19).

Equation (20) will give the exact belt length for a drive. However, since all belts and sheaves have reasonable manufacturing tolerances, an approximation to Equation (20) is normally used to calculate belt length. Equation (21) gives belt length values which are very close to the exact values and the equation is easier to use.

$$L_e = 2C + \frac{\pi}{2}(D_e + d_e) + \frac{(D_e - d_e)^2}{4C} \tag{21}$$

where

C = center distance between shafts (in.) and all other variables are as given in Equation (19)

Equation (21) is a series expansion of a form of Equation (20). It is terminated at the $(D_e - d_e)^2/4C$ term because adequate accuracy

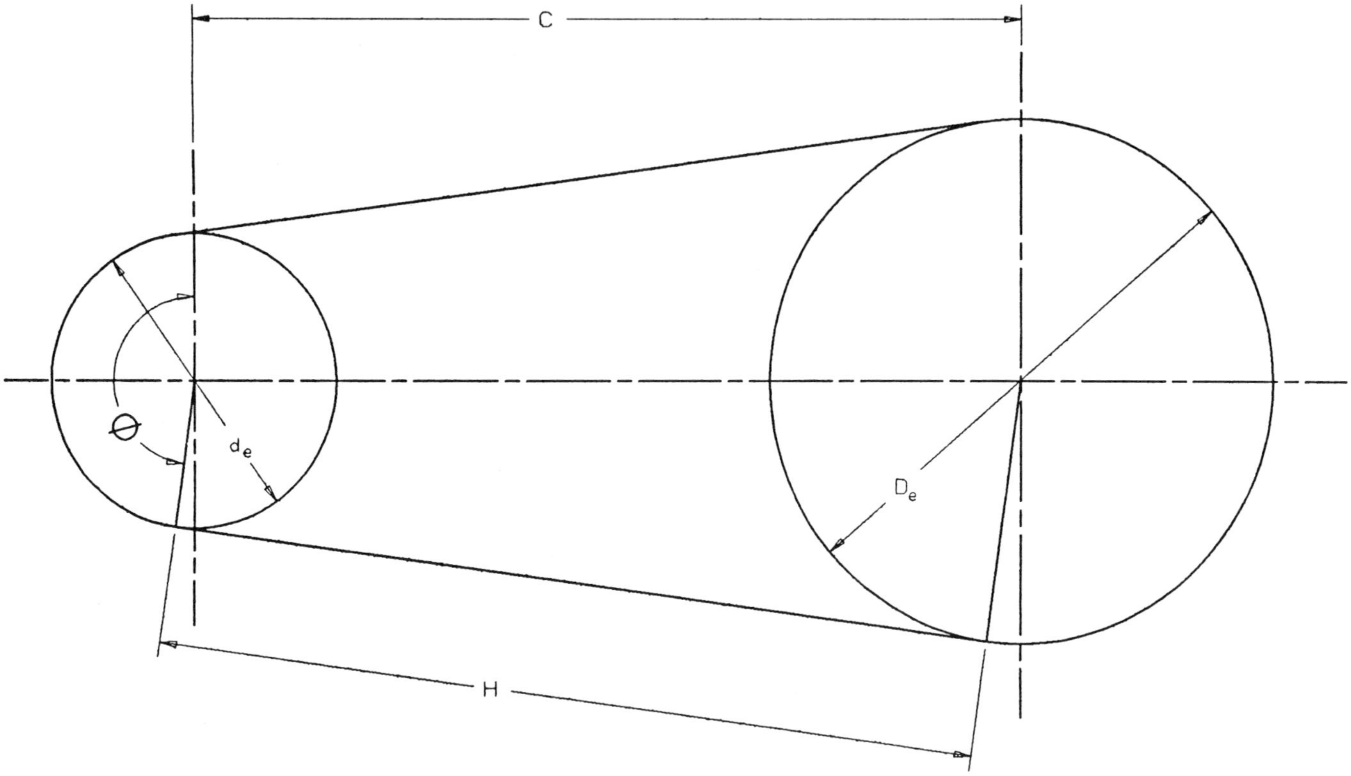

FIGURE 12 Geometric sketch of V-belt drive used in calculating belt length.

is obtained for V-belt drives. This equation is accepted by the industry as the standard formula for all power transmission belts except synchronous belts. It is published in the joint standards issued by the Rubber Manufacturers Association, the Mechanical Power Transmission Association, and the Rubber Association of Canada.

In industrial V-belt drive design, the designer is usually working with standard belt lengths, so he frequently needs to be able to calculate the center distance that is required for a given belt length and sheave diameters. A solution for using the belt length Equation (21) can be written in standard quadratic form:

$$8C^2 - \left[4L_e - \frac{\pi}{2}(D_e + d_e)\right] C - (D_e - d_e)^2 = 0 \qquad (22)$$

Solving for center distance, C:

$$C = \frac{K + \sqrt{K^2 - 32\,(D_e - d_e)^2}}{16} \qquad (23)$$

where

$K = 4L_e - \frac{\pi}{2}(D_e + d_e)$ and all other variables are as given in Equation (19)

This is the equation that is used by V-belt and sheave manufacturers in calculating the center distance values shown in the body of the drive selection tables as represented by Table 9.

5. Capacity Drive Design

In multiple industrial V-belts, capacity drive design involves determining the number of belts required for a drive that has been designed from a geometrical standpoint as discussed previously in this chapter.

If the number of belts required is more than the number of grooves available in stock sheaves or if the drive is too wide for the available space, it will be necessary to redesign the drive geometrically using large sheave diameters or perhaps even consider another cross section.

Using standard sheave diameters and motor speeds: The drive selection table shown in Table 9 from which the pair of sheave diameters and the belt length were selected is actually only the left hand portion of a typical selection table. The right hand portion of a drive selection table contains the HP rating for each specific pair of sheave diameters.

TABLE 10 Drive Selection Table For Use in Determining the Number of Belts Required For a Drive

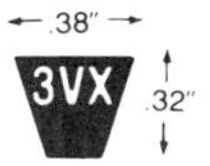

V-Belt Number and Center Distance														Sheave Outside Diameters		Rated HP per Belt (Including Allowance for Speed Ratio) RPM of Small Sheave			
3VX 670	3VX 710	3VX 750	3VX 800	3VX 850	3VX 900	3VX 950	3VX 1000	3VX 1060	3VX 1120	3VX 1180	3VX 1250	3VX 1320	3VX 1400	Small Sheave	Large Sheave	870 RPM	1160 RPM	1750 RPM	3450 RPM
2.8	24.8	26.8	29.3	31.9	34.4	36.9	39.4	42.4	45.4	48.4	51.9	55.5	59.5	2.65	10.60	1.30	1.68	2.40	4.23
					18.2	21.0	23.8	27.0	30.2	33.3	36.9	40.6	44.7	6.00	25.00	4.23	5.48	7.88	13.6
										22.9	26.9	30.8	35.1	8.00	33.50	5.90	7.63	10.9	†
.1	21.2	23.3	25.8	28.4	30.9	33.5	36.0	39.0	42.0	45.1	48.6	52.1	56.1	3.35	14.00	1.93	2.50	3.59	6.38
0	15.3	17.6	20.3	22.9	25.5	28.1	30.7	33.8	36.8	39.9	43.4	47.0	51.0	4.50	19.00	2.94	3.81	5.50	9.70
9	24.9	26.9	29.4	32.0	34.5	37.0	39.5	42.5	45.5	48.5	52.1	55.6	59.6	2.50	10.60	1.17	1.50	2.14	3.75
	21.3	23.4	26.0	28.5	31.1	33.6	36.1	39.2	42.2	45.2	48.7	52.3	56.3	3.15	14.00	1.75	2.26	3.26	5.77
					18.4	21.3	24.0	27.3	30.4	33.6	37.2	40.8	44.9	5.60	25.00	3.89	5.04	7.26	12.6
	25.0	27.0	29.5	32.1	34.6	37.1	39.6	42.6	45.6	48.7	52.2	55.7	59.7	2.35	10.60	1.03	1.32	1.88	3.27
	15.6	17.8	20.5	23.2	25.8	28.4	31.0	34.0	37.1	40.2	43.7	47.3	51.3	4.12	19.00	2.61	3.38	4.88	8.64
	21.4	23.5	26.1	28.6	31.2	33.7	36.2	39.3	42.3	45.3	48.8	52.4	56.4	3.00	14.00	1.62	2.09	3.00	5.31
					18.6	21.5	24.2	27.4	30.6	33.8	37.4	41.0	45.1	5.30	25.00	3.63	4.71	6.78	11.9
										23.5	27.6	31.5	35.8	6.90	33.50	4.99	6.46	9.26	15.7
	1	27.1	29.7	32.2	34.7	37.2	39.7	42.7	45.8	48.8	52.3	55.8	59.8	2.20	10.60	.89	1.14	1.62	2.79
				15.8	18.8	21.6	24.4	27.6	30.8	34.0	37.6	41.2	45.3	5.00	25.00	3.38	4.37	6.31	11.1
		23.6	26.2	28.8	31.3	33.8	36.4	39.4	42.4	45.5	49.0	52.5	56.5	2.80	14.00	1.44	1.85	2.66	4.70
										23.8	27.8	31.7	36.1	6.50	33.50	4.65	6.03	8.65	14.8
		8.1	20.8	23.5	26.1	28.7	31.3	34.4	37.4	40.5	44.1	47.6	51.6	3.65	19.00	2.20	2.84	4.10	7.27
				15.9	18.9	21.8	24.6	27.8	31.0	34.1	37.8	41.4	45.5	4.75	25.00	3.16	4.09	5.90	10.4
		[illegible]	26.3	28.9	31.4	34.0	36.5	39.5	42.6	45.6	49.1	52.6	56.6	2.65	14.00	1.30	1.68	2.40	4.23
				16.1	19.1	21.9	24.7	28.0	31.2	34.3	38.0	41.6	45.7	4.50	25.00	2.94	3.81	5.50	9.70
										24.1	28.1	32.0	36.4	6.00	33.50	4.23	5.48	7.88	13.6
				29.0	31.5	34.1	36.6	39.6	42.7	45.7	49.2	52.7	56.8	2.50	14.00	1.17	1.50	2.14	3.75
				[illegible]3.7	26.3	28.9	31.5	34.6	37.6	40.7	44.3	47.8	51.9	3.35	19.00	1.93	2.50	3.59	6.38
									20.6	24.3	28.4	32.3	36.7	5.60	33.50	3.89	5.04	7.26	12.6
						[illegible]4.2	36.7	39.7	42.8	45.8	49.3	52.8	56.9	2.35	14.00	1.03	1.32	1.88	3.27
							[illegible]1.6	34.7	37.8	40.8	44.4	48.0	52.0	3.15	19.00	1.75	2.26	3.26	5.77
								[illegible]	[illegible]1.4	34.6	38.2	41.8	46.0	4.12	25.00	2.61	3.38	4.88	8.64
														[illegible]	33.50	3.63	4.71	6.78	11.9
																1.62	2.09	3.00	5.31
																	1.14	1.62	2.79
																		[illegible]	11.1

Source: Courtesy of The Gates Rubber Company.

This right hand portion along with the center distance values for the longer belt lengths is shown in Table 10.

The drive selection tables illustrated in Table 10 can be used if the drive has a standard electric motor for the prime mover of an rpm shown in one of the column headings to the far right. If the rpm of the prime mover is not listed, the next section will have to be used.

The procedure in determining the HP rating per belt is easy to follow. It simply entails finding the pair of diameters in the center column and, following across to the proper standard motor speed, reading the HP rating per belt. For example, for the pair of diameters that was picked in the geometric design section (2.35 and 10.60 in.) the HP rating is 1.88.

Since this HP rating is for a specific pair of diameters, it includes the add-on HP for speed ratio. But it still needs to be corrected for belt length and for arc of contact on the small pulley. Both of these correction factors are a function of belt length as well as sheave diameters so this is handled most simply by coding the belt length and center distance section with a combined correction factor that includes belt length as well as arc of contact.

This coding is shown in Table 10 with different shading designating combined correction factors in increments of 0.1. Different V-belt and sheave manufacturers use various methods of coding but the end result is a combined length/arc correction factor by which the HP rating is multiplied. For the previous example (2.35 and 10.60 in. sheave diameters with a 3VX500 belt), the combined correction factor would be 0.9.

To find the corrected HP rating, multiply the HP rating of 1.88 by 0.9 to obtain 1.69 HP. This is the rating that is used to find the number of belts required for a specific design HP. Simply divide the design HP by the corrected HP per belt.

For example, if the design HP is 6, the number of belts required would be 6/1.69 = 3.55. The calculated number of belts is usually not an integer so it is recommended that the value be rounded upward to the next larger stock number of grooves which is shown in catalogs and design manuals published by belt and sheave manufacturers.

Nonstandard drives: If one or both of the sheave diameters is nonstandard, or if the rpm of the faster shaft is not shown in the standard rpm columns of the drive selection tables, the HP rating per belt must be determined in a somewhat different manner.

The drive selection manuals published by belt and sheave manufacturers contain HP rating tables that are a function of the small sheave diameter and the rpm of the faster shaft. These rating tables give the basic HP only because they are intended to be used with any speed ratio or any combination of sheave diameters.

It is common practice, however, to incorporate as part of the general HP rating tables a second table that gives the add-on HP for speed ratio. This table is typically arranged to the right of the basic HP table and since the add-on HP is a function of rpm of the faster shaft and speed ratio, the rows of the two tables can align on the same value of rpm. An example of this complete table giving both the basic HP and the add-on HP for speed ratio is given in Table 11.

TABLE 11 Horsepower Rating Table Showing Both Basic and Add-On Ratings

Rated Horsepower per Belt

5V (.62" × .54")

Basic Horsepower per Belt for Small Sheave Outside Diameter

RPM of Faster Shaft	7.00	7.20	7.40	7.60	7.80	8.00	8.20	8.40	8.60	8.80	9.00	9.50	10.00	10.50	11.00	11.50	12.00	12.50	13.00	13.50	14.00	14.50	15.00	15.50	16.00
435	4.86	5.11	5.36	5.61	5.86	6.10	6.35	6.60	6.84	7.09	7.33	7.94	8.55	9.15	9.76	10.4	11.0	11.5	12.1	12.7	13.3	13.9	14.5	15.0	15.6
485	5.34	5.62	5.89	6.17	6.44	6.71	6.99	7.26	7.53	7.80	8.07	8.75	9.42	10.1	10.7	11.4	12.1	12.7	13.4	14.0	14.7	15.3	15.9	16.6	17.2
575	6.19	6.51	6.83	7.15	7.47	7.79	8.11	8.42	8.74	9.06	9.37	10.2	10.9	11.7	12.5	13.2	14.0	14.8	15.5	16.3	17.0	17.7	18.5	19.2	19.9
690	7.23	7.61	7.99	8.36	8.74	9.12	9.49	9.86	10.2	10.6	11.0	11.9	12.8	13.7	14.6	15.5	16.4	17.3	18.2	19.0	19.9	20.8	21.6	22.5	23.3
725	7.54	7.94	8.33	8.73	9.12	9.51	9.90	10.3	10.7	11.1	11.5	12.4	13.4	14.3	15.3	16.2	17.1	18.0	18.9	19.9	20.8	21.6	22.5	23.4	24.3
870	8.79	9.25	9.72	10.2	10.6	11.1	11.6	12.0	12.5	12.9	13.4	14.5	15.6	16.7	17.8	18.9	20.0	21.0	22.1	23.1	24.1	25.2	26.2	27.2	28.1
950	9.45	9.95	10.5	11.0	11.5	11.9	12.4	12.9	13.4	13.9	14.4	15.6	16.8	18.0	19.2	20.3	21.5	22.6	23.7	24.8	25.9	27.0	28.0	29.1	30.1
1160	11.1	11.7	12.3	12.9	13.5	14.1	14.7	15.2	15.8	16.4	17.0	18.4	19.8	21.1	22.5	23.8	25.1	26.4	27.7	28.9	30.2	31.4	32.5	33.7	34.8
1425	13.1	13.8	14.5	15.2	15.9	16.5	17.2	17.9	18.6	19.2	19.9	21.5	23.1	24.7	26.2	27.7	29.2	30.6	32.0	33.4	34.7	36.0	37.2	38.4	39.5
1750	15.2	16.0	16.8	17.6	18.4	19.2	20.0	20.8	21.5	22.3	23.0	24.9	26.7	28.4	30.1	31.7	33.2	34.7	36.1	37.5	38.8	40.0	41.1		
2850	20.0	21.0	22.0	23.0	24.0	24.9	25.8	26.6	27.5	28.3	29.0														
3450	20.8	21.8	22.7	23.6																					
100	1.33	1.39	1.45	1.52	1.58	1.64	1.71	1.77	1.83	1.89	1.96	2.11	2.27	2.43	2.58	2.73	2.89	3.04	3.20	3.35	3.50	3.65	3.80	3.95	4.11
200	2.46	2.58	2.70	2.82	2.94	3.06	3.18	3.30	3.42	3.54	3.66	3.96	4.26	4.56	4.85	5.14	5.44	5.73	6.02	6.31	6.60	6.89	7.18	7.47	7.75
300	3.51	3.69	3.87	4.04	4.22	4.40	4.57	4.75	4.92	5.10	5.27	5.70	6.14	6.57	7.00	7.42	7.85	8.27	8.70	9.12	9.54	9.95	10.4	10.8	11.2
400	4.52	4.75	4.98	5.21	5.44	5.67	5.90	6.13	6.35	6.58	6.81	7.37	7.94	8.50	9.05	9.61	10.2	10.7	11.3	11.8	12.3	12.9	13.4	14.0	14.5
500	5.48	5.77	6.05	6.33	6.62	6.90	7.18	7.46	7.74	8.01	8.29	8.98	9.67	10.4	11.0	11.7	12.4	13.1	13.7	14.4	15.1	15.7	16.4	17.0	17.7
600	6.42	6.75	7.08	7.42	7.75	8.08	8.41	8.74	9.07	9.40	9.73	10.5	11.3	12.2	13.0	13.7	14.5	15.3	16.1	16.9	17.6	18.4	19.2	19.9	20.7
700	7.32	7.70	8.09	8.47	8.85	9.23	9.61	9.99	10.4	10.7	11.1	12.0	13.0	13.9	14.8	15.7	16.6	17.5	18.4	19.3	20.1	21.0	21.9	22.7	23.6
800	8.19	8.62	9.06	9.49	9.92	10.3	10.8	11.2	11.6	12.0	12.5	13.5	14.5	15.6	16.6	17.6	18.6	19.6	20.6	21.6	22.5	23.5	24.4	25.4	26.3
900	9.04	9.52	10.0	10.5	10.9	11.4	11.9	12.4	12.8	13.3	13.8	14.9	16.1	17.2	18.3	19.4	20.5	21.6	22.7	23.8	24.8	25.9	26.9	27.9	28.9
1000	9.86	10.4	10.9	11.4	12.0	12.5	13.0	13.5	14.0	14.5	15.0	16.3	17.5	18.8	20.0	21.2	22.4	23.5	24.7	25.8	27.0	28.1	29.2	30.3	31.3
1100	10.7	11.2	11.8	12.4	12.9	13.5	14.0	14.6	15.1	15.7	16.2	17.6	18.9	20.3	21.6	22.9	24.1	25.4	26.6	27.8	29.0	30.2	31.3	32.5	33.6
1200	11.4	12.0	12.6	13.3	13.9	14.5	15.1	15.7	16.2	16.8	17.4	18.9	20.3	21.7	23.1	24.5	25.8	27.1	28.4	29.7	30.9	32.1	33.3	34.5	35.6
1300	12.2	12.8	13.5	14.1	14.8	15.4	16.0	16.7	17.3	17.9	18.5	20.1	21.6	23.1	24.5	26.0	27.4	28.7	30.1	31.4	32.7	33.9	35.2	36.3	37.5
1400	12.9	13.6	14.3	15.0	15.6	16.3	17.0	17.7	18.3	19.0	19.6	21.2	22.8	24.4	25.9	27.4	28.9	30.3	31.7	33.0	34.3	35.6	36.8	38.0	39.2
1500	13.6	14.3	15.0	15.8	16.5	17.2	17.9	18.6	19.3	20.0	20.7	22.4	24.0	25.6	27.2	28.7	30.2	31.7	33.1	34.5	35.8	37.1	38.3	39.5	40.6
1600	14.2	15.0	15.8	16.5	17.3	18.0	18.8	19.5	20.2	20.9	21.7	23.4	25.1	26.8	28.4	30.0	31.5	33.0	34.4	35.8	37.1	38.4	39.6	40.7	41.8
1700	14.9	15.7	16.5	17.3	18.1	18.8	19.6	20.4	21.1	21.9	22.6	24.4	26.2	27.9	29.5	31.1	32.7	34.2	35.6	36.9	38.2	39.5	40.6	41.7	
1800	15.5	16.3	17.2	18.0	18.8	19.6	20.4	21.2	21.9	22.7	23.5	25.3	27.1	28.9	30.6	32.2	33.7	35.2	36.6	38.0	39.2	40.4			
1900	16.1	16.9	17.8	18.6	19.5	20.3	21.1	21.9	22.7	23.5	24.3	26.2	28.0	29.8	31.5	33.1	34.7	36.1	37.5	38.8	40.0				
2000	16.6	17.5	18.4	19.3	20.1	21.0	21.8	22.7	23.5	24.3	25.1	27.0	28.9	30.6	32.3	33.9	35.5	36.9	38.2						
2100	17.1	18.1	19.0	19.9	20.8	21.6	22.5	23.3	24.2	25.0	25.8	27.7	29.6	31.4	33.1	34.7	36.1	37.5							
2200	17.6	18.6	19.5	20.4	21.3	22.2	23.1	23.9	24.8	25.6	26.4	28.4	30.3	32.0	33.7	35.2	36.7								
2300	18.1	19.1	20.0	20.9	21.9	22.8	23.6	24.5	25.4	26.2	27.0	29.0	30.9	32.6	34.2	35.7									
2400	18.5	19.5	20.5	21.4	22.3	23.3	24.2	25.0	25.9	26.7	27.6	29.5	31.4	33.1	34.6										
2500	18.9	19.9	20.9	21.9	22.8	23.7	24.6	25.5	26.4	27.2	28.0	30.0	31.8	33.4											
2600	19.3	20.3	21.3	22.2	23.2	24.1	25.0	25.9	26.8	27.6	28.4	30.3	32.1												
2700	19.6	20.6	21.6	22.6	23.5	24.5	25.4	26.2	27.1	27.9	28.7	30.6													
2800	19.9	20.9	21.9	22.9	23.8	24.8	25.7	26.5	27.4	28.2	29.0	30.8													
2900	20.1	21.2	22.2	23.1	24.1	25.0	25.9	26.7	27.6	28.4	29.1														
3000	20.3	21.4	22.4	23.3	24.3	25.2	26.1	26.9	27.7	28.5															
3100	20.5	21.5	22.5	23.5	24.4	25.3	26.2	27.0	27.8																
3200	20.6	21.7	22.7	23.6	24.5	25.4	26.2																		
3300	20.7	21.8	22.7	23.7	24.6	25.4																			
3400	20.8	21.8	22.7	23.7	24.5																				
3500	20.8	21.8	22.7	23.6																					
3600	20.7	21.7	22.6																						
3700	20.7	21.6																							
3800	20.5																								

Drives for rpm-diameter combinations where no horsepower is shown may be practical if all conditions are known.

Additional Horsepower per Belt for Speed Ratio

RPM of Faster Shaft	1.00 to 1.01	1.02 to 1.05	1.06 to 1.11	1.12 to 1.18	1.19 to 1.26	1.27 to 1.38	1.39 to 1.57	1.58 to 1.94	1.95 to 3.38	3.39 and over
435	0.00	0.04	0.11	0.20	0.27	0.33	0.38	0.43	0.47	0.50
485	0.00	0.05	0.13	0.22	0.30	0.37	0.43	0.48	0.53	0.56
575	0.00	0.06	0.15	0.26	0.36	0.43	0.51	0.57	0.62	0.66
690	0.00	0.07	0.18	0.32	0.43	0.52	0.61	0.69	0.75	0.79
725	0.00	0.07	0.19	0.33	0.45	0.55	0.64	0.72	0.79	0.83
870	0.00	0.08	0.23	0.40	0.54	0.66	0.77	0.86	0.94	1.00
950	0.00	0.09	0.25	0.43	0.59	0.72	0.84	0.94	1.03	1.09
1160	0.00	0.11	0.30	0.53	0.72	0.87	1.02	1.15	1.26	1.33
1425	0.00	0.14	0.37	0.65	0.89	1.07	1.26	1.42	1.54	1.63
1750	0.00	0.17	0.46	0.80	1.09	1.32	1.55	1.74	1.90	2.01
2850	0.00	0.27	0.75	1.30	1.77	2.15	2.52	2.83	3.09	3.27
3450	0.00	0.33	0.90	1.58	2.15	2.60	3.05	3.43	3.74	3.96
100	0.00	0.01	0.03	0.05	0.06	0.08	0.09	0.10	0.11	0.11
200	0.00	0.02	0.05	0.09	0.12	0.15	0.18	0.20	0.22	0.23
300	0.00	0.03	0.08	0.14	0.19	0.23	0.26	0.30	0.32	0.34
400	0.00	0.04	0.10	0.18	0.25	0.30	0.35	0.40	0.43	0.46
500	0.00	0.05	0.13	0.23	0.31	0.38	0.44	0.50	0.54	0.57
600	0.00	0.06	0.16	0.27	0.37	0.45	0.53	0.60	0.65	0.69
700	0.00	0.07	0.18	0.32	0.44	0.53	0.62	0.70	0.76	0.80
800	0.00	0.08	0.21	0.37	0.50	0.60	0.71	0.80	0.87	0.92
900	0.00	0.09	0.24	0.41	0.56	0.68	0.79	0.89	0.97	1.03
1000	0.00	0.10	0.26	0.46	0.62	0.75	0.88	0.99	1.08	1.15
1100	0.00	0.11	0.29	0.50	0.68	0.83	0.97	1.09	1.19	1.26
1200	0.00	0.12	0.31	0.55	0.75	0.91	1.06	1.19	1.30	1.38
1300	0.00	0.13	0.34	0.59	0.81	0.98	1.15	1.29	1.41	1.49
1400	0.00	0.13	0.37	0.64	0.87	1.06	1.24	1.39	1.52	1.61
1500	0.00	0.14	0.39	0.69	0.93	1.13	1.32	1.49	1.62	1.72
1600	0.00	0.15	0.42	0.73	1.00	1.21	1.41	1.59	1.73	1.84
1700	0.00	0.16	0.45	0.78	1.06	1.28	1.50	1.69	1.84	1.95
1800	0.00	0.17	0.47	0.82	1.12	1.36	1.59	1.79	1.95	2.06
1900	0.00	0.18	0.50	0.87	1.18	1.43	1.68	1.89	2.06	2.18
2000	0.00	0.19	0.52	0.91	1.24	1.51	1.77	1.99	2.17	2.29
2100	0.00	0.20	0.55	0.96	1.31	1.58	1.85	2.09	2.27	2.41
2200	0.00	0.21	0.58	1.01	1.37	1.66	1.94	2.19	2.38	2.52
2300	0.00	0.22	0.60	1.05	1.43	1.73	2.03	2.29	2.49	2.64
2400	0.00	0.23	0.63	1.10	1.49	1.81	2.12	2.39	2.60	2.75
2500	0.00	0.24	0.66	1.14	1.56	1.89	2.21	2.49	2.71	2.87
2600	0.00	0.25	0.68	1.19	1.62	1.96	2.30	2.58	2.82	2.98
2700	0.00	0.26	0.71	1.23	1.68	2.04	2.38	2.68	2.92	3.10
2800	0.00	0.27	0.73	1.28	1.74	2.11	2.47	2.78	3.03	3.21
2900	0.00	0.28	0.76	1.33	1.80	2.19	2.56	2.88	3.14	3.33
3000	0.00	0.29	0.79	1.37	1.87	2.26	2.65	2.98	3.25	3.44
3100	0.00	0.30	0.81	1.42	1.93	2.34	2.74	3.08	3.36	3.56
3200	0.00	0.31	0.84	1.46	1.99	2.41	2.83	3.18	3.47	3.67
3300	0.00	0.32	0.87	1.51	2.05	2.49	2.91	3.28	3.57	3.79
3400	0.00	0.33	0.89	1.55	2.12	2.56	3.00	3.38	3.68	3.90
3500	0.00	0.34	0.92	1.60	2.18	2.64	3.09	3.48	3.79	4.01
3600	0.00	0.35	0.94	1.65	2.24	2.72	3.18	3.58	3.90	4.13
3700	0.00	0.36	0.97	1.69	2.30	2.79	3.27	3.68	4.01	4.24
3800	0.00	0.37	1.00	1.74	2.36	2.87	3.36	3.78	4.12	4.36

All sheaves are given a careful static balance for normal (shaft or operating) speeds; they will however, operate safely at rim speeds up to 6,500 FPM. Where vibration may be a problem, we recommend that sheaves be dynamically balanced.

Source: Courtesy of The Gates Rubber Company.

The speed ratio of the drive is known from the geometric design (either calculated or obtained from the speed ratio column of Table 9). It is a simple task to locate the rpm of the faster shaft in the left hand column and read across to the proper smaller diameter column for the basic HP rating. Then one continues across to the proper speed ratio column to obtain the add-on HP for speed ratio. These two ratings are added together.

The next step is to determine the correction factors for belt length and arc of contact. Again tables are provided in design manuals published by the V-belt and sheave manufacturers. These tables are calculated from the equations discussed earlier in this chapter.

Table 12 is an example of the belt length correction tables which show the belt length correction factor for each standard narrow cross section belt.

Table 7 gives the arc of contact correction factor as a function of $(D_e - d_e)/C$. Since the sheave diameters and center distance are known from the geometric design, this calculation can easily be made. Table 7 also gives the arc of contact for a given value of $(D_e - d_e)/C$.

The total HP rating (the sum of the basic rating plus the add-on) from Table 11 is multiplied by both the arc of contact correction factor (Factor G) and the belt length correction factor (C_L). The product is the corrected HP rating which can be used to find the required number of belts in the same manner as for the section using standard sheave diameters and standard motor rpm.

6. Summary Comments on Industrial Multiple Belt Drive Design

The drive selection process for heavy duty industrial drives described in this chapter does not necessarily give the best drive design for a given application on the first try. There are usually many options open to the designer depending on what his specific needs are. If he needs a drive with small diameters to fit into a limited space, then he should select a pair of diameters that are toward the small end of the range of a specific cross section, or possibly explore the next smaller cross section where even smaller diameters would be available.

If the width of the drive seems to be a problem or if bearing or shaft loads need to be minimized, then the designer should pick diameters toward the large end of the range for his particular cross section. Of course, drive cost is always an important factor.

Many V-belt and sheave manufacturers have computerized drive design methods by which all possible V-belt drives from standard or stock components can be evaluated for space requirements, shaft pull (bearing loads), initial cost, 10 year cost, and other requirements that a designer may consider important. More detail is given on computerized drive design in Chapter 15.

TABLE 12 Belt Length Correction Factors

3VX		5VX & 5V		8V	
V-Belt Number	Correction Factor	V-Belt Number	Correction Factor	V-Belt Number	Correction Factor
3VX250	0.83	5VX500	0.85	8V1000	0.87
3VX265	0.84	5VX530	0.86	8V1060	0.88
3VX280	0.85	5VX560	0.87	8V1120	0.88
3VX300	0.86	5VX600	0.88	8V1180	0.89
3VX315	0.87	5VX630	0.89	8V1250	0.90
3VX335	0.88	5VX670	0.90	8V1320	0.91
3VX355	0.90	5VX710	0.91	8V1400	0.92
3VX375	0.91	5VX750	0.92	8V1500	0.93
3VX400	0.92	5VX800	0.93	8V1600	0.94
3VX425	0.93	5VX850	0.94	8V1700	0.94
3VX450	0.94	5VX900	0.95	8V1800	0.95
3VX475	0.95	5VX950	0.96	8V1900	0.96
3VX500	0.96	5VX1000	0.96	8V2000	0.97
3VX530	0.97	5VX1060	0.97	8V2120	0.98
3VX560	0.98	5VX1120	0.98	8V2240	0.98
3VX600	0.99	5VX1180	0.99	8V2360	0.99
3VX630	1.00	5VX1250	1.00	8V2500	1.00
3VX670	1.01	5VX1320	1.01	8V2650	1.01
3VX710	1.02	5VX1400	1.02	8V2800	1.02
3VX750	1.03	5VX1500	1.03	8V3000	1.03
3VX800	1.04	5VX1600	1.04	8V3150	1.03
3VX850	1.06	5VX1700	1.05	8V3350	1.04
3VX900	1.07	5VX1800	1.06	8V3550	1.05
3VX950	1.08	5VX1900	1.07	8V3750	1.06
3VX1000	1.09	5VX2000	1.08	8V4000	1.07
3VX1060	1.10	5V2120	1.09	8V4250	1.08
3VX1120	1.11	5V2240	1.09	8V4500	1.09
3VX1180	1.12	5V2360	1.10	8V4750	1.09
3VX1250	1.13	5V2500	1.11	8V5000	1.10
3VX1320	1.14	5V2650	1.12	8V5600	1.12
3VX1400	1.15	5V2800	1.13		
		5V3000	1.14		
		5V3150	1.15		
		5V3350	1.16		
		5V3550	1.17		

Source: Courtesy of The Gates Rubber Company.

II. LIGHT-DUTY INDUSTRIAL DRIVES

A. Definition

1. Type of Service

Light-duty industrial drives are applied to machines where the HP requirements are low – usually fractional HP which is basically defined as less than 1 HP at 1750 rpm.

These belts are also usually used on machines such as heavy-duty drives that are subject to intermittent service rather than continuous service. This is because light-duty drives are usually designed to give much shorter service life than heavy-duty drives in order to be compatible with machinery that is run on an intermittent basis in an industrial operation. There are methods, which will be discussed later, by which some V-belt manufacturers can design light-duty drives for continuous service on fractional HP applications.

2. Single Belt Operation

Since light-duty drives are designed for fractional HP service, they are single belt drives rather than multiple belt drives as are the heavy duty drives. Therefore, the V-belts which are stocked by the various V-belt and pulley manufacturers are not marked with an exact length which would allow them to be matched into sets. (See Chapter 12 for more details in operating V-belt drives in multiple.)

3. Standard Cross Sections and Construction

Light-duty or single belts come in four standard cross sections. Figure 13 shows these four cross sections along with the nominal dimensions of each.

The typical construction for light-duty belts is shown in Figure 14. Note that single-unit tensile cord is used and that the belt has a cover of rubber-impregnated fabric. Occasionally, raw edge belts (belts without a cover) are used for light applications in the standard cross sections but these are usually designed for specific applications.

Light duty belts of standard construction will withstand a reasonable amount of oil and heat, but if the normal environment is consistently at elevated temperatures or subject to oil, a special construction will probably be necessary.

4. Special Snap-On Type Belts

For very lightly loaded applications, there are some all-rubber or thermoplastic light-duty belts on the market. These belts are usually of the 2L cross section or a special cross section which is even smaller. These belts are manufactured without the usual tensile member and are

	Nominal Top Width x Thickness, Inches	Cross Section Designation Gates	Industry
	1/4 x 5/32 (0.25 x 0.16)	0	2L
	3/8 x 7/32 (0.38 x 0.22)	1	3L
	1/2 x 5/16 (0.50 x 0.31)	2	4L
	21/32 x 3/8 (0.66 x 0.38)	3	5L

FIGURE 13 Light-duty V-belt cross sections.

elongated during installation and snapped on the drive rather than adjusting the center distance for installation and tensioning.

These belts have the advantage of being able to be used on drives where there is no center distance adjustment (fixed center drives). However, in order to maintain adequate tension in the drive for a reasonable service life, the initial tension must be very high. This can cause high bearing loads and sometimes bearing or shaft failures.

If significant HP is being transmitted with this type of belt, the belt will tend to grow or stretch until it will slip and burn. So use of this type of belt is usually confined to lower power transmission applications which involve motion transmission rather than power transmission. In applying this type of belt, it is important to follow the procedures specified by each individual manufacturer.

B. Nomenclature

1. V-Belts

Light-duty industrial V-belts are on the effective system as described in Chapter 1. The nomenclature consists of one or two numerals that represent the width of the belt, followed by a letter which designates the type of belt. Then the effective length is given by three or four digits.

FIGURE 14 Banded light-duty industrial V-belt with single unit cord.

Conventional USA units: The cross section designation and belt type is shown in Figure 13. The belt length in tenths of an inch follows the belt type letter "L." For example:

4 L 450

Nominal width of belt (4/8 or 1/2 in.)
light duty
effective length designation (45.0 in.)

Metric-SI units: The same system is used in metric units. The first one or two digits is the nominal belt width in mm followed by the letter "R" to designate the belt type. The effective length in mm comes next. For example:

13 R 1200

Nominal width of belt (mm) — 13; R — light duty; 1200 — effective length (mm)

2. Pulleys

Nomenclature for pulleys is not included in the industry standard. When ordering pulleys, the following information needs to be included:

Cross section designation
Outside pulley diameter
Bore of hub or bushing
Keyseat dimensions if not standard as specified in Table 1

C. Stock Components

1. V-Belts

The standard belt lengths for the four cross sections mentioned earlier in this chapter are usually in 1 in. increments up through 80 in. Above 80 in., the standard lengths are in 2-in. increments. See Table 13 for a listing of standard lengths.

Most V-belt manufacturers stock the standard lengths in the three largest cross sections (3L, 4L, and 5L). Stocking of the 2L section is inconsistent with several manufacturers not stocking this particular cross section.

The light-duty belts are on the effective system as described in Chapter 1.

2. Pulleys

The industry standard on light-duty or single belt drives does not specify standard pulley type or diameter. Most manufacturers do stock essentially the same diameters, however,

Light-duty pulleys are made of a wide variety of materials such as cast iron, formed steel, and zinc die cast. Most of the pulleys are made with a combination groove that will accommodate two different cross sections. For example, Table 14 shows a typical series of sheave diameters for pulleys with a combination groove that will accept either a 3L or a 4L cross section. There is another series of pulleys in common use that will accept 4L and 5L sections.

Figure 15 shows the various types of pulley designs available in a cast iron construction using a bushing for easy mounting to the shaft.

TABLE 13 Standard Light-Duty Industrial V-Belt Lengths

Standard Effective Outside Length (Inches)				Permissible Deviation From Standard Effective Length (Inches)	Standard Effective Outside Length (Inches)				Permissible Deviation From Standard Effective Length (Inches)
Cross Section					Cross Section				
2L	3L	4L	5L		2L	3L	4L	5L	
8				+0.12, −0.38		48	48	48	+0.25, −0.62
9				+0.12, −0.38		49	49	49	+0.25, −0.62
10				+0.12, −0.38		50	50	50	+0.25, −0.62
11				+0.12, −0.38			51	51	+0.25, −0.62
12				+0.12, −0.38		52	52	52	+0.25, −0.62
13				+0.12, −0.38			53	53	+0.25, −0.62
14	14			+0.12, −0.38		54	54	54	+0.25, −0.62
15	15			+0.12, −0.38			55	55	+0.25, −0.62
16	16			+0.12, −0.38		56	56	56	+0.25, −0.62
17	17			+0.12, −0.38			57	57	+0.25, −0.62
18	18	18		+0.12, −0.38		58	58	58	+0.25, −0.62
19	19	19		+0.12, −0.38			59	59	+0.25, −0.62
20	20	20		+0.12, −0.38		60	60	60	+0.25, −0.62
	21	21		+0.25, −0.62			61	61	+0.31, −0.69
	22	22		+0.25, −0.62			62	62	+0.31, −0.69
	23	23		+0.25, −0.62			63	63	+0.31, −0.69
	24	24		+0.25, −0.62			64	64	+0.31, −0.69
	25	25	25	+0.25, −0.62			65	65	+0.31, −0.69
	26	26	26	+0.25, −0.62			66	66	+0.31, −0.69
	27	27	27	+0.25, −0.62			67	67	+0.31, −0.69
	28	28	28	+0.25, −0.62			68	68	+0.31, −0.69
	29	29	29	+0.25, −0.62			69	69	+0.31, −0.69
	30	30	30	+0.25, −0.62			70	70	+0.31, −0.69
	31	31	31	+0.25, −0.62			71	71	+0.31, −0.69
	32	32	32	+0.25, −0.62			72	72	+0.31, −0.69
	33	33	33	+0.25, −0.62			73	73	+0.31, −0.69
	34	34	34	+0.25, −0.62			74	74	+0.31, −0.69
	35	35	35	+0.25, −0.62			75	75	+0.31, −0.69
	36	36	36	+0.25, −0.62			76	76	+0.31, −0.69
	37	37	37	+0.25, −0.62			77	77	+0.31, −0.69
	38	38	38	+0.25, −0.62			78	78	+0.31, −0.69
	39	39	39	+0.25, −0.62			79	79	+0.31, −0.69
	40	40	40	+0.25, −0.62			80	80	+0.62, −0.88
	41	41	41	+0.25, −0.62			82	82	+0.62, −0.88
	42	42	42	+0.25, −0.62			84	84	+0.62, −0.88
	43	43	43	+0.25, −0.62			86	86	+0.62, −0.88
	44	44	44	+0.25, −0.62			88	88	+0.62, −0.88
	45	45	45	+0.25, −0.62			90	90	+0.62, −0.88
	46	46	46	+0.25, −0.62			92	92	+0.62, −0.88
	47	47	47	+0.25, −0.62			94	94	+0.62, −0.88
							96	96	+0.62, −0.88
							98	98	+0.62, −0.88
							100	100	+0.62, −0.88

Source: Standard IP-23 published jointly by the Rubber Manufacturers Association, the Mechanical Power Transmission Association, and the Rubber Association of Canada.

TABLE 14 Commonly Stocked Pulley Diameters in a Combination Groove for 3L and 4L Cross Sections

O.D. (In.)	P.D. (In.)		Catalog No.	Product No.	Type	Dimensions (In.)								Wt. (Lbs.) Rim Only
	No. 1	No. 2				O.L.	F	L	P	C	X	G	E	
3.05	2.46	2.80	AG30D	7801-0030	1	1 13/16	3/4	1 1/4	7/8	—	13/16	7/16	3/16	1.1
3.25	2.66	3.00	AG32D	7801-0032	1	1 13/16	3/4	1 1/4	7/8	—	13/16	7/16	3/16	1.2
3.45	2.86	3.20	AG34D	7801-0034	2	1 1/2	3/4	1 1/4	9/16	—	7/8	7/16	3/16	1.0
3.75	3.16	3.50	AG39D	7801-0100	2	1 1/2	3/4	1 1/4	9/16	—	7/8	7/16	3/16	1.4
3.95	3.36	3.70	AG41D	7801-0101	2	1 1/2	3/4	1 1/4	9/16	—	7/8	7/16	3/16	1.6
4.25	3.66	4.00	AG44D	7801-0102	2	1 1/2	3/4	1 1/4	9/16	—	7/8	7/16	3/16	1.9
4.45	3.86	4.20	AG46D	7801-0103	2	1 1/2	3/4	1 1/4	9/16	—	7/8	7/16	3/16	1.9
4.75	4.16	4.50	AG49D	7801-0104	2	1 1/2	3/4	1 1/4	9/16	—	7/8	7/16	3/16	2.1
4.95	4.36	4.70	AG51D	7801-0105	2	1 1/2	3/4	1 1/4	9/16	—	7/8	7/16	3/16	2.3
5.25	4.66	5.00	AG54D	7801-0106	2	1 1/2	3/4	1 1/4	9/16	—	7/8	7/16	3/16	2.0
5.45	4.86	5.20	AG56D	7801-0107	2	1 1/2	3/4	1 1/4	9/16	—	7/8	7/16	3/16	2.3
5.75	5.16	5.50	AG59D	7801-0108	2	1 1/2	3/4	1 1/4	9/16	—	7/8	7/16	3/16	2.4
5.95	5.36	5.70	AG61D	7801-0109	3	1 1/2	3/4	1 1/4	9/16	0	7/8	7/16	3/16	2.5
6.25	5.66	6.00	AG64D	7801-0110	3	1 1/2	3/4	1 1/4	9/16	0	7/8	7/16	3/16	2.7
6.45	5.86	6.20	AG66D	7801-0111	3	1 1/2	3/4	1 1/4	9/16	0	7/8	7/16	3/16	2.8
6.75	6.16	6.50	AG69D	7801-0112	3	1 1/2	3/4	1 1/4	9/16	0	7/8	7/16	3/16	3.2
6.95	6.36	6.70	AG71D	7801-0113	3	1 1/2	3/4	1 1/4	9/16	0	7/8	7/16	3/16	3.1
7.25	6.66	7.00	AG74D	7801-0114	3	1 1/2	3/4	1 1/4	9/16	0	7/8	7/16	3/16	3.3
7.75	7.16	7.50	AG79D	7801-0115	3	1 1/2	3/4	1 1/4	9/16	0	7/8	7/16	3/16	3.5
8.25	7.66	8.00	AG84D	7801-0118	3	1 1/2	3/4	1 1/4	9/16	0	7/8	7/16	3/16	3.6
8.75	8.16	8.50	AG89D	7801-0116	3	1 1/2	3/4	1 1/4	9/16	0	7/8	7/16	3/16	4.0
9.25	8.66	9.00	AG94D	7801-0120	3	1 1/2	3/4	1 1/4	9/16	0	7/8	7/16	3/16	4.4
9.75	9.16	9.50	AG99D	7801-0121	3	1 1/2	3/4	1 1/4	9/16	0	7/8	7/16	3/16	4.7
10.25	9.66	10.00	AG104D	7801-0122	3	1 1/2	3/4	1 1/4	9/16	0	7/8	7/16	3/16	4.5
10.75	10.16	10.50	AG109D	7801-0123	3	1 1/2	3/4	1 1/4	9/16	0	7/8	7/16	3/16	5.1
11.25	10.66	11.00	AG114D	7801-0124	3	1 1/2	3/4	1 1/4	9/16	0	7/8	7/16	3/16	5.5
12.25	11.66	12.00	AG124D	7801-0126	3	1 1/2	3/4	1 1/4	9/16	0	7/8	7/16	3/16	6.1
13.25	12.66	13.00	AG134D	7801-0128	3	1 1/2	3/4	1 1/4	9/16	0	7/8	7/16	3/16	7.4
14.25	13.66	14.00	AG144D	7801-0130	3	1 1/2	3/4	1 1/4	9/16	0	7/8	7/16	3/16	7.8
15.25	14.66	15.00	AG154D	7801-0131	3	1 1/2	3/4	1 1/4	9/16	0	7/8	7/16	3/16	8.8
18.25	17.66	18.00	AG184D	7801-0133	3	1 1/2	3/4	1 1/4	9/16	0	7/8	7/16	3/16	11.3

Source: Courtesy of The Gates Rubber Company.

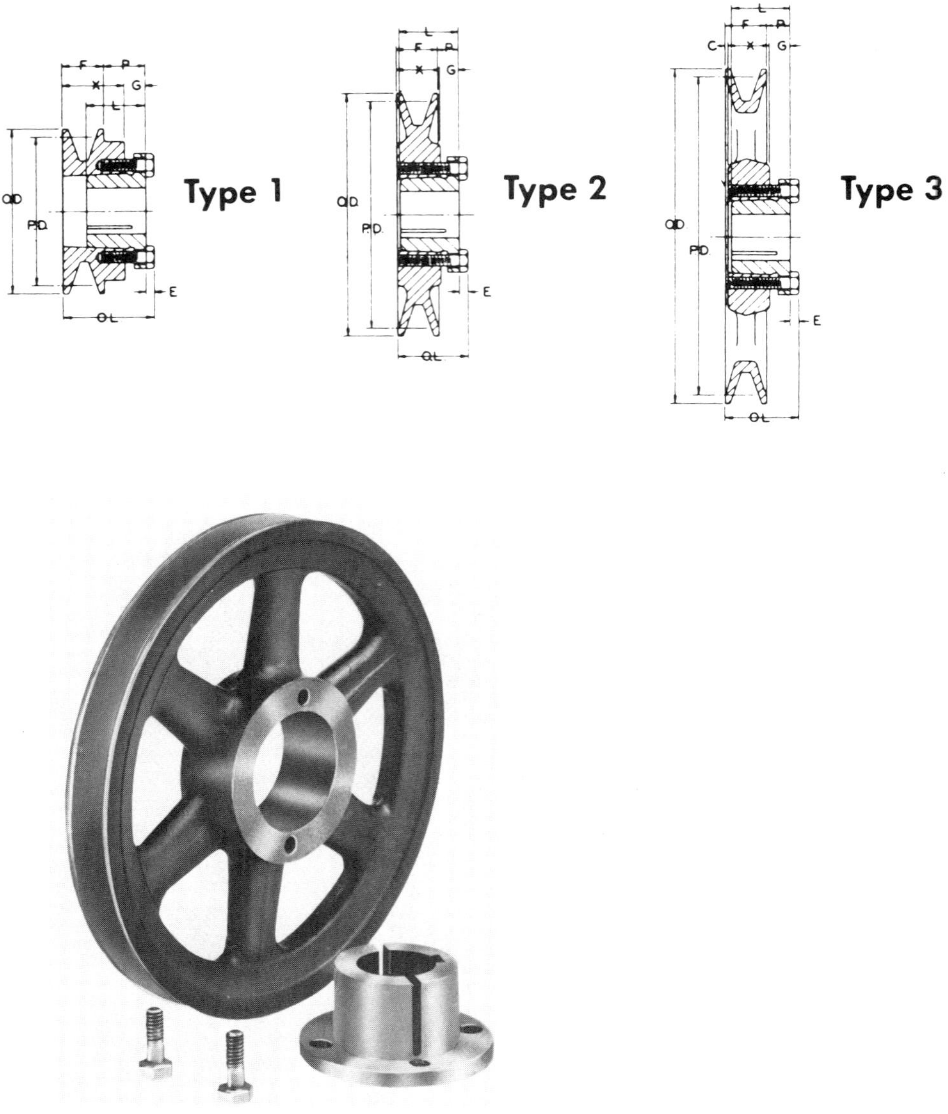

FIGURE 15 Typical cast iron light-duty sheaves in three types of construction.

D. Drive Design Procedure

The drive design procedure for light-duty drives is not as detailed as it is for the classical or narrow industrial drives. This is because the light-duty drives are intended for light-duty intermittent service and because they are basically single belt drives which eliminates the choice as to how many belts are needed to transmit the required HP.

As with the heavy-duty design section of this chapter, it is not the intent to give a detailed drive design procedure. This is available in design manuals published by V-belt and pulley manufacturers. The discussion that follows will describe in general how light-duty industrial drives are designed and why the procedure is different from heavy-duty drive design.

1. Capacity Drive Design

Since light-duty drives are basically one-belt drives, the designer does not have the choice of adding belts to increase the drive capacity to satisfy his needs. This means he must select the belt cross section as well as the small sheave diameter that is adequate to satisfy his power transmission needs with one belt.

To aid the designer in making this initial selection of cross section and small sheave diameter, V-belt or pulley manufacturers publish information like that shown in Table 15.

The information for Table 15 is based on a HP rating formula similar to the ones used for heavy-duty drive design. But rather than showing the HP rating for a specific rpm and pulley diameter, this table shows the minimum pulley diameter that is required to handle the power transmission needs of a specific drive. For example, if a drive is being designed for a 1/3 HP, 1750 rpm motor, the designer would locate the 1750 rpm line in the left hand column of Table 15. Moving across the motor HP columns of the 2L section of the table, he finds that a 1/4 HP motor is the largest that a 2L belt will handle.

Going to the 3L section of the table, the designer finds he must use a 2.75 or larger diameter to satisfy his 1/3 HP requirements. Exploring further, he finds that a 4L section requires a diameter at least as large as 3.00 in.

The belt length and arc of contact correction factors as well as the add-on HP for speed ratio are simplified and combined with the service factor to give one correction factor as shown in Table 16. This combined factor is divided into the HP rating shown in Table 17 for the 3L belt to get the corrected HP rating. The industry standard IP-23 shows rating tables for other sections as do design manuals published by V-belt and pulley manufacturers.

TABLE 15 Cross Section and Small Pulley Diameter Selection Tables For Light-Duty Drive Design

#0 (2L) Section—Small Sheave Outside Diameter (Inches)

Small Sheave RPM	Motor Or Engine Horsepower						
	1/20	1/15	1/12	1/10	1/8	1/6	1/4
200				Use Table 9 or			
400	2.25	2.75			Table 10 For		
600	1.75	2.25	2.50	3.00	This Range		
1000	1.38	1.62	1.75	2.00	2.50	3.00	
1160	1.25	1.50	1.62	1.88	2.25	2.75	
1400	1.25	1.38	1.50	1.62	1.88	2.50	
1750	1.12	1.25	1.38	1.50	1.75	2.25	3.00
1800	1.12	1.25	1.38	1.50	1.75	2.00	2.75
2200	1.00	1.12	1.25	1.38	1.50	1.88	2.50
2600	1.00	1.12	1.25	1.25	1.50	1.75	2.25
3000	1.00	1.12	1.12	1.25	1.38	1.62	2.00
3450	1.00	1.00	1.12	1.25	1.25	1.50	1.88

#1 (3L) Section—Small Sheave Outside Diameter (Inches)

Small Sheave RPM	Motor Or Engine Horsepower							
	1/12	1/10	1/8	1/6	1/4	1/3	1/2	3/4
200	3.50	3.75			Use Table 10 or			
400	2.50	2.75	3.00	3.75		Table 11 For		
600	2.25	2.40	2.60	3.00	3.75		This Range	
1000	2.00	2.25	2.25	2.50	3.00	3.50		
1160	2.00	2.25	2.25	2.40	2.75	3.25	4.00	
1400	2.00	2.00	2.25	2.25	2.60	3.00	3.75	
1750	2.00	2.00	2.25	2.25	2.50	2.75	3.50	4.25
1800	2.00	2.00	2.25	2.25	2.50	2.75	3.25	4.25
2200	2.00	2.00	2.00	2.25	2.40	2.60	3.25	4.00
2600	2.00	2.00	2.00	2.25	2.25	2.50	3.00	3.75
3000	2.00	2.00	2.00	2.25	2.25	2.40	2.75	3.50
3450	2.00	2.00	2.00	2.25	2.25	2.40	2.75	3.25

#2 (4L) Section—Small Sheave Outside Diameter (Inches)

Small Sheave RPM	Motor Or Engine Horsepower								
	1/10	1/8	1/6	1/4	1/3	1/2	3/4	1	1 1/2
200	3.25	3.50	4.00	4.75	Use Table 11 For This Range				
400	2.60	3.00	3.25	3.50	4.00				
600	2.50	2.60	3.00	3.25	3.50	4.25			
1000	2.50	2.50	2.60	3.00	3.25	3.50	4.25	5.00	
1160	2.50	2.50	2.60	3.00	3.00	3.50	4.00	4.75	
1400	2.50	2.50	2.60	2.75	3.00	3.25	3.75	4.25	
1750	2.50	2.50	2.60	2.75	3.00	3.25	3.50	4.00	5.00
1800	2.50	2.50	2.60	2.75	3.00	3.25	3.50	4.00	4.75
2200	2.50	2.50	2.60	2.75	2.75	3.00	3.50	3.75	4.50
2600	2.50	2.50	2.60	2.75	2.75	3.00	3.25	3.75	4.25
3000	2.50	2.60	2.60	2.75	2.75	3.00	3.25	3.50	4.25
3450	2.60	2.60	2.60	2.75	2.75	3.00	3.25	3.50	4.00

Source: Courtesy of The Gates Rubber Company.

TABLE 16 Combined Correction Factor For HP Ratings Used in Light-Duty Drive Design

Type Of DriveN Unit	Speed Ratio	
	Less Than 1.5	1.5 And Over
Fans and Blowers	1.0	0.9
Domestic Laundry Machines	1.1	1.0
Centrifugal Pumps	1.1	1.0
Generators	1.2	1.1
Rotary Compressors	1.2	1.1
Machine Tools	1.3	1.2
Reciprocating Pumps	1.4	1.3
Reciprocating Compressors	1.4	1.3
Woodworking Machines	1.4	1.3

Source: Courtesy of The Gates Rubber Company.

This corrected HP rating can then be compared to the power requirements of the drive to be sure the initial selection of the cross section and minimum diameter was satisfactory.

This simplified design procedure is adequate for light-duty drives because the type of service they are subjected to is not as demanding as heavy-duty drives.

TABLE 17 HP Ratings For 3L Section Light-Duty Belts

RPM of Faster Shaft	Rated Horsepower Per Belt For Small Sheave Outside Diameter						
	1.50	1.75	2	2.25	2.50	2.75	3
1160	0.07	0.13	0.18	0.23	0.28	0.34	0.39
1750	0.09	0.17	0.25	0.32	0.39	0.46	0.54
3450	0.11	0.25	0.38	0.50	0.62	0.73	0.83
500	0.04	0.07	0.09	0.12	0.14	0.17	0.19
1000	0.07	0.12	0.16	0.21	0.25	0.30	0.34
1500	0.09	0.15	0.22	0.29	0.35	0.41	0.47
2000	0.10	0.19	0.27	0.35	0.43	0.51	0.59
2500	0.11	0.21	0.31	0.41	0.51	0.60	0.69
3000	0.11	0.23	0.35	0.46	0.57	0.68	0.78
3500	0.11	0.25	0.38	0.50	0.62	0.74	0.84
4000	0.11	0.26	0.40	0.54	0.66	0.78	0.88
4500	0.10	0.26	0.42	0.56	0.68	0.80	0.90
5000	0.09	0.26	0.42	0.57	0.69	0.80	0.89

Source: Courtesy of The Gates Rubber Company.

2. Geometric Drive Design

Knowing the cross section and the minimum small pulley diameter that can be used, the designer can select the large sheave diameter using the speed ratio of the drive in the same manner described in the heavy duty section of this chapter. It is important to remember that the combination pulleys usually used for light-duty drives will have a different pitch diameter depending on the cross section of the belt being used. Table 14 shows the pitch diameter for a 3L belt for 4.75 in. outside diameter pulley to be 4.16 in. whereas the 4L belt has a pitch diameter of 4.50 in. in the same combination pulley.

When the pulley diameters have been selected to give the desired speed ratio, the belt length/center distance relationship can be calculated using the same formulas as given in the heavy-duty section of this manual.

III. OTHER DRIVE DESIGN CONSIDERATIONS

There are other design considerations which are covered in separate chapters since they are similar for all types of belts. These are belt tensioning and bearing loads covered in Chapter 11 and installation and maintenance provisions which are covered in Chapter 12. Chapter 14, which gives more detail on sheaves and pulleys, will also be of interest to the industrial drive designer.

4

Agricultural Drive Design Using V-Belts

WALTER T. NAWROCKI *Industrial Product Application Department, The Gates Rubber Company, Denver, Colorado*

I. V-BELT DRIVES ON AGRICULTURAL MACHINERY

Agricultural belt drives are unique in both belt construction and application. They are subjected to extremes in environmental conditions, vibration, and shock loading. They are used either as single belts or multiples depending on the load-life conditions. The drives can be of a fixed speed ratio or variable ratio. The design procedure for both is given below.

Because of the conditions that a V-belt experiences, it becomes imperative that sound drive design procedures be followed and thorough field testing be carried out.

V-belt drives on agricultural equipment are considered limited-service drives and a life-in-hours approach is used. With this in mind, the life of a V-belt could be calculated giving some assurance of drive success before field testing.

II. DRIVE CONSIDERATIONS

A. Acceptable Belt Life

Design engineers, with the experience gained from product testing, are in the best position to establish the acceptable belt life which typically ranges from 1000 to 3000 hr. When determining what would be acceptable life, it is important to consider the following factors:

1. The operating time of the drive per day, week, month, as compared to the total time desired between belt replacements
2. Probability of longer and more severe usage than normal in some geographical areas or in some specific applications
3. The warranty period

B. DriveN Loads

If the drive is to operate over a range of speeds and loads it is essential that more than one operating condition be selected for analysis. This is because load-life relationships are not linear, and it is extremely unlikely that a single condition could be selected which would represent the average. It is particularly important to realize that checking a drive for only relatively high or a relatively low load is unsatisfactory.

When available, load-speed curves or graphs of each of the driveN units should be used to establish the driveN HP loads. The torque trace is the most usable data since loading and percentage of operating time are directly indicated. When torque values are known, the following formulas can be used to convert to horsepower:

$$HP = \frac{Q_f\ (rpm)}{5,252}$$

$$HP = \frac{Q_i\ (rpm)}{63,025}$$

where

Q_f = torque, pound-feet
Q_i = torque, pound-inches
rpm = revolutions per minute of sheave

When a drive uses a slip clutch and will operate a significant portion of time at the slip load, it is desirable to use this maximum load condition as one of the operating conditions.

C. Component Placement

Placement of the different shafts with respect to one another makes a great difference in belt life. The main factors to consider when determining the position of the different shafts with respect to one another, are as follows:

1. As a general rule, try to arrange the path of the belt so that, after leaving the driveR sheave, it enters the driveN sheave with the lowest horsepower load first, then enters the other driveN sheaves in order of increasing HP loads. By such an arrangement, the tight side tension will be relatively low at the sheaves that are more lightly loaded and will increase gradually as the belt moves on to the more highly-loaded sheaves. This is not always possible, but since this sequence has an important effect on belt life, it should be given as much consideration as possible.
2. Exceptionally small diameter sheaves should be in spans of lesser tension. This prevents the combination of a high tight-side tension with a high bending tension. In the event this contradicts the placement of the sheaves when following No. 1 (above), several designs should be made and checked for life to pick the optimum placement.
3. If an idler is used, it should be placed in the span with the least tension.

4. Smaller arcs of contact require higher adjacent-span tensions to transmit the same load. These higher tensions decrease belt life. It is generally advantageous to increase the arc of contact of a given sheave.

D. Belt Type and Cross Section

Many factors must be considered in selecting the proper belt construction and cross section. See the Appendix (Figures 1A–1D) for description of Agbelt type. A given drive may have loads such that one belt type and cross section looks like a logical choice, but the diameters may be such that the drive will perform more satisfactorily with another belt type or cross section. NOTE: Agbelts have basic construction differences than industrial belts.

A guide to the proper cross section to use for any combination of design HP and speed is given in Figures 2 and 3 in the Appendix. When the intersection of the design HP and speed of the faster shaft falls near a line between two areas on the chart, it is always desirable to investigate the possibilities in both areas.

Note that the normal diameter range for each of the various cross sections is also given in the charts. A diameter within this normal range may be used as a preliminary diameter for the drive analysis. In the process of designing the drive, it is useful to verify this initial selection. Special circumstances such as space limitations may lead to a choice of belt cross sections or diameters different from those indicated in the chart.

III. DESIGN PROCEDURE

Step 1 Establish driveN speed for a given driveR speed.

Step 2 Determine acceptable belt life as described above.

Step 3 Select several speed and load conditions if the machine operates at different conditions. Approximate as closely as possible the actual operating conditions and the percentage of operating time at each condition.

Step 4 Determine the driveN load for each condition.

Step 5 Find the service factor in Table 1 in the Appendix.

Step 6 Multiply the load at each shaft for each driveN shaft by the service factor to obtain the design HP.

Step 7 Select type of Agbelt and cross section as described abcve.

Step 8 Select a driveR pitch diameter and calculate the driveN pitch diameter using the following formula.

$$\text{driveN pitch diameter} = \text{driveR pitch diameter}\ \frac{(\text{driveR rpm})}{(\text{driveN rpm})}$$

Step 9 Calculate belt length using the following formula.

$$L = 2C + 1.57(D + d) + \frac{(D - d)^2}{4C}$$

where

L = effective belt length (in.)
C = center distance (in.)
D = effective diameter of large sheave (in.)
d = effective diameter of small sheave (in.)

Step 10 Find the allowable tension ratio for each loaded driveN sheave, using the arcs of contact and the graph given in Figure 4 in the Appendix. Find Factor G (arc of contact correction factor) for the driveR, using its arc of contact and Table 3 and graph in Figure 5 in the Appendix.

Step 11 Calculate belt speed using the following formula.

$$\text{belt speed} = \frac{(\text{rpm})\ \text{pitch diameter}}{3.82}$$

(Belt speed is in ft/min; diameter is in in.)

Step 12 Calculate the effective pull for the drive at each condition using the following formula.

$$\text{effective pull} = \frac{(\text{design HP})\ (33{,}000)}{\text{belt speed}}$$

(Effective pull is in lb; belt speed is in ft/min)

Step 13 Calculate the tight side tension for each condition using the formula below. This is the total tight side tension. Calculate tensions in the remaining spans by subtracting the effective pull of each of the driveN sheaves from the tension in the preceding span.

$$T_1 = (1.25)\ \frac{(\text{effective pull})}{G}$$

where

T_1 = tight side span tension (lb)

Step 14 If a multiple belt drive is being considered, divide the total tight side tension by the number of belts.

Step 15 Calculate the tension ratio at each driveN sheave. Compare these tension ratios with the allowable tension ratio found in

Step 10 for each sheave. If the tension ratio is higher than the allowable tension ratio for any loaded driveN sheave, the tension in the belt must be increased using the formula below to prevent slip. If more than one driveN sheave has a higher tension ratio than the allowable tension ratio, calculate the increased tension for the greatest difference. The tension for each span for a given drive condition must be increased by this amount.

$$T_A = \frac{T_T - (R_A \times T_S)}{R_A - 1}$$

where

T_A = additional tension
T_T = tight side tension at sheave
T_S = slack side tension at sheave
R_A = allowable tension ratio at sheave

Step 16 Find the fatigue rate for each sheave or pulley at each condition by using the formula below. Note: The pitch diameters of all flat pulleys which contact the back of an Agbelt are multiplied by a flex factor 0.75 for the purpose of finding the fatigue rates for those pulleys. Also, use this same 0.75 multiplier for all sheaves or pulleys of a double V-belt drive.

$$\log FR = K_1\left[T_T + \frac{K_2}{d} + K_3 \frac{(V)^2}{(1000)}\right] + \log \frac{V}{1000} - K_4$$

where

V = belt speed (ft/min)
FR = fatigue rate
T_1 = tight side tension (lb)
d = pitch diameter (in.)
K_1, K_2, K_3, K_4 = constants depending on belt cross section from Table 2 given in the Appendix.

Step 17 Find the total fatigue rate for each condition by adding the fatigue rate for each sheave for that condition.

Step 18 Calculate belt life from the information now recorded. Multiply the percentage of the total operating time at each condition by the total fatigue rate for each condition. Add the results and use with the following formula.

$$\text{life} = \frac{100 \times \text{belt length}}{\text{sum of} \left(\text{fatigue rate} \times \frac{\%}{100}\right)}$$

(Belt length is in in.)

Step 19 If the calculated belt life is significantly above or below the acceptable belt life, a redesign would be in order.

IV. DRIVE EVALUATION

Consider the following questions when redesigning a drive for greater or lesser belt life:

1. Can the drive be redesigned with different diameter sheaves? Belt life is affected by sheave diameter; as diameter is increased, belt life is increased.
2. Can the sequence of the driveN shafts be changed?
3. Can another cross section be used to advantage? If the fatigue rate for the larger sheave is approximately the same as that for the smaller sheave, repeat analysis for a larger cross section. If, however, the smaller sheave has fatigue rates that are 5–10 times greater than the large sheave, repeat analysis for a smaller cross section.
4. Can the drive be broken into separate drives, thus assigning a smaller HP load to each separate belt?
5. Can the drive be designed with more belts, thus splitting the HP load between two or more belts?
6. Can the drive be redesigned with a longer belt? Belt life is directly proportional to belt length.

A. Field Testing

Even though all the calculations may have shown that the belt drive designed will give satisfactory service, it is imperative that the drive be tested on the actual application. The calculated belt life does not take into account the effects of the environmental conditions, misalignment, maintenance, and the machine operator. By taking the machine to the field, the designer may find that the load conditions are different than anticipated in the calculations.

V. AGRICULTURAL ADJUSTABLE SPEED DRIVE DESIGN

Typically, Ag adjustable speed belts are used where speed variation is required. (See Chapter 6 for details of speed variation.) On a self-propelled combine, for example, speed change can be a necessity for the cylinder drive, cleaning fan drive, header drive, or traction drive. These can take the form of one adjustable pitch sheave and one fixed pitch companion sheave, two adjustable pitch sheaves or the variator type drive. Normally, the amount of speed variation required will determine the type of drive that will be used.

VI. DESIGN PROCEDURE

The following is the step-by-step procedure for designing Ag adjustable speed drives:

Step 1 DriveN speeds. Establish the maximum, minimum, and average driveN speeds for a given driveR speed and the percentage of operating time at each condition.

Step 2 Acceptable belt life. Determine the acceptable life of the belt.

Step 3 DriveN load. Determine the average and peak driveN loads at each speed condition and the percentage of operating time at each load condition.

Step 4 Service factor and design horsepower. Find the drive service factor and multiply the driveN load for each speed condition by the service factor to obtain the design HP.

Step 5 Belt type and cross section. Select the cross section to be checked using the appropriate chart.

Step 6 Required speed variation. Calculate the required variation using the following formula.

$$\text{speed variation} = \frac{\text{maximum dN speed (rpm)}}{\text{minimum dN speed (rpm)}}$$

Step 7 Determine if one or two adjustable sheaves are required. The maximum change in pitch diameter that is obtainable with adjustable speed belts is given by the following formula.

$$\text{maximum PD change (in.)} = \frac{\text{(groove top width, minimum (in.))}}{\tan\ (1/2\ \text{groove angle})} - 2 \times \text{nominal belt thickness (in.)}$$

where

PD = pitch diameter

Using this maximum pitch diameter change, it is possible to calculate the maximum speed variation in a drive using one or two adjustable sheaves. This maximum possible variation is only slightly different with different cross sections.

If the drive requires a speed variation of 2.00 or below (from the speed variation formula), that variation can be obtained with one adjustable sheave. If it requires a speed variation between 2.00 and 4.24, two adjustable sheaves will be required. If a speed variation over 4.24 is required, a single agricultural adjustable drive cannot be used. (This is assuming the use of American Society of Automotive Engineers (ASAE) standard groove dimensions.) Two or more successive drives could be used to increase the total speed variation.

If a larger speed variation forces the use of smaller-than-recommended sheave diameters, severe bending stresses will be imposed and shorter calculated belt life will result.

Another consideration in deciding whether one or two adjustable sheaves will be used is whether a shaft can be moved when a speed change is made. When only one adjustable sheave is used, the change in pitch diameter must be offset by a change in shaft position (because the belt length stays the same). If centers cannot be changed (and an automatic tensioning idler is not being used), two adjustable sheaves might possibly be used.

Step 8 Speed ratios. Calculate the required speed-up and speed-down ratios using the driveR rpm as the base using the following formula.

$$\text{speed-up ratio} = \frac{\text{maximum dN speed (rpm)}}{\text{dR speed (rpm)}}$$

$$\text{speed-down ratio} = \frac{\text{dR speed (rpm)}}{\text{minimum dN speed (rpm)}}$$

Step 9 Pitch diameters.

One adjustable sheave. First calculate the pitch diameter of the fixed sheave using the following formula.

$$\text{PD, fixed sheave} = \frac{\text{maximum PD change (in.)} \times \text{speed-up ratio}}{\text{speed variation} - 1}$$

(Note: Maximum PD change from Step 7, speed-up ratio from Step 8, and speed variation from Step 6.)

Then calculate the maximum and minimum pitch diameters of the adjustable sheave from the following formulas.

$$\text{maximum PD, adjustable sheave} = \text{PD (in.), fixed sheave} \times \text{speed-down ratio}$$

$$\text{minimum PD, adjustable sheave} = \frac{\text{PD (in.), fixed sheave}}{\text{speed-up ratio}}$$

Check calculations by making certain that the difference between the maximum and minimum pitch diameters of the adjustable sheave equals the maximum PD change. Find the outside diameters by adding the PD to the outside diameter (OD) value. If these outside diameters fit the available space satisfactorily, proceed to Step 10. If the outside diameter of the adjustable sheave is too large for the space and is larger than the minimum recommended diameter, it can be reduced, but not below the recommended minimum. If the adjustable sheave diameter is reduced, recalculate using the above formulas to find the new diameter of the fixed sheave.

Two adjustable sheaves. Calculate the maximum and minimum pitch diameters of both adjustable sheaves using the following formulas:

maximum PD, driveR =

$$\text{maximum PD change } \frac{(\text{speed-up ratio} + \text{speed variation})}{\text{speed variation} - 1}$$

minimum PD, driveR =

$$\text{maximum PD change } \frac{(\text{speed-up ratio} + 1)}{\text{speed variation} - 1}$$

maximum PD, driveN =

$$\text{maximum PD change } \frac{(\text{speed-down ratio} + \text{speed variation})}{\text{speed variation} - 1}$$

minimum PD, driveN =

$$\text{maximum PD change } \frac{(\text{speed-down ratio} + 1)}{\text{speed variation} - 1}$$

Calculations can be checked by making sure that the difference between the maximum and minimum pitch diameters of the adjustable sheaves equals the maximum PD change.

Find the outside diameters by adding the PD to OD value. If these outside diameters fit the available space satisfactorily, proceed to Step 10. If the outside diameters of the adjustable sheaves are too large for the space available, and are larger than the minimum recommended diameter, they can be reduced but not below the recommended minimum. If diameters are reduced, recalculate using the above formulas to find the new diameter of the other sheave.

Variator drive. Since the variator drive has a fixed pitch driveR and driveN sheave with an adjustable pitch sheave using a floating center flange between them, the procedure for finding speed variation is much like the one "adjustable sheave" (see Chapter 6).

When determining the speed variation for the variator-type drive, it must be kept in mind that the jackshaft sheave (variator) must have the same PD change for the primary and secondary drives.

Step 10 Approximate belt length (in.). Details of determining required belt length will be discussed separately. At this point, only an approximation is necessary. Calculate the approximate belt length using the following formula:

$$L_A = 2C + 1.57\,(D + d)$$

where

L_A = approximate belt length (in.)
C = desired center distance (in.)
D = diameter of large sheave (in.)
d = diameter of small sheave (in.)

Step 11 Factor G. Find three values of Factor G, one for each of the three driveN speed conditions established in Step 1. For a two-sheave drive, first calculate (D-d)/c using the values of pitch diameter established in Step 9 and the desired center distance used in Step 10. Then find Factor G from the appropriate table.

Find the values of Factor G for drives with inside idlers by using the smallest arcs of contact for either the driveR or driveN sheave, for each of the three speed conditions.

Step 12 Belt speed. Calculate belt speed for each of the three driveN speed conditions established in Step 1. Use the belt speed formula given previously.

Step 13 Effective pull. Calculate the effective pull for the drive at each speed condition.

Step 14 Tight side tension. Calculate the tight side tension for each condition using the appropriate formula and the value of Factor G found in Step 11.

Step 15 Fatigue rate. Find the fatigue rate for each sheave at each condition by using the fatigue rate formula and the constants for the cross section chosen in Step 5.

Step 16 Total fatigue rate. Add the fatigue rates for each condition.

Step 17 Belt life. Using the life formula, calculate the belt life.

Step 18 Drive evaluation. If the calculated belt life is significantly above or below the acceptable belt life, the drive evaluation discussion given previously can be helpful. If the calculated life is too low, and if one adjustable speed sheave of the minimum recommended diameter was used, consider using two adjustable speed sheaves of larger diameter, even though the additional speed variation may not be required.

VII. FITTING THE DRIVE

The effective belt length, installation, and take-up requirements for adjustable-speed Ag belt drives involve the same basic principles as used with other Ag belt drives. There is, however, the additional consideration of providing for speed variation.

When the position of a V-belt in a sheave changes, the diameter at which it is operating also changes. This changes the effective belt length. The difference must be accounted for in some manner. Fitting of adjustable speed drives requires finding the proper effective belt length and providing for the following:

Installation allowance (sometimes an adjustable sheave can be blocked open when installing the belt, eliminating the need for installation allowance)
Take-up allowance
Speed variation allowance

These allowances can be accomplished mechanically by combinations of the following:

Diameter change
Movement of driveR or driveN
Movement of an inside idler

The most satisfactory method, and the one preferred when it is practical, is movement of driveR or driveN shafts by increasing or decreasing the center distance.

Several specific examples of fitting adjustable speed drives are given below.

A. One Adjustable Sheave

The following procedure is for a drive using one adjustable, spring-loaded sheave and one fixed-pitch companion sheave, where the required allowances are provided by movement of the driveR or driveN shafts.

Step 1 Establish the effective belt length. Use the maximum diameter of the adjustable speed sheave. For center distance, use the minimum value desired, not counting any installation allowance that might be found necessary in Step 2.

Step 2 It is not necessary to provide an installation allowance on this type of drive when the adjustable sheave can be blocked open for belt installation. If this cannot be done, for any reason, then the center distance used in Step 1 is decreased by 1/2 the amount shown in the installation allowance table (because the table is based on length rather than center distance). The result is the minimum center distance for the drive.

Step 3 Provide the allowance for speed variation by calculating a maximum center distance using the following formula.

$$C = A - \frac{K(D_o - d_o)}{2}$$

where

C = center distance calculated (in.)
A = L_o - 1.57 (D_o - d_o)
D_o = effective diameter of large sheave (in.)
d_o = effective diameter of small sheave (in.)
K = center distance factor from table in the Appendix
L_o = effective belt length found in Step 1 (in.)

Use the minimum diameter of the adjustable speed sheave as D_o or d_o — whichever is applicable. The difference between the center distance found here and that used in Step 1 is the center distance change allowed for speed variation.

Step 4 Provide the allowance for take-up by adding 1/2 the amount shown in the take-up allowance table to the center distance found in Step 3. The result is the maximum center distance for the drive. The total center distance movement required is usually provided for in one adjustment. However, the best arrangement is to provide the allowances for speed variation and take-up separately. If the speed variation adjustment is an integral part of the movable assembly, then the desired speed range can be achieved without the possibility of moving the belt into the flats of the adjustable sheave.

If the above drive illustration were changed to one where neither shaft could be moved to provide the allowances, an inside idler would have to be used. In this case, scale layouts are made to achieve the same results as the above procedure.

If the above drive were not automatically tensioned, then it would be necessary to retension after each speed change is made.

B. Two Adjustable Sheaves

When two equal-diameter adjustable sheaves are used (with one sheave spring-loaded and with no inside idler involved) the fitting procedure is simplified. Use only Steps 1 and 4 above. To establish the effective belt length, use the maximum diameter for only one of the sheaves.

Steps 2 and 3 are not necessary. In this case, installation can usually be accomplished by blocking open the spring-loaded sheave, and a speed variation allowance is not necessary because the calculated belt length is the same in all speed conditions. Take-up allowance is provided by adding 1/2 the amount shown in the take-up allowance table to the center distance used in Step 1.

C. The Variator

The variator-type drive, which involves two adjustable-speed Ag belts in tandem, is treated the same as a one adjustable-sheave drive, except that belt length and center distance allowance calculations must be performed concurrently on both drives.

VIII. CHANGE IN SPEED VARIATION WITH TIME

During the life of any V-belt, wear of both the sheave and belt sidewalls takes place. Also, the length of the belt increases slightly. In a regular Ag belt drive, these factors create a need for occasional tension maintenance. In an adjustable speed Ag belt drive, these factors can also cause a slight shift in the speed variation range.

Both the amount and direction of this speed-variation shift will vary with the application. These factors in the application may include the number of adjustable sheaves, the use of a spring-loaded sheave, the speed ratio of the drive, and the center distance adjustment for take-up.

The adjustable speed Ag belt procedure given previously is based on using the total available range of pitch-diameter variation of an adjustable sheave. Therefore, the wear and length change factors will change, slightly, the actual speed attainment from that calculated. Therefore, if the driveN speed requirements are very close to those calculated, the wear and length factors must be taken into account.

IX. CONCLUSION

The design procedure given here is to be used as a guide. This procedure is used by the Gates Rubber Company and may be different from those of other V-belt manufacturers. The fatigue constants given for the fatigue-rate formula are periodically updated. It is highly recommended that before finalization of any V-belt drive design, the belt manufacturer be contacted to insure all up-to-date procedures and formulas are used.

APPENDIX

TABLE 1 Service Factors, Farm Implement Drives

Machine or operating unit	Service factor
Ammonia applicator	1.0
Auger, delivery	1.3
Bean thresher, stationary	1.3
Beet harvester	1.3
Cane harvester	1.3
Chopping, combine	1.5
Cleaning shoe	1.0
Corn head	1.5
Corn husker	1.3
Corn picker	1.3
Corn sheller cylinder	1.5
Corn shredder	1.5
Cotton harvester, stripper type	1.3
Cotton picker	1.3
Cutting, reel	1.2
Cutting, sickle bar	1.5
Cutting, sickle bar and counter weight	1.3
Cylinder, combine	1.5
Draper, canvas	1.3
Elevator, delivery	1.1
Elevator, portable and stationary	1.2
Ensilage blower	1.5
Ensilage cutter	1.5
Ensilage field harvester	1.5
Expelling, combine	1.2
Fan, cleaning	1.0
Fan over 10 HP	1.3
Feeder roll	1.3
Feeding, combine	1.3
Flail harvester	1.5
Front cylinder beater	1.3
Grain blower	1.3
Hammermill rotor and mixer	1.5
Hay bale loader, field	1.2
Hay baler, pickup or stationary	1.5
Hay chopper, field	1.5

(continued)

TABLE 1 (Continued)

Machine or operating unit	Service factor
Hay conditioner	1.5
Hay hoist and conveyor	1.2
Hay mower	1.5
Hay rake, side delivery	1.2
Hay waferer and cuber	1.5
Header, combine	1.5
Husk blower	1.3
Hydraulic system oil pump	1.3
Hydrostatic pump and motor	1.2
Loader, tractor front end	1.3
Main drive, combine	1.3
Manure loader	1.5
Manure spreader	1.5
Nut and fruit shaker harvester	1.5
Orchard sprayer and duster	1.3
Pea thresher, stationary	1.3
Pea viner, stationary and self-propelled	1.3
Peanut combine	1.3
Peanut digger	1.3
Pickup attachment, combine	1.2
Pitman, combine	1.3
Potato harvester	1.3
Potato loader	1.2
Propulsion, combine	1.3
Rear cylinder beater	1.0
Reel, pickup	1.3
Reel, bat	1.0
Rotary cutter	1.5
Separation, combine	1.0
Sieve, combine	1.0
Silo unloader, thrower	2.0
Stock shredder	1.5
Straw chopper, combine	1.5
Straw walker	1.0
Threshing, combine	1.2
Traction, self-propelled machine	1.3
Unloader, combine	1.5
Vegetable harvester	1.2
Wagon unloader	1.3
Windrower and swather	1.2

TABLE 2 Constants for Fatigue Rate Formula

Agbelt type and cross section	K_1	K_2	K_3	K_4
Agbelt HA	0.02541	536.51	0.6676	6.14267
Agbelt HB	0.01668	1132.10	1.1003	6.14267
Agbelt HC	0.01068	2500.00	1.8695	6.14267
Agbelt HD	0.00614	6666.60	3.6036	6.14267
Agbelt HA-P	0.020817	647.06	0.6676	6.14267
Agbelt HB-P	0.013654	1422.20	1.1003	6.14267
Agbelt HC-P	0.0087291	3279.20	1.8695	6.14267
Agbelt HD-P	0.0050166	9225.10	3.6037	6.14267
Adjustable speed Agbelt, HI	0.01507	993.01	1.9419	5.1423
Adjustable speed Agbelt, HJ	0.00964	1939.50	3.0342	5.1423
Adjustable speed Agbelt, HK	0.00670	3351.40	4.3692	5.1423
Adjustable speed Agbelt, HL	0.00492	5322.10	5.9470	5.1423
Adjustable speed Agbelt, HM	0.00377	7943.90	7.7674	5.1423

TABLE 3 Center Distance Factor (K)

$\frac{D_o - d_o}{A}$	Factor K	$\frac{D_o - d_o}{A}$	Factor K
0.00	0.00	0.29	0.15
0.02	0.01	0.30	0.16
0.04	0.02	0.32	0.17
0.06	0.03	0.34	0.18
0.08	0.04	0.35	0.19
0.10	0.05	0.37	0.20
0.12	0.06	0.39	0.21
0.14	0.07	0.40	0.22
0.16	0.08	0.41	0.23
0.18	0.09	0.43	0.24
0.20	0.10	0.44	0.25
0.21	0.11	0.46	0.26
0.23	0.12	0.47	0.27
0.25	0.13	0.48	0.28
0.27	0.14	0.50	0.29
		0.51	0.30

Conventional Agbelt

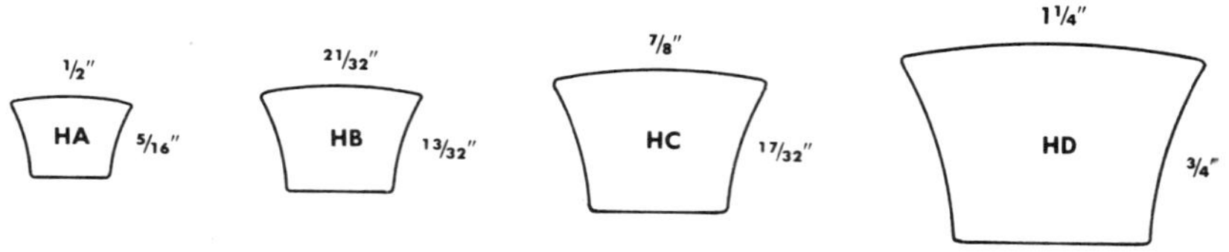

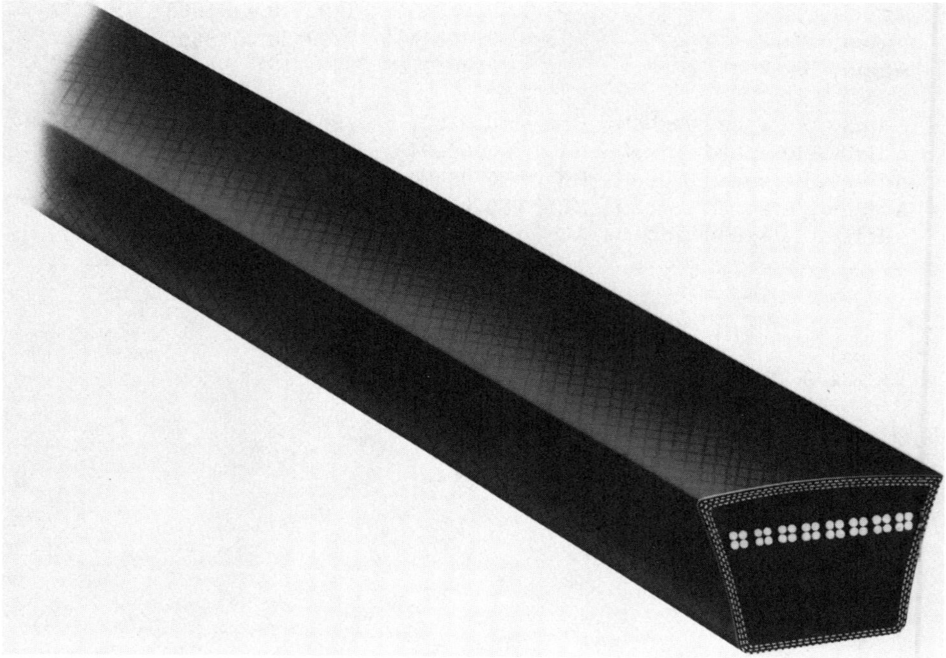

FIGURE 1A

Dubl-V Agbelt

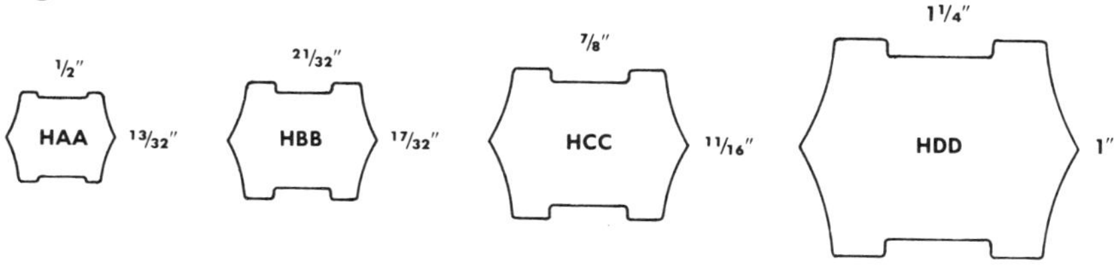

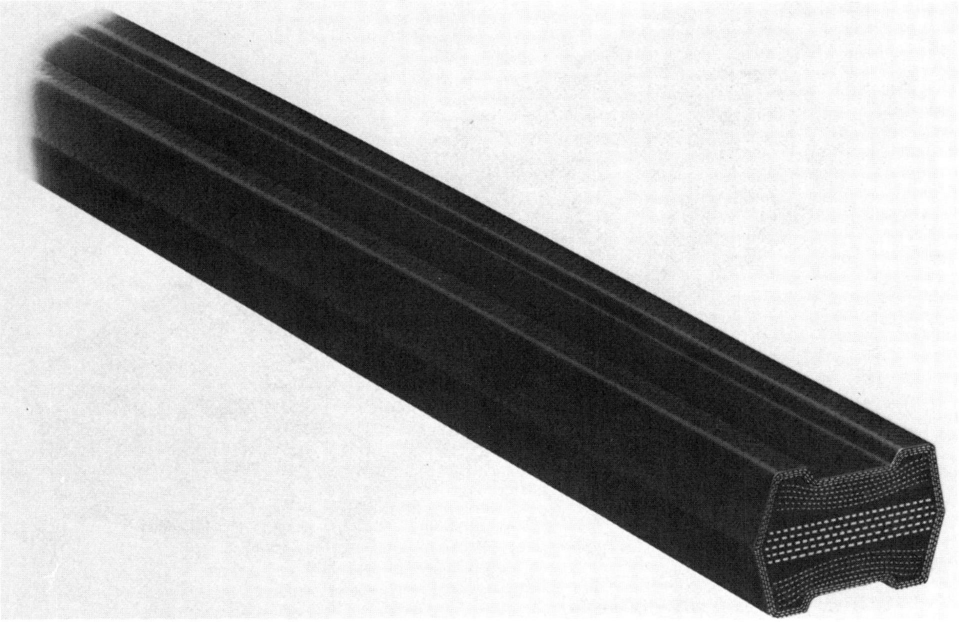

FIGURE 1B

Adjustable Speed Agbelt

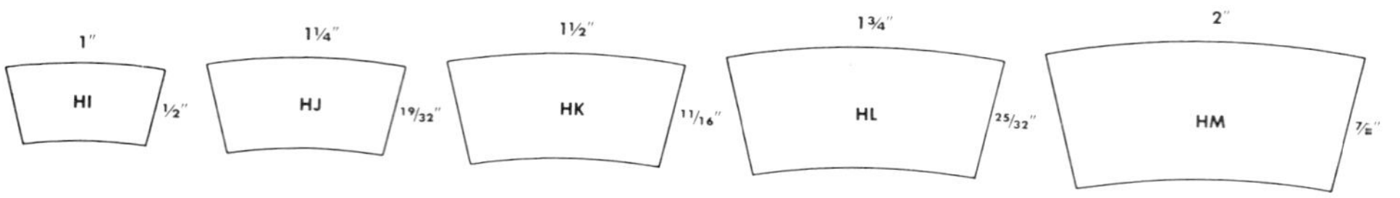

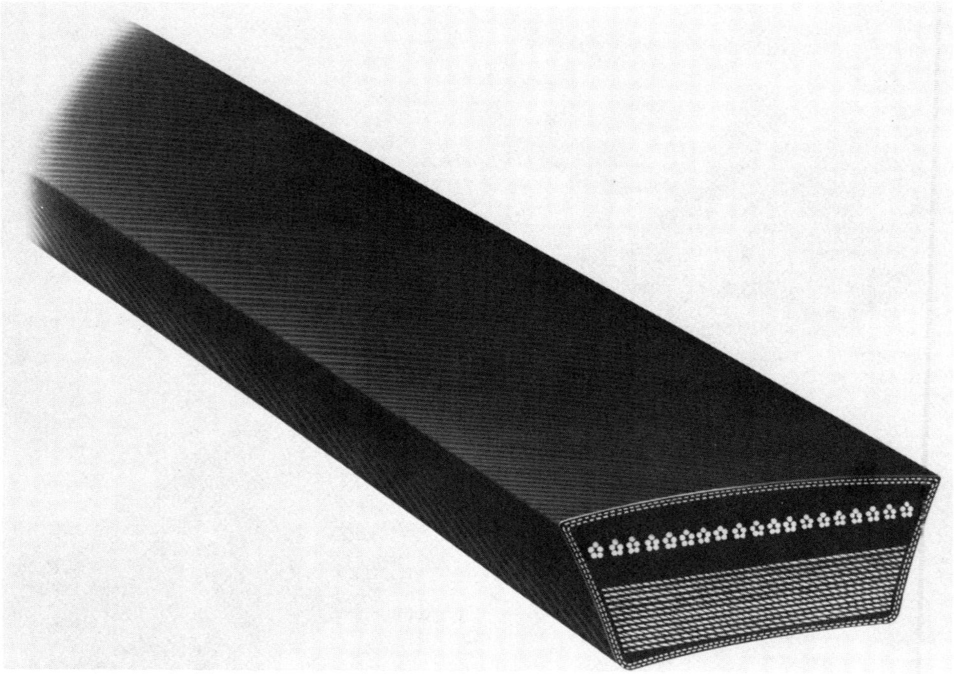

FIGURE 1C (top and bottom)

Joined Agbelt

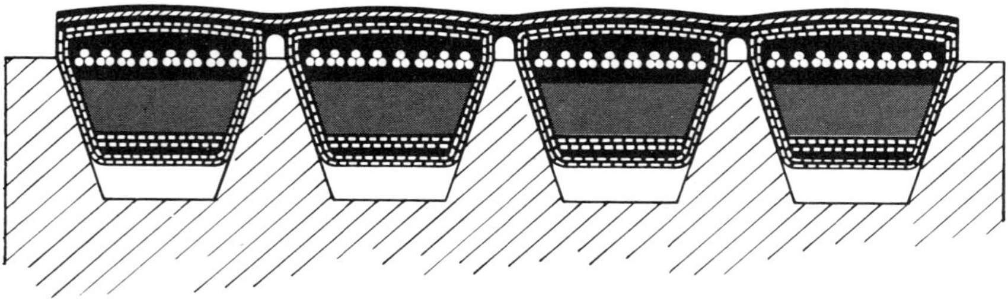

FIGURE 1D

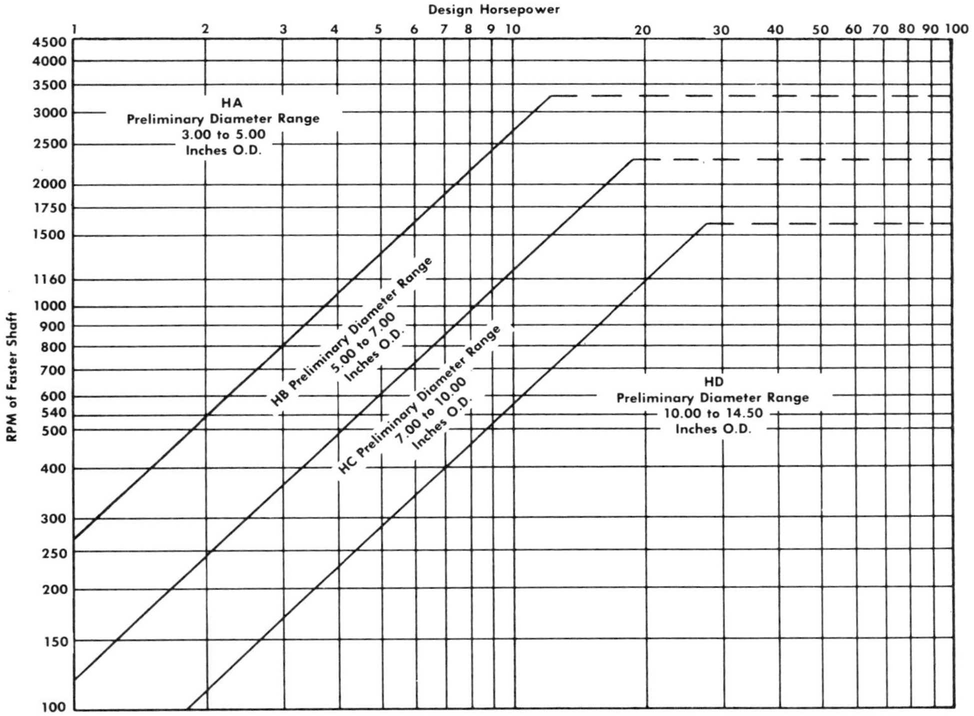
Design Horsepower
1
2
3
4
5
6
7
8
9
10
20
30
40
50
60
70
80
90
100
RPM of Faster Shaft
4500
4000
3500
3000
2500
2000
1750
1500
1160
1000
900
800
700
600
540
500
400
300
250
200
150
100
HA
Preliminary Diameter Range
3.00 to 5.00
Inches O.D.
HB Preliminary Diameter Range
5.00 to 7.00
Inches O.D.
HC Preliminary Diameter Range
7.00 to 10.00
Inches O.D.
HD
Preliminary Diameter Range
10.00 to 14.50
Inches O.D.

FIGURE 2

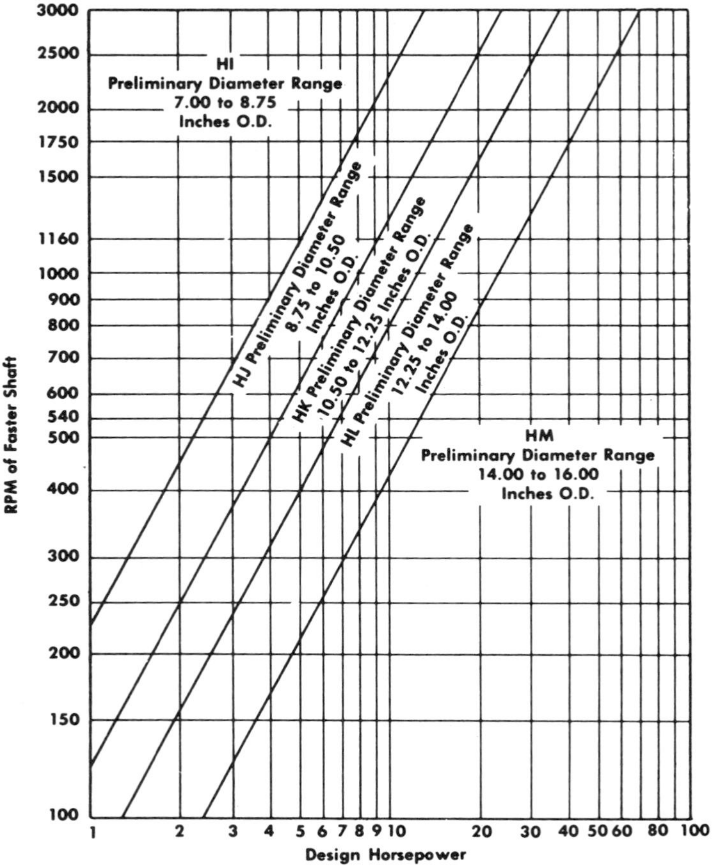
HI
Preliminary Diameter Range
7.00 to 8.75
Inches O.D.
HJ Preliminary Diameter Range
8.75 to 10.50
Inches O.D.
HK Preliminary Diameter Range
10.50 to 12.25 Inches O.D.
HL Preliminary Diameter Range
12.25 to 14.00
Inches O.D.
HM
Preliminary Diameter Range
14.00 to 16.00
Inches O.D.
RPM of Faster Shaft
3000
2500
2000
1750
1500
1160
1000
900
800
700
600
540
500
400
300
250
200
150
100
1
2
3
4
5
6
7
8
9
10
20
30
40
50
60
80
100
Design Horsepower

FIGURE 3

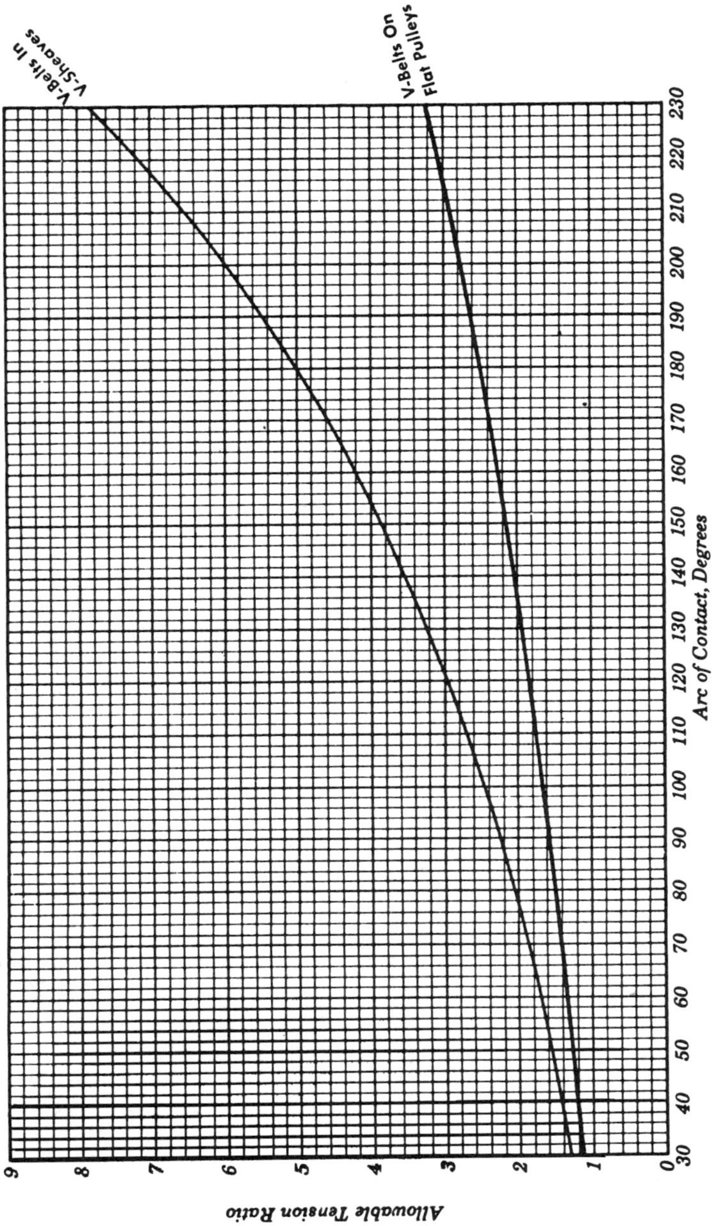
V-Belts In
V-Sheaves
V-Belts On
Flat Pulleys
Allowable Tension Ratio
0
1
2
3
4
5
6
7
8
9
Arc of Contact, Degrees
30
40
50
60
70
80
90
100
110
120
130
140
150
160
170
180
190
200
210
220
230

FIGURE 4

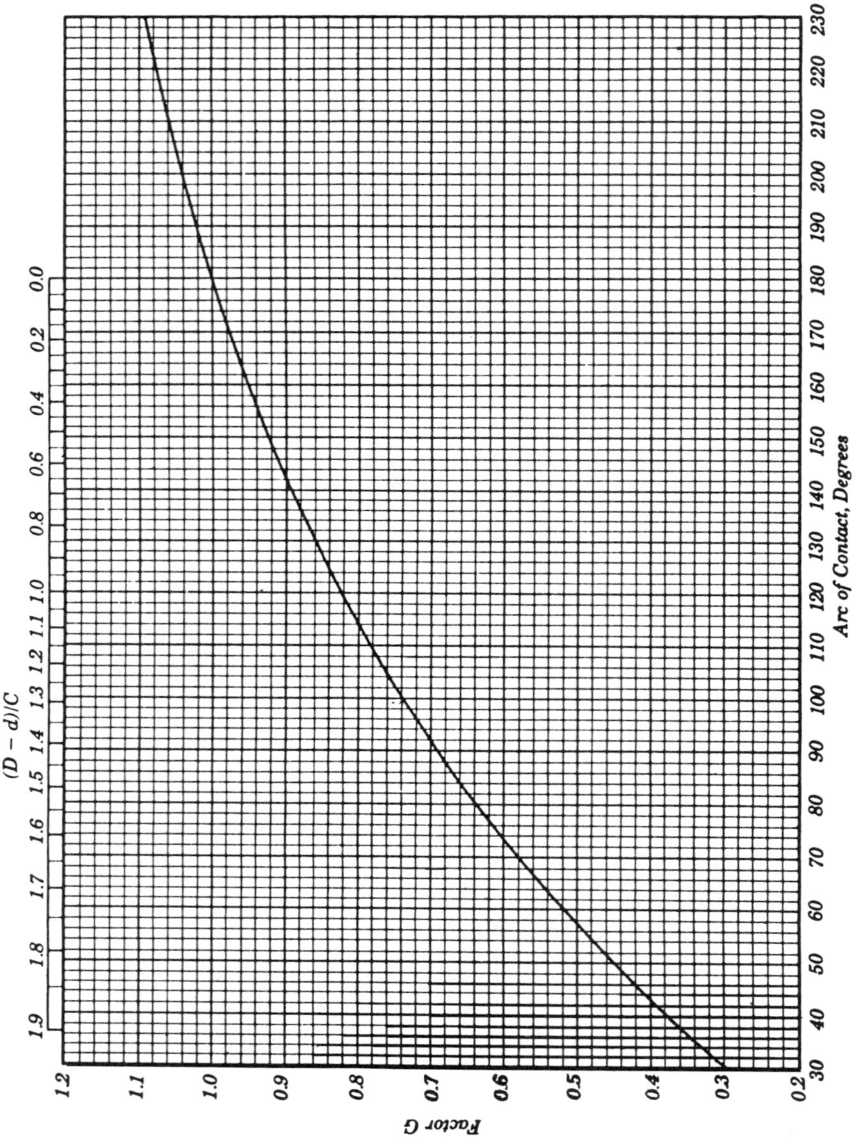
(D − d)/C
0.0
0.2
0.4
0.6
0.8
1.0
1.1
1.2
1.3
1.4
1.5
1.6
1.7
1.8
1.9
Factor G
1.2
1.1
1.0
0.9
0.8
0.7
0.6
0.5
0.4
0.3
0.2
Arc of Contact, Degrees
30
40
50
60
70
80
90
100
110
120
130
140
150
160
170
180
190
200
210
220
230

FIGURE 5

5

Automotive Accessory Drives Using V-Belts

ROBERT E. STETSON *Automotive Application Engineering, The Gates Rubber Company, Denver, Colorado*

I. V-BELTS ON AUTOMOTIVE ACCESSORY DRIVES

A. Types of Drives and General Description

V-belts are widely used to transmit power to the cooling fan, water pump, air conditioning compressor, and exhaust emission air pump on passenger car engines. V-belts also drive the air brake compressor, vacuum pump, hydraulic pump, alternator, and fan water pump on heavy-duty truck engines. Wherever an engine is used, its accessories usually obtain their operating power through V-belts.

1. Fixed Speed Ratio

The majority of drives are normally designed using a fixed speed ratio compatible with the characteristics of the driven accessory. These drives are usually single-belt drives on passenger car engines. Two or more V-belts, in matched sets, are used on heavy-duty applications when the accessory power requirements dictate.

2. Variable Speed Ratio

The current trend toward more fuel-efficient engines has fostered the development of variable-ratio accessory drives. These drives are designed to limit accessory speed range and are sometimes called *constant speed accessory drives*. Two variable-pitch sheaves and a variable-speed V-belt are used to increase accessory speeds at idle engine speed and to decrease the accessory speeds at high engine speed. The driven sheave is usually the driver of all other engine accessories so the advantage of the constant speed range can be realized. Refer to Chapter 6 for the design principles involved.

Two-speed drives are also designed to shift between two basic speed ratios. A controlling magnetic clutch and an overrunning clutch on two shafts coupled with V-belts to the accessories constitute a two-speed drive system. This system is not a true variable-speed drive, but it does satisfy the concept of reducing accessory speed and thus reducing the accessory power required.

II. DRIVE DESIGN CONSIDERATIONS

A. Acceptable Belt Life

The engine designers and vehicle engineers are best able to judge acceptable belt life from their experience gained from track and field testing. A guide has been developed that indicates an acceptable life range for accessory V-belts.

Type of vehicle	Acceptable belt life (hr)
Passenger cars, light trucks	600 – 1000
Heavy trucks (not transports)	1000 – 1500
Buses, small tractors, transports	1500 – 4000

Considerations of operating time, environment, accuracy of accessory load requirements, and possible addition of optional equipment to the engine or vehicle must be made when using the above guide.

B. Engine Duty Cycle

A basic selection of engine speeds needs to be determined before an accessory drive is designed. This includes assigning a percentage of time for each engine speed. A typical passenger car engine duty cycle is as follows:

Condition	% Operating time	C/S rpm
A	5	600
B	20	1000
C	45	2000
D	25	3000
E	5	4000

It is particularly important not to select the highest speed for a one-condition check. A drive may appear to give very short life under such a condition, and yet be satisfactory when all conditions are evaluated.

C. Accessory Loads

Horsepower load curves for each accessory are needed to calculate the tensions required for each condition of operation in the engine duty cycle. Load curves are usually plotted against accessory speed. Accessory speeds therefore are computed for each engine speed, using chosen speed ratios, before the accessory loads are determined.

D. Service Factor

To compensate for extremes in environment, a service factor may be used. The normal service factor to use with most automobile accessories is as follows:

Accessory	Service factor
Fans, water pumps, centrifugal compressors	1.0
Generators, small alternators	1.0
Large alternators (over 50 amp)	1.3
Piston compressors and hydraulic pumps	1.3

However, if unusually high temperatures or extremely oily or dusty conditions will be encountered by the drive, the normal service factor may be increased to compensate for these V-belt life-reducing factors.

E. Belt Type and Cross Section

Initial selection of a belt type and cross section must consider the number of pulleys in the drive, their diameters, and the accessory loads. Perhaps the best reference for belt cross section selection is the Society of Automotive Engineers (SAE) standard J636c "V-Belts and Pulleys." Recommended minimum effective diameters for the SAE belt sizes are shown in Table 1A of the standard. See discussion on belt types and constructions (Section IV) for a guide to belt type selection.

F. Component Placement

Discussions with the engine and vehicle designers will usually provide the location of the various accessories. Space limitations are numerous in today's engine compartments. Generally, the small diameter pulleys should be located on the slack side (low tension) of the driver pulley. Heavier loaded accessories (larger diameter pulleys) should then be positioned around the crankshaft in ascending order. Most accessory drives are designed after the accessories are positioned about the engine in the allowable space provided.

III. DESIGN PROCEDURE

Use the following steps in designing and checking out an automotive accessory V-belt drive. In these steps, calculate the necessary factors for determining the fatigue rate and thus the expected life of a belt on a drive. In this procedure, it is assumed that the drive has three or more pulleys and is being checked using a typical five- or six-condition engine duty cycle.

1. Establish all driveN shaft speeds using selected speed ratios for a given driveR speed.

2. Select a driveR pulley pitch diameter and calculate all driveN pulley pitch diameters using the following formula.

$$\text{d of DN} = \frac{(\text{rpm of DR})\ (\text{d of DR})}{\text{rpm of DN}}$$

where

d = pitch diameter (in.)
DN = driveN
DR = driveR
rpm = revolutions per minute

3. Determine acceptable belt life.

4. Select the driveN loads from the accessory HP curves for each condition.

5. Find the service factors for each driveN accessory then multiply each load to obtain the design HP.

6. Total the design HP for each accessory and each condition to obtain the driveR design HP.

7. Select the type of belt and cross section. At this point, more than one selection could be made to run a comparison, if desired.

8. Lay out the drive to scale, using pulley pitch diameters. Show the movable pulley in its normal operating position as well as its two extreme positions. Number the belt spans and driveN pulleys in a direction opposite to belt travel. The first driveN pulley is that pulley that is adjacent to span 1.

9. Calculate approximate belt length. Measure the span lengths on the layout. Measure the arc of contact on each sheave. Calculate the arc lengths by:

$$\text{AL} = \pi \times d \times \frac{\theta}{360}$$

where

AL = arc length (in.)
θ = arc of contact (deg)
d = pitch diameter of pulley (in.)

Belt length is the sum of the span lengths and arc lengths.

10. Find allowable tension ratio and Factor G (see Figure 1). Locate the arc of contact for each of the loaded driveN pulleys. Locate the corresponding value of allowable tension ratio. From Figure 2, locate the arc of contact for the driveR along the bottom of the chart. Find the corresponding value of Factor G.

11. Calculate the belt speed for each condition using the following formula:

$$V = \frac{(rpm)\ (d)}{3.82}$$

where

V = belt speed (ft/min)
rpm = revolutions per minute of pulley
d = pitch diameter of same pulley (in.)

12. Calculate the effective pull for each pulley, and for each drive condition by use of the following formula. When calculating belt life for a drive with more than one belt, divide the design HP by the number of belts.

$$T_T - T_S = \text{effective pull} = \frac{(HP_D \text{ per belt})\ (33{,}000)}{V}$$

where

$T_T - T_S$ = effective pull per belt (lb)
V = belt speed (ft/min)
d = pitch diameter (in.)
HP_D = design horsepower

Note: The total of the effective pulls for each driveN pulley should equal the effective pull of the driveR pulley.

13. Calculate the tension in span 1 for each condition as follows:

$$T_1 = \frac{1.25 \times (T_T - T_S)}{G}$$

where

T_1 = tension of span 1 (lb)
$T_T - T_S$ = effective pull per belt of driveR pulley (lb)
G = Factor G of driveR

Calculate tensions in the remaining spans by subtracting the effective pull of each of the driveN sheaves from the tension in the preceding span, proceeding around the drive opposite to belt travel.

14. Calculate the tension ratio at each loaded driveN pulley for each condition. Compare these tension ratios with the allowable tension ratio for each pulley. If the tension ratio is higher than the allowable tension ratio for any driveN pulley, the tension must be increased for this condition. When more than one driveN pulley meets the above, calculate the increased tension for the greatest difference. The additional tension required can be calculated by the formula below:

$$T_A = \frac{T_T - (R_A \times T_A)}{R_A - 1}$$

where

T_A = additional tension (lb)
T_T = tight side tension (lb)
T_S = slack side tension (lb)
R_A = allowable tension ratio

When it is necessary to add this tension, the tension for each span for a given drive condition must be increased by this amount.

16. Record the total span tensions as calculated above.

17. Select the tight side tensions for each loaded pulley.

18. Find the fatigue rate for each pulley at each condition by using the formula below:

$$\log FR = K_1 \left(T_T + \frac{K_2}{d} + K_3 S^2\right) + \log S - K_4$$

where

K_1, K_2, K_3, and K_4 = cross section constants (see Table 1)
FR = fatigue rate
S = belt speed/1,000
T_T = tight side tension (lb)
d = pitch diameter of pulley (in.)

19. Add the fatigue rates for each pulley to find the total fatigue rate for each condition.

20. Calculate belt life from the information recorded.

a. Multiply the percentage of the total operating time at each condition by the total fatigue rate for each condition and add the results.

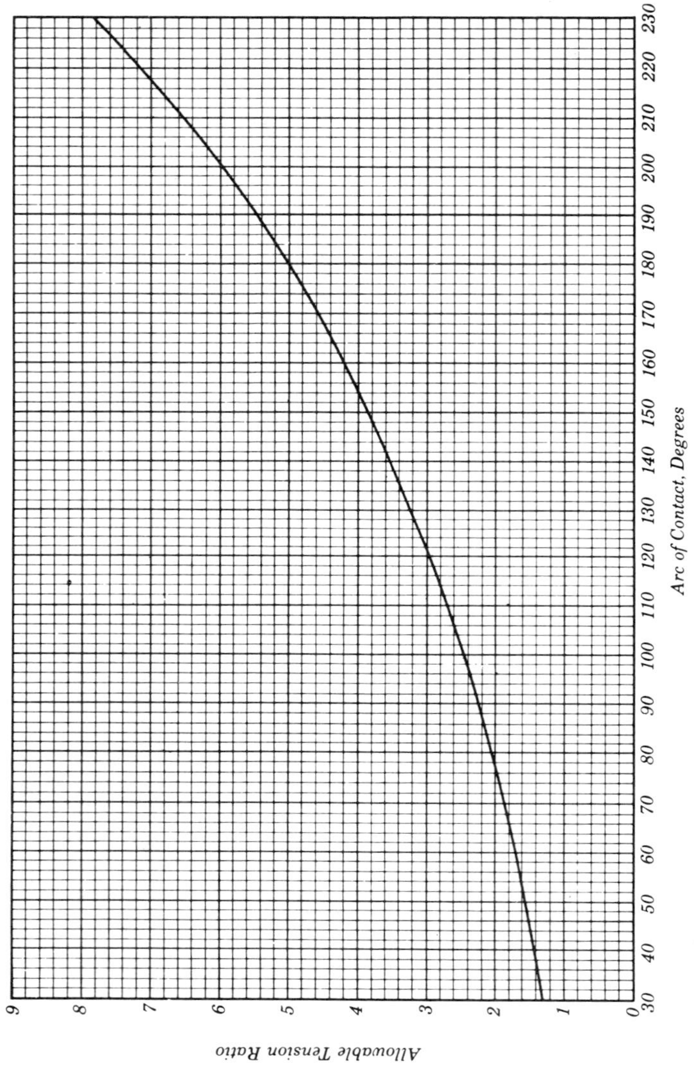

FIGURE 1 Allowable tension ratio.

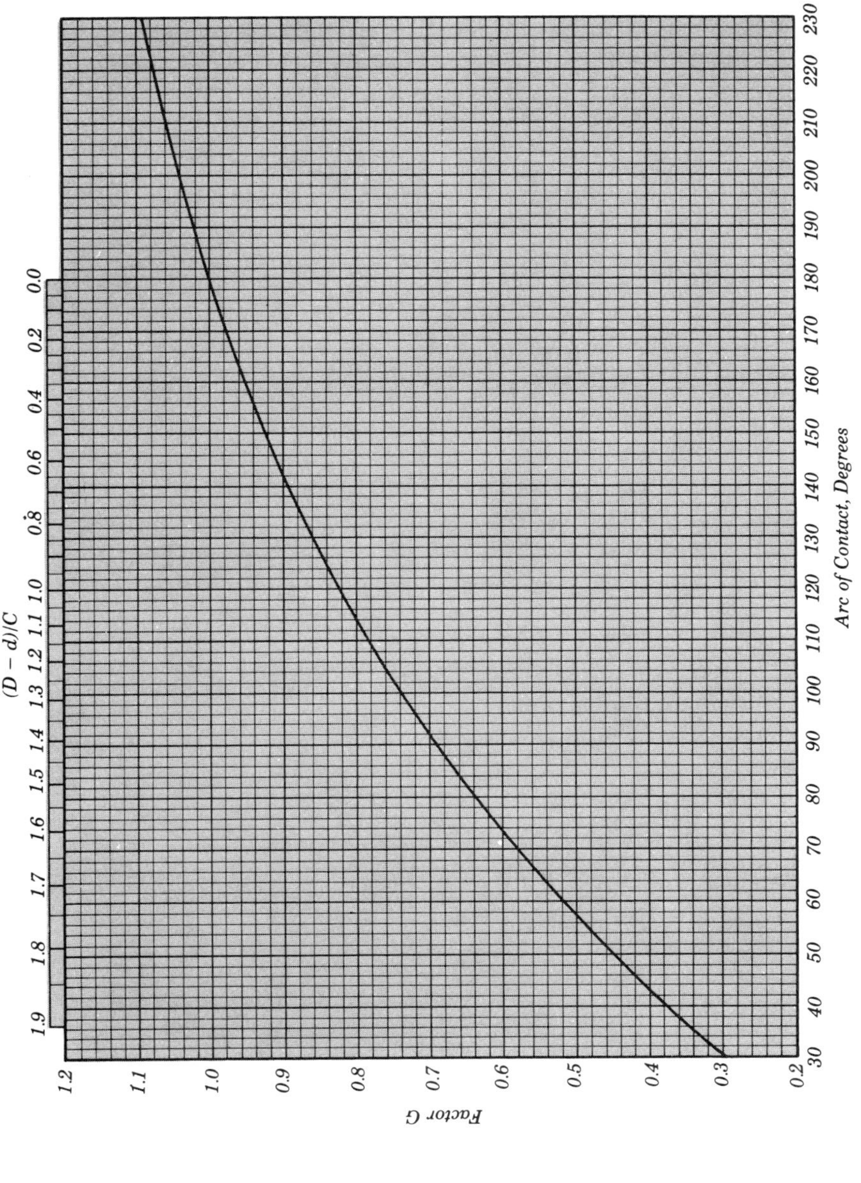

FIGURE 2 Factor G.

TABLE 1 Fatigue Rate Formula Constants

SAE size	K_1	K_2	K_3	K_4
0.250	0.048787	149.17	0.21521	6.1427
0.315	0.034608	265.36	0.31156	6.1427
0.380	0.025495	443.08	0.43306	6.1427
0.440	0.020312	648.68	0.55320	6.1427
0.500	0.016661	904.42	0.68485	6.1427
11/16 (0.600)	0.022722	1144.00	1.13400	9.1427
3/4 (0.660)	0.018968	1530.30	1.36390	9.1427
7/8 (0.790)	0.013882	2530.60	1.87660	9.1427
1 (0.910)	0.010640	3884.40	2.46300	9.1427

Source: Courtesy of the Gates Rubber Company.

b. Calculate belt life using the following formula:

$$h = \frac{100\ La}{\text{sum of } (FR\text{'s} \times \%/100)}$$

where

h = belt life (hr)
La = approximate belt length (in.)
FR = fatigue rate
% = percent of total operating time

Round the belt life to the nearest 100 hours.

21. Drive evaluation. Compare the calculated belt life to the acceptable belt life range selected. When the life is compatible with the selected life, no further design effort is needed. A wide variation, greater or less than the acceptable life range, requires a study for redesign. Points to consider include the following:

a. Change pulley diameters? Increasing pulley diameters (maintaining speed ratios) will increase belt life.
b. Change component placement? Changes in arcs of contact may lower required tensions and increase belt life.

TABLE 2 PD to Nominal OD Values

SAE size	PD to OD (in.)
0.250	0.04
0.315	0.05
0.380	0.06
0.440	0.07
0.500	0.08
11/16 (0.600)	0.00
3/4 (0.660)	0.02
7/8 (0.790)	0.04
1 (0.910)	0.06

Source: Courtesy of the Gates Rubber Company.

c. Change in belt cross sections (SAE size)? Perhaps a larger or smaller SAE size is indicated. Check the fatigue rates on the small pulleys.
d. Change in drive geometry? Two separate drives, each using one belt, may be the best solution.
e. Change in number of belts? Maybe the use of two belts on the original drive, splitting the total power requirements, is indicated.

22. Determine effective belt length. Add the PD to the nominal OD value from Table 2 to the pulley pitch diameters found in part 2 of the design procedure (Section III) to obtain nominal outside diameters. Draw the nominal outside diameters on the layout with the movable pulley in its two extreme positions. Then draw in the belt for both positions of the movable pulley. Measure the span lengths and calculate the nominal OD arc lengths for both positions of the movable pulley.

From the longest effective length, subtract the applicable take-up allowance from Table 3. The result is the effective length of the longest belt that can be used on the drive. Select the next shorter standard effective length from Table 4.

Subtract from this length the shortest effective length calculated above. The result is the amount of installation allowance on the drive. Compare this value with the applicable value in Table 3. If the allowance is not sufficient, arrange for additional movement of the movable pulley.

TABLE 3 Installation and Take-Up Allowances. S = Single Belt, M = Multiple Belt.

Installation allowances: belt length (in.)

Length range	SAE							
	0.250		0.315		0.380		0.440	
	S	M	S	M	S	M	S	M
50 and less	5/8	---	5/8	---	11/16	15/16	3/4	1
50.01 thru 60.00	11/16	---	11/16	---	3/4	1	13/16	1 1/16
60.01 thru 80.00	3/4	---	3/4	---	13/16	1 1/16	7/8	1 1/8
80.01 thru 100.00	13/16	---	13/16	---	7/8	1 1/8	15/16	1 3/16

Take-up allowances: belt length (in.)

All SAE belt sizes	
Length range	Allowance for tensioning and take-up
50 and less	1
50.01 thru 60.00	1 1/4
60.01 thru 80.00	1 5/8
80.01 thru 100.00	1 15/16

Source: Courtesy of the Gates Rubber Company.

IV. BELT TYPES AND CONSTRUCTIONS

Automotive V-belts are produced in a number of types and constructions. The two basic types are the SAE sizes of 0.500 and smaller and the SAE sizes of 11/16 (0.600) and larger. Passenger cars and light trucks make use of the small SAE sizes which are sometimes referred to the high capacity sizes. The conventional SAE sizes of 11/16 (0.600) and larger are used in certain classes of heavy duty trucks, buses, and stationary power units.

belt size									
0.500		11/16 (0.600)		3/4 (0.660)		7/8 (0.790)		1 (0.910)	
S	M	S	M	S	M	S	M	S	M
3/4	1	7/8	1 3/16	7/8	1 3/16	15/16	1 5/16	1	1 3/8
13/16	1 1/16	15/16	1 1/4	15/16	1 1/4	1	1 3/8	1 1/16	1 7/16
7/8	1 1/8	1	1 5/16	1	1 5/16	1 1/16	1 7/16	1 1/8	1 1/2
15/16	1 3/16	1 1/16	1 3/8	1 1/16	1 3/8	1 1/8	1 1/2	1 3/16	1 9/16

All SAE-size belts are available in two basic constructions: banded or bandless. V-belts are either made with a band or a cover around the belt or they are of a cut-edge bandless construction. Both basic constructions are also made in a notched or cogged construction. Figure 3 illustrates a plain banded V-belt. A cover or band is applied around the belt to help prevent the entrance of dust, dirt, or moisture to the inner parts of the belt. A cut-notch banded V-belt is shown in Figure 4. The construction is the same as the plain belt except that notches are cut in the underside of the belt. This makes

TABLE 4 Standard Belt Lengths and Tolerances (in.)

SAE size	Effective length range	Standard length	Tolerance on center distance
All belt sizes	50.00 and less	Every exact 1/2	± 1/8
	50.50 thru 60.00	Every exact 1/2	± 5/32
	60.50 thru 80.00	Every exact 1/2	± 3/16
	81.00 thru 100.00	Every exact 1	± 7/32

Source: Courtesy of The Gates Rubber Company.

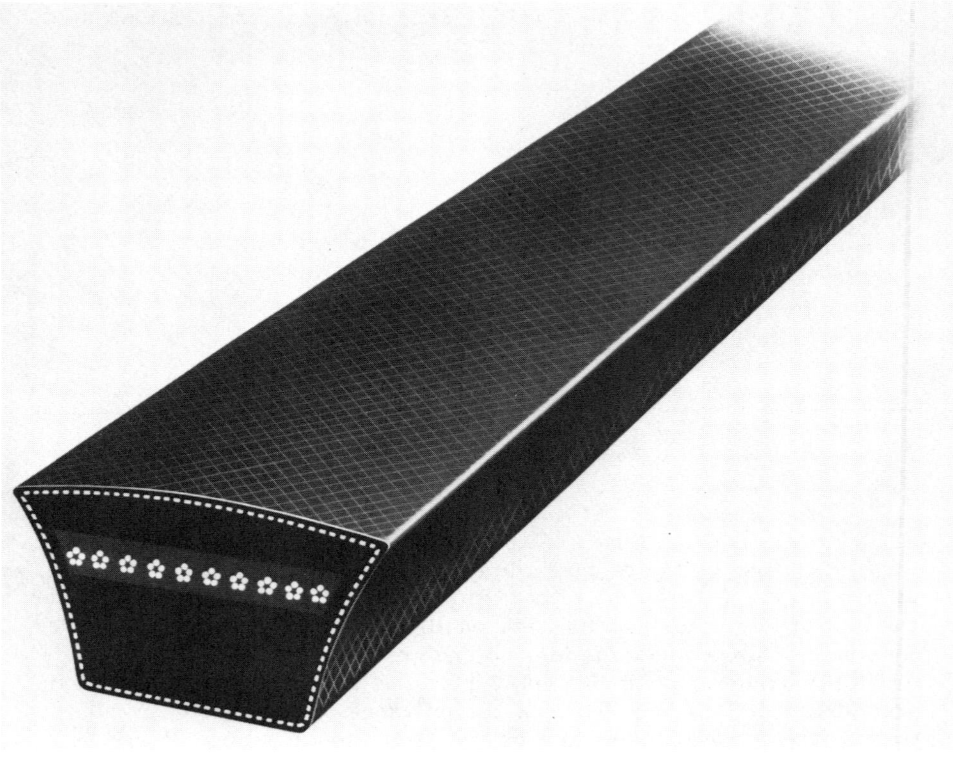

FIGURE 3 Banded construction.

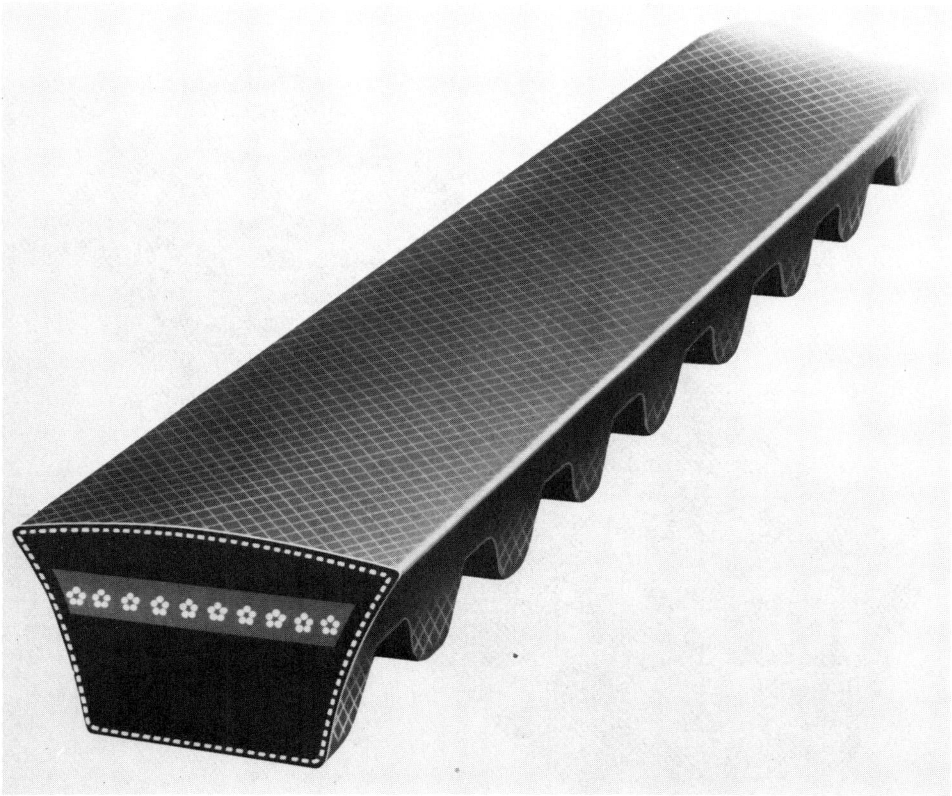

FIGURE 4 Banded cut notch construction.

the belt more flexible and better able to run on smaller subminimum diameters.

Figures 5 and 6 show the bandless and molded-notch bandless constructions. The plain bandless belts use a laminated fabric undercord. The molded-notch bandless belts are more flexible and are also used on drives requiring extremely small pulley diameters. Other special constructions are also available for low temperature applications. The most popular constructions today are the bandless constructions.

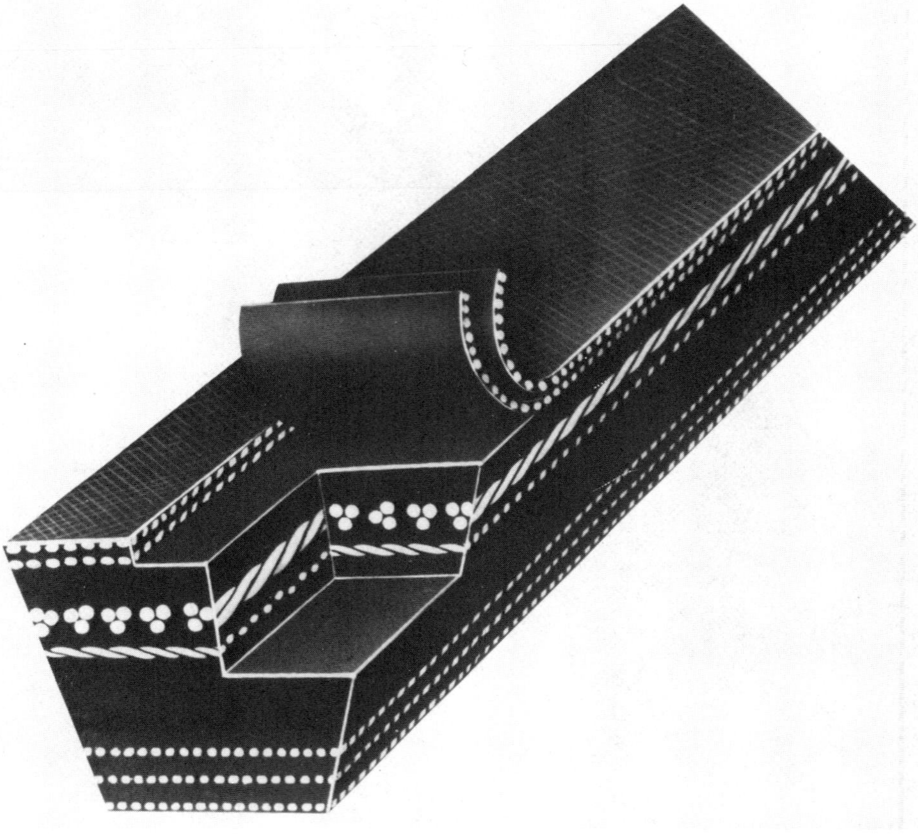

FIGURE 5 Bandless construction.

FIGURE 6 Bandless molded notch construction.

6

Variable-Speed Drive Design Using V-Belts

DAVID E. ROOS *Synchronous Application Engineering, The Gates Rubber Company, Denver, Colorado*

I. DEFINITION AND DESCRIPTION OF VARIABLE-SPEED DRIVES

Many driven machines require a change in speed. Some speed variation requirements are quite minimal, such as in a ventilating fan that is adjusted only once after installation to give the proper amount of air flow. Other machines, such as drill presses and similar machine tools, may require changes in driveN speed for different kinds of work. Still others require speed changes while operating.

In addition to the typical industrial drives, variable speed at the driveN shaft is often required on various types of agricultural and lawn-and-garden equipment, recreational vehicles, and automotive transmission applications.

Many types of V-belts and sheave combinations are used by all industries to accomplish driveN speed variation.

A. Principles of Operation

Within certain limits, V-belts are well suited to drives that must run at varying input and output speeds. Speed ratio on these drives is controlled by moving one sheave sidewall relative to the other so the belt rides at different pitch diameters.

Unlike other means of power transmission where speed variation is limited to finite steps, variable-speed V-belt drives offer an infinite selection of speed ratios within a certain range. In simple applications, speed is varied manually by adjusting the sheave sidewalls while the drive is stopped. On other, more complex drives, speed ratio is varied while the drive is operating – either by manual control or by an automatic control system sensitive to the speed and torque requirements of the machine. This simplified concept is illustrated in Figure 1.

The change in diameter at which a V-belt operates in one or more sheaves of a drive gives a speed variation to the driveN sheave. Speed variation can be best expressed as a range, such as variation from 1750 rpm to 875 rpm of the driveN shaft. Most often, however, it will be expressed as a ratio, with the terminology "speed ratio" applied. For example, the above variation can be expressed as a 2.0 speed ratio. This can be confusing because the normal meaning of speed ratio indicates a relationship of driveR and driveN shaft speeds. However, in this case, it is a relationship between the highest and lowest output speeds of the driveN shaft. Care must be taken to properly express the variation being considered.

B. Various Types of Drives

There are four standard types of variable-speed V-belt drives for industrial, agricultural, and automotive-related applications. Perhaps

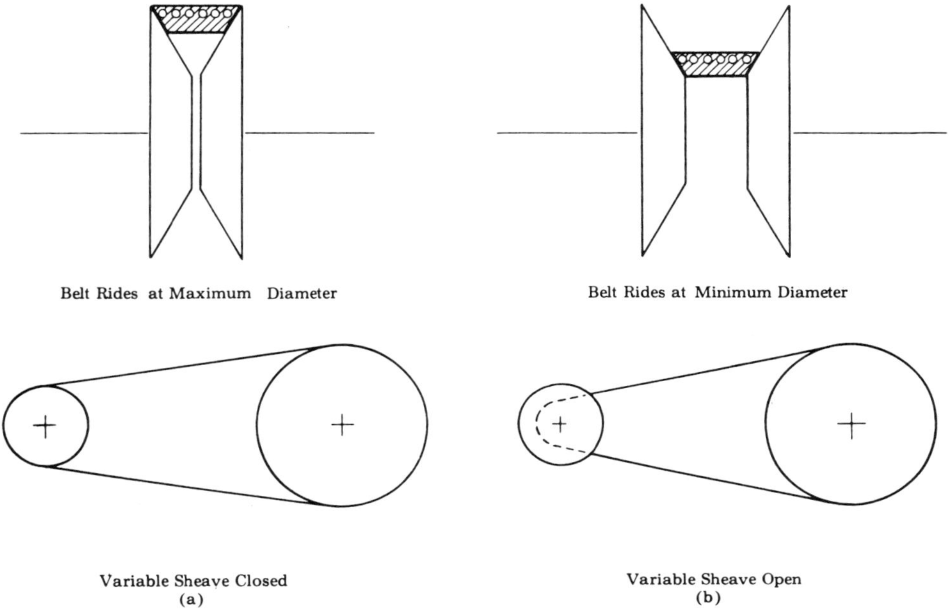

FIGURE 1 Speed variation is dependent on belt position in the sheave groove.

the most common of these are the standard variable-speed drives using classical, narrow, and light-duty belts. These variable-speed drives are used where required speed variation is not great. Wide-range variable-speed drives are another, and are used for greater speed variation. Variator drives incorporate two fixed-pitch sheaves with a single variator sheave. They typically exhibit lower speed variation.

1. Single Variable-Sheave Drives

In designing variable-speed drives, either the driveR or driveN can be the variable-pitch sheave, or, in cases requiring more speed variation, both can be. Typically, one variable-pitch sheave gives about 20–40% speed variation, depending on cross section and diameter. Two variable-pitch sheaves give from 40–100% speed variation.

Standard lines of variable-pitch sheaves are available for heavy-duty industrial belt sections. The most common include the classical belts (A, B, C, and D). Less common are the narrow belts (5V and 8V). They are also available for the standard light-duty belts (3L,

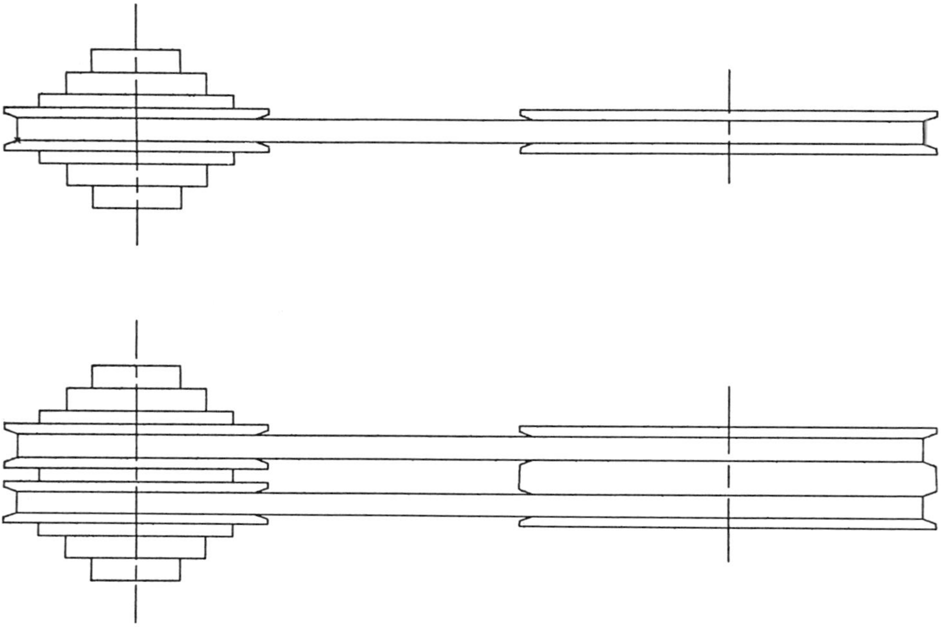

FIGURE 2 Single and multiple variable-pitch belt drives.

4L, and 5L). All of these belt types and the related industry standards are discussed in Chapter 3. They are also described in the Rubber Manufacturers Association RMA/MPTA Engineering Standards. For classical belts see Publication IP-20; for narrow belts see Publication IP-22; and for single belts (light duty) see Publication IP-23.

These drives can use either single or multiple belts. Typically, the sheaves available for variable-speed light-duty drives incorporate either one or two grooves. Industrial variable-speed drives incorporating the classical and narrow belt sections are available with as many as ten grooves. These drive variations are shown in Figure 2.

2. Double Variable-Sheave Drives

Typically, variable-speed drives incorporating light-duty belts use only one variable-pitch sheave and a fixed-pitch companion sheave. However, variable-speed drives for larger industrial applications incorporating the classical and narrow belt sections often use double variable sheaves. In this case, a variable-pitch sheave is installed on both the driveR and driveN shafts, which in turn allows a greater speed change at the driveN shaft. As with single variable-sheave drives, either single or multiple belts can be used on these applications.

3. Variator-Type Variable-Speed Drives

One of the simplest and most durable shifting systems is the variator drive, in which neither driveR nor driveN sheaves are adjustable. Speed variation is provided entirely by a set of intermediate sheaves called the variator. The centerline of the variator is adjustable toward or away from the driveR and driveN sheaves. The variator consists of two fixed outer-sheave halves separated by a one-piece floating element serving as the mate to both fixed halves. As the variator is moved toward the driveR sheave, increasing tension on the belt to the driveN sheave pulls that belt deeper into the variator sheave. This motion forces the floating element against the belt to the driveR sheave, causing the belt to ride on a larger pitch diameter. The net effect is speed reduction from driveR to driveN. Motion of the variator in the opposite direction produces an increase in driveN speed (see Figure 3). The disadvantage of this type of drive is the large base required compared to the more compact two-sheave arrangements. Variator drives are widely used on agricultural machinery, particularly lawn-and-garden equipment. Variator drives have also been known to be used on various types of industrial equipment.

4. Wide-Range Variable-Speed Drives

For industrial applications requiring significant speed variation, special variable-speed V-belts are available. These belts have wide, thin cross sections to provide a wide range of speed variation. These belts transmit up to 100 HP per belt and provide speed variations up to 10:1.

An RMA/MPTA Engineering Standard entitled, "Specifications for Drives Using Variable Speed V-Belts," IP-25, gives sheave-groove and belt dimensions for 12 of the more commonly used cross sections. These belts have top widths ranging from 0.875 in. to 3.0 in.

The simplest of these variable-speed drives includes one variable-pitch sheave and a fixed-pitch companion sheave. In this case, the fixed-pitch companion sheave can be either flat or V-groove (see Figure 4).

In this type of application, the variable-pitch pulley can be located on either the driveR or the driveN shaft. The pulley is typically spring-loaded and is infinitely variable while running under full load. Belt position in the variable-pitch pulley is usually controlled in one of three ways. The most common incorporates a sliding motor base

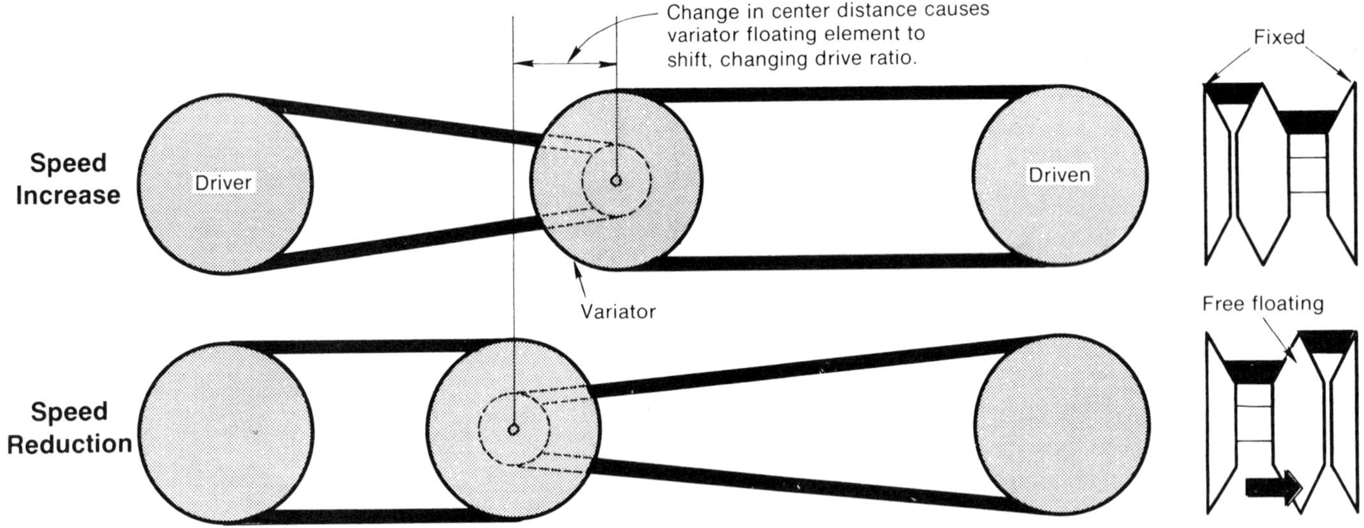

FIGURE 3 Variable-speed drive consisting of two fixed-pitch sheaves and a variator sheave.

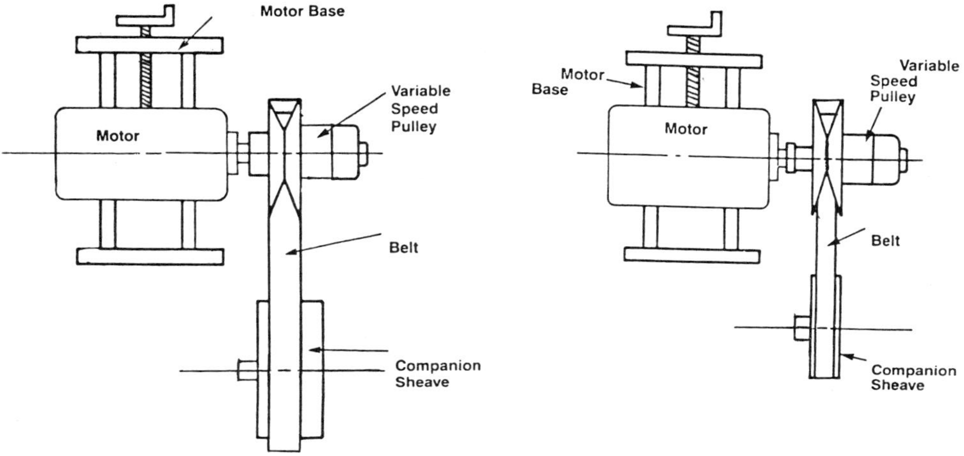

FIGURE 4 Single variable-pitch sheave shown with a flat companion and a fixed-pitch V-groove sheave.

that changes drive center distance, thus affecting the position of the belt in the variable-pitch sheave as the pulley halves open and close. Pivoting motor bases can also be used to accomplish the same objective, and inside idlers are sometimes used to accomplish belt shifting in the variable-pitch sheave. The use of a single variable-pitch sheave typically limits the maximum amount of driveN speed variation to less than 4:1.

The widest speed range is available if both driveR and driveN sheaves are adjustable. In systems where both driveR and driveN sheaves are adjustable, there are two ways to regulate sheave movement. In one system the shifting mechanisms of both sheaves are interconnected through a mechanical link. As one sheave opens, the other closes by the same amount. This system provides maximum ratio range, but also has a pronounced drawback. Namely, belt tension is too low at mid-speed range if properly tensioned for extreme ratios. Thus, this system is seldom used. A better arrangement avoids direct interconnection and has one sheave spring-loaded. The spring automatically provides correct belt tension and causes its sheave to respond properly to any shift in the controlled sheave. Belt location in the control sheave can be accomplished using various mechanical, hydraulic, or electronic arrangements (see Figure 5). It is also important that the fixed and adjustable sheave halves be on opposite sides of the belt to maintain correct belt alignment (see Figure 6).

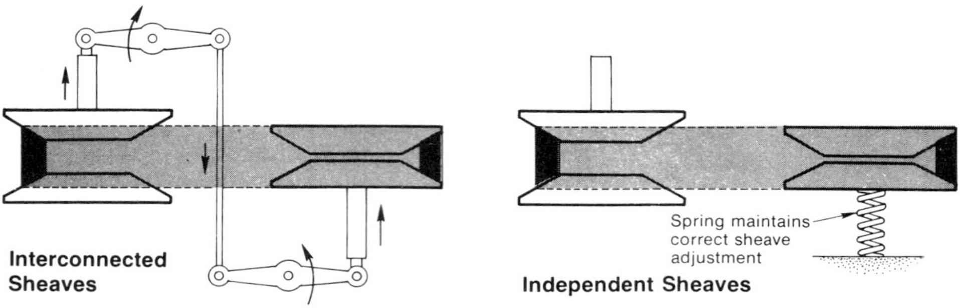

FIGURE 5 Variable-speed drives using interconnected and independent sheaves.

C. Geometric Variations Available

The change in pitch diameter and the subsequent speed variation is controlled by belt width and thickness and by the included sheave-groove angle. These three dimensions must be balanced with load-carrying capacity to produce a belt with wide speed variation and adequate lateral rigidity to prevent belt collapse.

1. Effect on Speed Ratio Range

The amount of speed ratio range, or speed variation, possible in a drive is controlled by the amount that the pitch diameter of the sheave can change. Pitch diameter variation (P_d) is given by:

$$P_d = \frac{B_b}{2 \tan \alpha/2} - H_b$$

where

B_b = belt top width
H_b = belt thickness
α = sheave groove angle

This relationship shows that the wider the belt top width and the thinner the belt section, the greater the pitch diameter change and speed variation. Also, as groove angle is reduced, the amount of allowable speed variation increases. For the greatest variation, the belt should be as wide as possible and the groove angle should be as

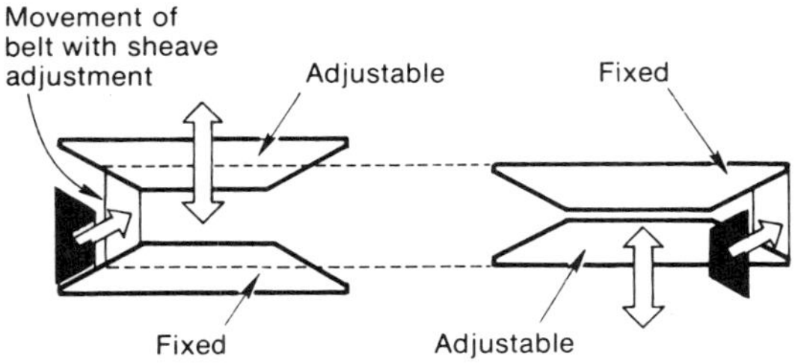

FIGURE 6 Proper sheave design is important in maintaining belt alignment. The adjustable and fixed sheaves must be on opposite sides of the belt to maintain correct alignment. In the illustration, for example, as the left sheave opens, the belt moves upward and to the right. To maintain proper tracking, the belt must also move upward as the right sheave closes. This condition is provided if the top sheave half remains fixed so that the belt "climbs" upward and to the right.

shallow as possible. The belt should also be as thin as possible. Because the available change in drive ratio depends upon the percent change in drive pitch diameter, the pitch diameter should be as small as possible if changes are to produce maximum ratio changes. However, to permit use of small sheaves, the belt must be thin (see Figure 7).

2. Effect on Load-Life Performance

The design of a variable-speed system involves trade-offs among belt width, belt thickness, and groove angle. While a variable-speed drive with a smaller groove angle tends to provide the greatest speed variation, the belt also has a greater tendency to collapse inward under the load tension. In basic belt terminology, the center tensile cords are said to have inadequate support. This factor limits the amount of power that can be transmitted. Therefore, compromises must be made away from this ideal belt if significant power is to be transmitted. As the belt is made thicker, it cannot operate on small-diameter sheaves, so allowable ratio variation drops.

Because of these complex interactions, there are very few stock variable-speed belts. Most mass-production drives are custom-engineered for the optimum balance of belt variables. Optimization of these variables must include both required speed variations for the drive and the maximum loads under which the belt must operate. Both of these requirements will affect those decisions involving minimum allowable pitch diameters.

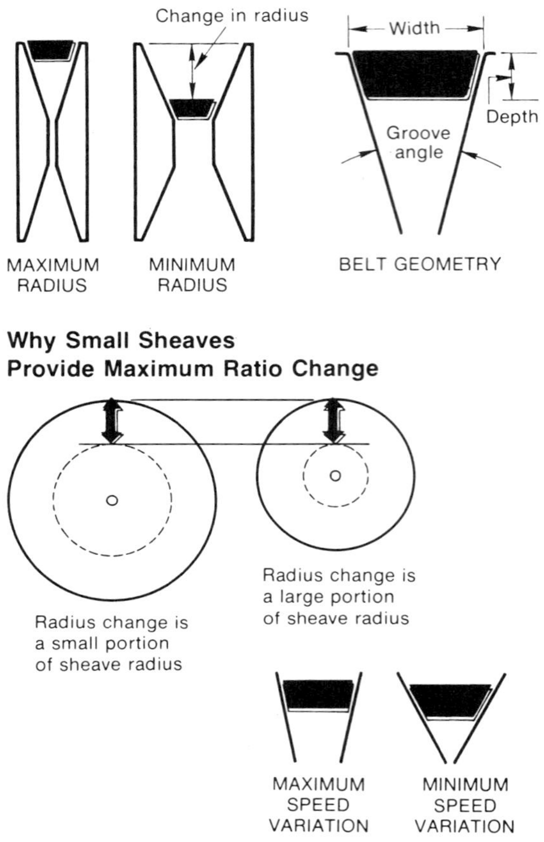

FIGURE 7 Sheave and belt geometry control allowable speed variation.

II. INDUSTRIAL VARIABLE-SPEED DRIVE DESIGN

Variable-speed belt drives have a long history in industry, but with the development of the modern high-capacity belts, significant changes have taken place in the design of variable-speed belt drives. The first variable-speed drives simply used conventional V-belts and the sheaves were not adjustable while the drive was operating. Sheave separation was adjusted manually while the drive was at rest. This type of system is referred to as stationary control. Systems where the ratio can be controlled while the drive is running are called motion-control drives.

Variable-pitch drives incorporating classical, narrow, and light-duty V-belts can typically be designed using existing industry standards and manufacturers drive-design catalogs. Variator drives are very much similar in design procedure to the variable-pitch belt drives. The wide range variable-speed drives using either one variable-speed sheave or two variable-speed sheaves can be designed using the RMA/MPTA Publication IP-25.

A. Typical Usage of Variable-Speed Industrial Drives

In addition to machine tools mentioned earlier in this chapter, industrial variable-speed drives can be found in almost every major industry. Variable-speed drives are used on machine tools, pumps, fans and blowers, conveyor systems, mixers and agitators, and textile processing equipment, to name only a few.

Design procedures for each of the above-described industrial variable speed systems will be discussed.

B. Variable-Pitch Drive Design Using Industrial Belts

Both classical and narrow section V-belts are used on variable-speed drives. Because of their wider and thinner cross sections, classical belts tend to be far more common. The same design procedures and practices apply to both the classical and narrow industrial V-belts and also to the light-duty or fractional-HP V-belts. These drives are typically used when only a limited range of speed variation is required. Common practice is to use a variable-pitch sheave on the motor and a fixed-pitch sheave on the driveN unit. For maximum speed change, the variable-pitch sheave should be mounted on the faster shaft. Because the spacing between belts must be wider on the variable-pitch sheave to allow for axial sidewall movement, a special fixed-pitch sheave with wider groove spacing is sometimes needed.

If only one variable-pitch sheave is used on a drive, speed variation is limited to 1:1.2–1:1.4, depending on the type of belt and the sheave diameters. However, if variable-pitch sheaves are used on both the driveR and driveN, speed variations from 1:1.4–1:2.0 are possible.

In these applications, speed changes are made when the drive is stopped. These sheaves are often referred to as stationary control. Typically, only the classical A, B, and C cross sections are used.

In addition to the stationary control variable-pitch sheaves, there also exist motion-control variable-pitch sheaves. Although the design practices are the same for either system, the motion-control variable-

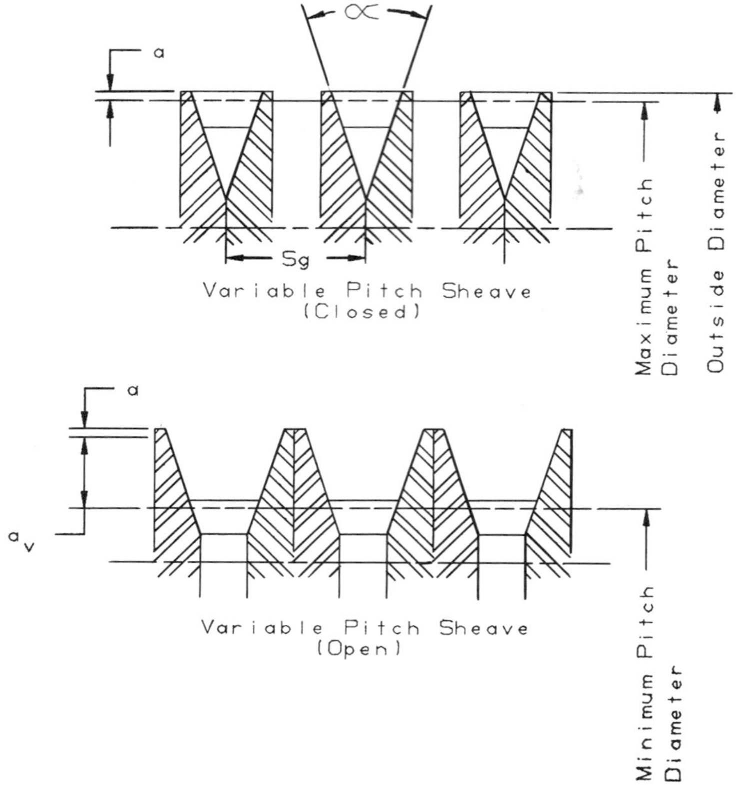

FIGURE 8 Variable-pitch sheave groove shown in full open and full closed position.

pitch sheaves will allow for speed changes being made while the belt drive is moving. This is typically accomplished through the use of sliding and pivoting motor bases, or with the use of inside belt idlers.

1. Design Procedures Using Stock Components

A variable-pitch sheave is designed so that one sidewall of each groove can move axially. This causes a radial movement of the belt in the groove (see Figure 8).

The dimension 2a is the same as shown in RMA/MPTA Publication IP-20. The dimensions a_V and α, as well as S_g, must be obtained from the manufacturer of the variable-pitch sheave.

The fixed-pitch sheave can often be of the standard type described in RMA/MPTA Publication IP-20. Sometimes the groove spacing (S_g) must be increased to the value of S_g for the variable-pitch sheave for satisfactory belt alignment. Rules governing this choice will be discussed later.

The selection of a service factor and appropriate cross section are as described in RMA/MPTA Publication IP-20 or in various manufacturers' V-belt drive-design catalogs. These publications also include all the HP ratings and appropriate correction factors for arc of contact and belt length. For additional discussion on HP ratings, see Chapter 3. Also see RMA Technical Bulletin IP-3-14.

Step 1 is to determine the minimum acceptable pitch diameter for the belt cross section selected. When applicable, be sure to consider the minimum recommended sheave diameter for the specified electric motor being used as established by the National Electric Motors Association (NEMA). This data can be found in most belt manufacturers' drive design catalogs.

Select an appropriate driveR or driveN sheave (variable-pitch or fixed-pitch) using the recommended minimum allowable pitch diameter dictated by NEMA and/or the belt cross section as a guide. Placement of the variable-pitch sheave can be on either the driveR or the driveN shaft. Selection of the other sheave, fixed-pitch or variable-pitch (dependent on which was earlier selected), can be accomplished using Table 1

where

D_F = pitch diameter, fixed pitch sheave (in. or mm)
D_{VP} ave = median pitch diameter, variable-pitch sheave (in. or mm)
DN rpm ave = median rpm for driveN sheave

Select the closest available stock sheave to meet the above requirements.

Step 2 involves selecting the type of fixed-pitch sheave to be used on any specific drive. The companion sheaves with wider groove spacing may not always be necessary. A standard multiple-groove fixed-pitch sheave described in RMA/MPTA Publication IP-20 or manufacturers' catalogs can be used if the angle of offset (misalignment) as shown in Figure 9, does not exceed 2°. The angle of offset (g) can be calculated as follows:

TABLE 1 Calculating Variable-Pitch Sheave Diameters

Given:	Fixed driveR (D_F)	Variable driveN (D_{VP} ave)
Determine:		
Variable driveN (D_{VP} ave)	$= \frac{(D_F)(\text{DR rpm})}{(\text{DN rpm ave})}$	
Fixed driveR (D_F)		$= \frac{(D_{VP}\text{ ave})(\text{DN rpm ave})}{(\text{DR rpm})}$
Given:	Fixed driveN (D_F)	Variable driveR (D_{VP} ave)
Determine:		
Variable driveR (D_{VP} ave)	$= \frac{(D_F)(\text{DN rpm ave})}{(\text{DR rpm})}$	
Fixed driveN (D_F)		$= \frac{(D_{VP}\text{ ave})(\text{DR rpm})}{(\text{DN rpm ave})}$

$$g = \tan^{-1} \frac{(F_a - F_f)}{2t} \tag{2}$$

where

F_a = adjustable sheave overall face width at minimum pitch diameter (in. or mm)
F_f = fixed sheave overall face width (in. or mm)
t = span length between sheaves (in. or mm)

The above formula is based on the center belt of the drive, or the two center-most belts on an even number belt drives being aligned with the variable-pitch sheave in mid-position.

Commercially available fixed-pitch sheaves with standard groove spacing can be used in two ways (see Figure 9). A drive can use

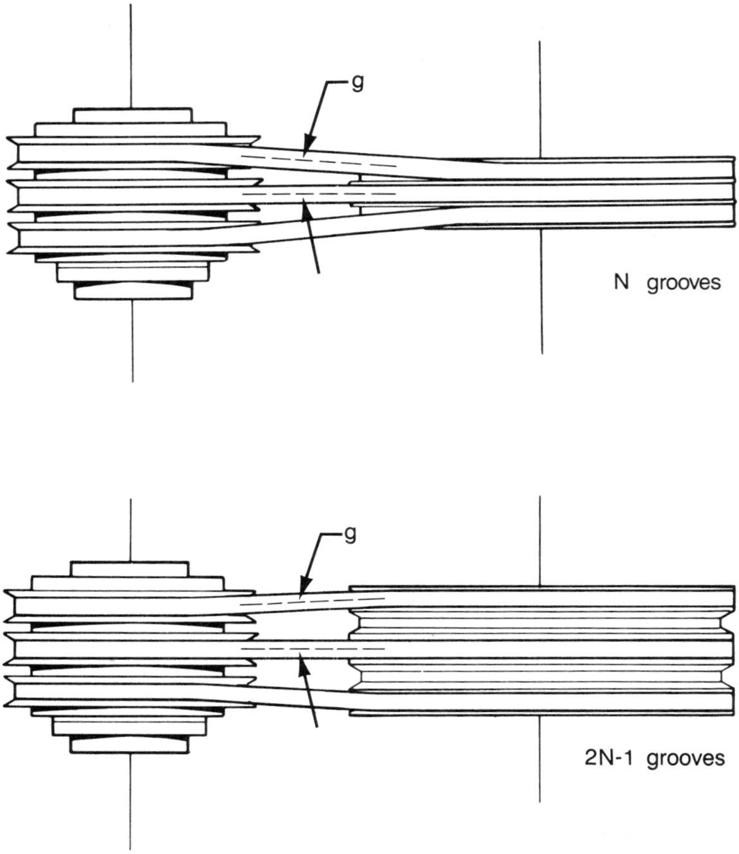

FIGURE 9 Misalignment with "N" and "2N-1" groove standard industrial sheaves.

every groove or every other groove on the sheave (number of grooves on sheave equals 2N-1).

If a flat pulley is used as the fixed-pitch sheave, it must be wide enough to accommodate the axial movement of the belts as the speed is changed in the variable-pitch sheave.

Step 3 involves determining the number of belts required for the application. This is obtained by dividing design HP (or design K_W) by the corrected HP rating (or K_W rating) for one belt.

The corrected HP rating (or K_W rating) can be obtained from the Design Factors Section of RMA/MPTA Publication IP-20 or manufacturers' catalogs. This rating depends on the sheave diameter and the rpm of the faster shaft. The rating should be determined for both the minimum and maximum speeds using the corresponding sheave diameter. The smaller value of the corrected HP rating (or K_W rating) should be used for determining the number of belts required on the drive.

Most commercially available variable-pitch sheaves have an included groove angle of 30° to obtain a greater speed variation. The V-belts specified in RMA/MPTA Publication IP-20 are designed to operate in sheave grooves of 34° or greater, depending on diameter. When these belts are operated in 30° sheaves, a reduction in belt life can be expected. On those drives using fixed-pitch sheaves yielding more than 1/2° misalignment, a resultant reduction in belt life can also be expected.

Rim speeds must also be evaluated, since special sheaves may be required if the rim speed exceeds 6,500 ft/min (33 m/s). The rim speed applies to the actual outside diameter of the variable-pitch sheave and the companion sheave and should be calculated for each with the variable pitch sheave in the position that gives the maximum rpm at the fixed-sheave.

The formula for calculating rim speed is:

$$\text{rim speed (ft/min)} = \frac{(\pi)(D_O)(\text{rpm}_{max})}{12} \tag{3}$$

where

D_O = outside diameter of sheave (in.)
rpm_{max} = maximum rpm of driveR or driveN shaft

$$\text{rim speed (m/s)} = \frac{(\pi)(D_O)(\text{rpm}_{max})}{60{,}000} \tag{4}$$

where

D_O = outside diameter of sheave (mm)
rpm_{max} = maximum rpm of driveR or driveN shaft

Center distance and belt length can be calculated using those formulas defined in Chapter 3. The required center distance should be calculated with the belt at the minimum and the maximum diameter of the variable-pitch sheave using a standard length belt. Sufficient center distance adjustment must be provided to accommodate the shortest center distance minus the installation allowance shown in RMA/MPTA Publication IP-20 in order to allow for damage-free installation of the belts. In the other direction, the center distance adjustment must be enough for the maximum center distance plus the take-up allowance shown in RMA/MPTA Publication IP-20 so that tension can be maintained throughout the life of the belt. Center distance installation and take-up allowances are shown in Table 2.

C. Wide-Range Variable-Speed Drive Design

For drives that require more speed variation than can be obtained with conventional industrial V-belts, standard-line variable-speed drives are available. These drives use special wide, thin belts. Package units of standard-line variable-speed belts and sheaves, combined with the motor and output gear box (if specified or required), are available in HP ranges from approximately 1/2 through 100 HP. These are either single- or dual-belt drives, dependent upon HP and manufacturer.

The speed range of wide variable-speed drives can be much greater than those variable-speed drives using classical V-belts. Speed ranges up to 10:1 can be obtained on lower HP units. Spring-loaded or hand-adjustable motor sheaves are available for the standard variable-speed belts in a package containing the motor, sheaves, belt, and sometimes an adjustable motor base.

RMA/MPTA Publication IP-25 covers specifications for drives using variable-speed V-belts. Specifically, 12 of the more common V-belt cross sections are covered (see Table 3).

1. Speed Variation

The speed variation capabilities for a single variable-sheave drive can be determined as follows:

$$\text{speed variation} = \frac{\text{PD maximum}}{\text{PD minimum}} \tag{5}$$

where

PD = pitch diameter

The actual output speeds will be established by the selection of the companion sheave and will be a function of the ratio of the pitch diameter of the companion sheave to the maximum and minimum pitch diameters of the variable sheave.

TABLE 2 Center Distance Allowances For Belt Installation and Take-Up

Standard Length Designation	Minimum Allowance Below Standard Center Distance for Installation of Belts										Minimum Allowance Above Standard Center Distance for Maintaining Tension All Cross Sections
	A	A Joined	B	B Joined	C	C Joined	D	D Joined	E	E Joined	
Up to and Incl. 35	0.75	1.20	1.00	1.50							1.00
Over 35 to and Incl. 55	0.75	1.20	1.00	1.50	1.50	2.00					1.50
Over 55 to and Incl. 85	0.75	1.30	1.25	1.60	1.50	2.00					2.00
Over 85 to and Incl. 112	1.00	1.30	1.25	1.60	1.50	2.00					2.50
Over 112 to and Incl. 144	1.00	1.50	1.25	1.80	1.50	2.10	2.00	2.90			3.00
Over 144 to and Incl. 180			1.25	1.80	2.00	2.20	2.00	3.00	2.50	3.40	3.50
Over 180 to and Incl. 210			1.50	1.90	2.00	2.30	2.00	3.20	2.50	3.50	4.00
Over 210 to and Incl. 240			1.50	2.00	2.00	2.50	2.50	3.20	2.50	3.60	4.50
Over 240 to and Incl. 300			1.50	2.20	2.00	2.50	2.50	3.50	3.00	3.90	5.00
Over 300 to and Incl. 390					2.00	2.70	2.50	3.60	3.00	4.00	6.00
Over 390					2.50	2.90	3.00	4.10	3.50	4.40	1.5% of belt length

TABLE 3 Industry Standard Cross Sections Defined by RMA/MPTA Publication IP-25. [Nominal dimensions of V-belt cross sections (in.)].

Cross Section	b_b	h_b	h_b/b_b
1422V	0.88	0.31	0.35
1922V	1.19	0.38	0.32
2322V	1.44	0.44	0.31
1926V	1.19	0.44	0.37
2926V	1.81	0.50	0.28
3226V	2.00	0.53	0.27
2530V	1.56	0.59	0.38
3230V	2.00	0.62	0.31
4430V	2.75	0.69	0.25
4036V	2.50	0.69	0.28
4436V	2.75	0.72	0.26
4836V	3.00	0.75	0.25

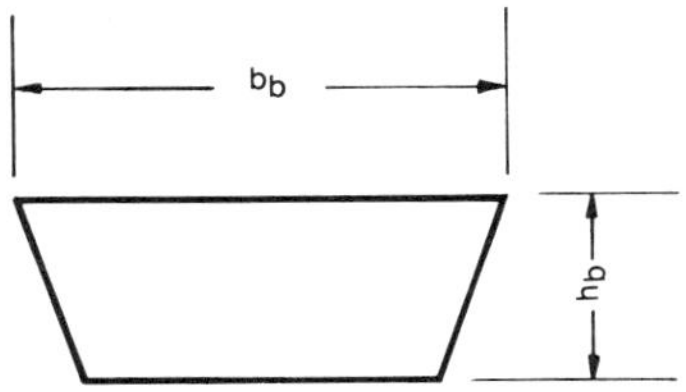

Speed variation with a double variable-sheave drive can be determined as follows:

$$\text{speed variation} = \frac{DR(DN)}{dR(dN)} \qquad (6)$$

where

DR = Maximum driveR pitch diameter (in. or mm)
DN = Maximum driveN pitch diameter (in. or mm)
dR = Minimum driveR pitch diameter (in. or mm)
dN = Minimum driveN pitch diameter (in. or mm)

Speed variation with a single variable-sheave drive is usually obtained by varying the center distance between the two sheaves. With the use of dual variable-sheaves, the center distance is generally fixed and speed variation is accomplished by mechanically changing the pitch diameter of one sheave. In this type of drive the other sheave is typically spring-loaded to make an opposite change in the pitch diameter and to provide the correct belt tension (see Figure 10). (*Note*: All speed ratio changes should be made while the drive is running. Any attempt to make speed ratio adjustments while the unit is stopped can create unnecessary and, possibly, destructive forces on either the belt or the sheaves. In the case of stationary controlled type drives, which are quite uncommon, the belt tension should be released so that the movable flanges are easily adjustable without belt force interference.)

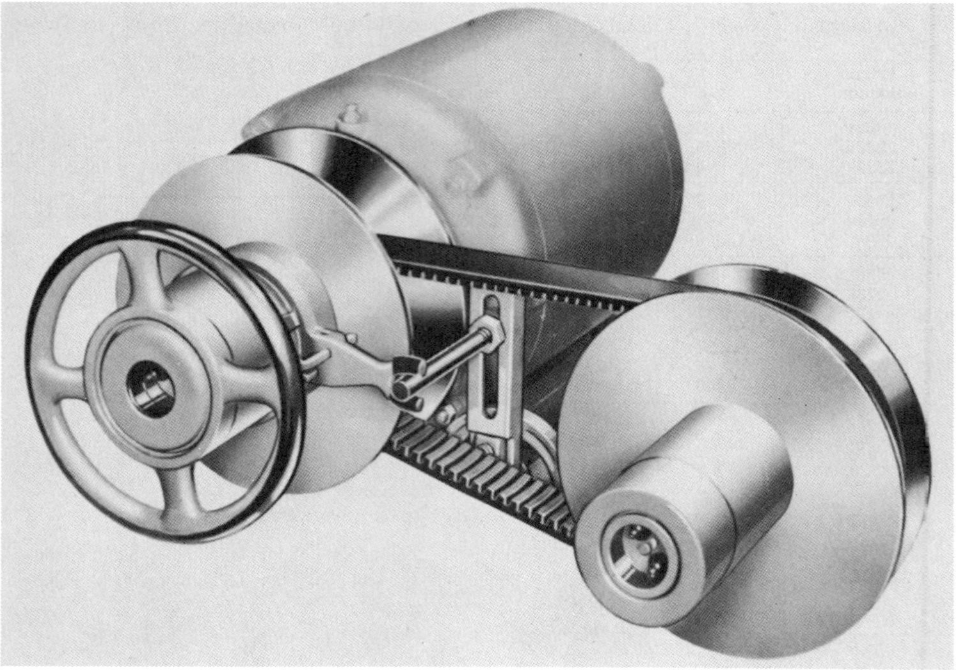

FIGURE 10 Mechanically controlled variable-pitch driveR with spring-loaded variable-pitch driveN on fixed centers.

2. Cross Section Selection

Selection of a variable-speed belt cross section is based on the drive design HP and speed variation requirements. Table 4 shows the maximum pitch diameter variation that each cross section can attain. Table 5 shows a typical HP-rating table from RMA/MPTA Publication IP-25. Although HP values are shown for some diameters that are less than the minimum recommended pitch diameters given in Table 4, the values are shown to inform the user of the potential capabilities of the cross section when other considerations may force the use of smaller diameters. The HP ratings are for standard belts of average length and with 180° arc of contact. The basic V-belt HP rating formula is discussed in Chapter 3. The specific constants for the wide-range variable-speed belts are identified in RMA/MPTA Publication IP-25. Corrected HP ratings are obtained by using the HP-rating tables and adjusting those values for the arc of contact correction factor

TABLE 4 Standard Groove Dimensions and Drive Design Factors for the 12 Basic Cross Sections

	STANDARD GROOVE DIMENSIONS									DRIVE DESIGN FACTORS			
	VARIABLE					COMPANION							
CROSS SECTION	α Groove Angle ±0.67 Degrees	b_g † Closed +0.000 −0.030	b_{go} Open max.	h_{gv} min.	S_g ±0.03	α Groove Angle ±0.33 Degrees	b_g ±0.010	h_g min.	S_g ±0.03	Minimum Recom-mended Pitch Diam.	2a	2av max.	CL min.
1422V	22	0.875	1.63	2.33	1.82	22	0.875	0.500	1.82	2.0	0.20	3.88	0.08
1922V	22	1.188	2.23	3.14	2.42	22	1.188	0.562	2.42	3.0	0.22	5.36	0.08
2322V	22	1.438	2.71	3.78	2.89	22	1.438	0.625	2.89	3.5	0.25	6.52	0.08
1926V	26	1.188	2.17	2.65	2.36	26	1.188	0.625	2.36	3.0	0.25	4.26	0.08
2926V	26	1.812	3.39	4.00	3.58	26	1.812	0.750	3.58	3.5	0.30	6.84	0.08
3226V	26	2.000	3.75	4.41	3.96	26	2.000	0.781	3.96	4.0	0.30	7.60	0.08
2530V	30	1.562	2.81	3.01	2.98	30	1.562	0.844	2.98	4.0	0.30	4.64	0.10
3230V	30	2.000	3.67	3.83	3.85	30	2.000	0.875	3.85	4.5	0.35	6.22	0.10
4430V	30	2.750	5.13	5.23	5.38	30	2.750	0.938	5.38	5.0	0.40	8.88	0.10
4036V	36	2.500	4.55	3.95	4.80	36	2.500	0.938	4.80	4.5	0.40	6.32	0.10
4436V	36	2.750	5.03	4.33	5.30	36	2.750	0.969	5.30	5.0	0.40	7.02	0.10
4836V	36	3.000	5.51	4.72	5.76	36	3.000	1.000	5.76	6.0	0.45	7.74	0.10

TABLE 5 Portion of HP Rating Tables for Variable-Speed V-Belts

HORSEPOWER RATING FOR 1422V SECTION VARIABLE SPEED BELT

RPM of Faster Shaft	Rated Horsepower Per Belt For Small Sheave Pitch Diameter										
	1.8	2.0	2.5	3.0	3.5	4.0	5.0	6.0	7.0	8.0	9.0
430	0.03	0.07	0.16	0.26	0.35	0.44	0.62	0.80	0.97	1.14	1.30
580	0.02	0.08	0.20	0.32	0.45	0.57	0.80	1.03	1.25	1.46	1.67
870	0.01	0.08	0.26	0.44	0.61	0.78	1.11	1.43	1.73	2.03	2.30
1160		0.08	0.31	0.54	0.76	0.98	1.39	1.78	2.15	2.49	2.79
1750		0.05	0.39	0.70	1.01	1.30	1.84	2.31	2.71	3.03	3.25
2000		0.04	0.40	0.76	1.09	1.41	1.98	2.45	2.82	3.06	3.17
2500			0.42	0.83	1.21	1.56	2.13	2.54	2.74	2.70	2.41
3000			0.41	0.86	1.27	1.61	2.12	2.33	2.20	1.68	0.70
3500			0.37	0.85	1.25	1.56	1.90	1.78	1.13		

RPM of Faster Shaft	Additional Horsepower per Belt for Speed Ratio							
	1.00 to 1.01	1.02 to 1.04	1.05 to 1.08	1.09 to 1.12	1.13 to 1.18	1.19 to 1.24	1.25 to 1.34	1.35 to [illegible]
430	0.00	0.00	0.01	0.01	0.02	0.02	0.0[illegible]	
580	0.00	0.01	0.01	0.02	0.03	0.03	[illegible]	
870	0.00	0.01	0.02	0.03	0.04	0.05		
1160	0.00	0.01	0.03	0.04	0.05	0.[illegible]		
1750	0.00	0.02	0.04	0.06	0.08	[illegible]		
2000	0.00	0.02	0.05	0.07	0.0[illegible]			
2500	0.00	0.03	0.06	0.08	[illegible]			
3000	0.00	0.03	0.07	0.10				
3500	0.00	0.04	0.08	0.1[illegible]				

HORSEPOWER RATING FOR 1922V SECTION VARIABLE SPEED

RPM of Faster Shaft	Rated Horsepower Per Belt For Small Sheave Pitch Diameter											RPM of Fas[illegible]
	2.5	3.0	3.5	4.0	4.5	5.0	6.0	7.0	8.0	9.0	10.0	
430	0.28	0.45	0.61	0.77	0.92	1.08	1.38	1.68	1.97	2.26	2.5[illegible]	
580	0.35	0.56	0.77	0.98	1.18	1.38	1.78	2.16	2.54	2.91	[illegible]	
870	0.45	0.76	1.06	1.36	1.65	1.93	2.48	3.02	3.52	4.01		
1160	0.54	0.94	1.32	1.70	2.06	2.42	3.10	3.74	4.34	4.8[illegible]		
1750	0.67	1.22	1.76	2.27	2.75	3.22	4.06	4.78	5.37	[illegible]		
2000	0.70	1.32	1.90	2.46	2.98	3.47	4.32	5.00	5.4[illegible]			
2500	0.74	1.46	2.13	2.74	3.29	3.78	4.55	4.98	5[illegible]			
3000	0.73	1.53	2.24	2.87	3.40	3.82	4.30	4.23				
3500	0.67	1.52	2.24	2.82	3.26	3.53	3.51	2.6[illegible]				

(see Chapter 3, Table 7) and by the length correction factor from Table 6.

3. Service Factors

The selection of a V-belt drive for any application should be based on the nature of the load and the type of driving unit. Service factors for different kinds of driveN machines combined with different types of driving units are shown in Chapter 3, Table 8.

4. Optimizing Drive Design

While RMA/MPTA Publication IP-25 shows 12 specific variable-speed V-belt cross sections, a considerably greater number of V-belt cross sections are available through various belt manufacturers.

The very nature of the wide, thin belt sections used on wide-range variable-speed drives dictates that some optimization must be

TABLE 6 Length Correction Factors For The 12 Basic Cross Sections

Standard Pitch Length Designation	Cross Section											
	1422V	1922V	2322V	1926V	2926V	3226V	2530V	3230V	4430V	4036V	4436V	4836V
315	0.93											
335	0.94											
355	0.95	0.90		0.90								
375	0.96	0.91		0.91								
400	0.97	0.92	0.90	0.92								
425	0.98	0.93	0.91	0.93								
450	0.99	0.94	0.92	0.94								
475	1.00	0.95	0.93	0.95								
500	1.01	0.95	0.94	0.95			0.90					
530	1.02	0.96	0.95	0.96	0.92		0.92					
560	1.03	0.97	0.96	0.97	0.93	0.92	0.93	0.91	0.90	0.91	0.91	0.92
600	1.04	0.98	0.97	0.98	0.94	0.93	0.94	0.93	0.92	0.93	0.92	0.93
630	1.05	0.99	0.98	0.99	0.95	0.94	0.95	0.94	0.93	0.94	0.93	0.94
670	1.06	1.00	0.99	1.00	0.97	0.95	0.96	0.95	0.94	0.95	0.95	0.95
710	1.07	1.01	1.00	1.01	0.98	0.96	0.98	0.96	0.96	0.96	0.96	0.96
750	1.08	1.02	1.01	1.02	0.99	0.98	0.99	0.97	0.97	0.97	0.97	0.98
800		1.03	1.02	1.03	1.00	0.99	1.00	0.99	0.99	0.99	0.99	0.99
850		1.04	1.03	1.04	1.01	1.00	1.01	1.00	1.00	1.00	1.00	1.00
900		1.05	1.04	1.05	1.02	1.01	1.02	1.01	1.01	1.01	1.01	1.01
950		1.06	1.05	1.06	1.03	1.02	1.04	1.02	1.03	1.02	1.02	1.02
1000		1.07	1.06	1.07	1.04	1.03	1.05	1.03	1.04	1.03	1.04	1.03
1060		1.08	1.07	1.07	1.06	1.04	1.06	1.05	1.06	1.05	1.05	1.04
1120		1.09	1.08	1.08	1.07	1.06	1.07	1.06	1.07	1.06	1.06	1.06
1180		1.09	1.09	1.09	1.08	1.07	1.08	1.07	1.08	1.07	1.07	1.07
1250					1.09	1.08	1.10	1.08	1.10	1.08	1.09	1.08
1320						1.09		1.09	1.11	1.09	1.10	1.09

achieved in drive design. The variables involved in this optimization are sheave diameters, belt top width, belt thickness, and belt (sheave) angle. In designing new variable-speed drives, it is helpful, if not essential, that a device manufacturer or V-belt manufacturer be contacted for drive design assistance. When standard, readily-available drive components (V-belts and sheaves) cannot be incorporated into a new drive design, special design considerations must be applied. Because of the limited drive design data available, the V-belt manufacturers must be involved in the drive selection process.

III. AGRICULTURAL ADJUSTABLE-SPEED DRIVE DESIGN

Agricultural adjustable-speed belts are generally thicker than industrial variable-speed belts so that they can transmit the higher HP required by large agricultural machinery. The standard cross sections cover a range of top widths from 1.00 in. to 2.50 in. in 0.25 in. increments. These belts provide a maximum speed variation of approximately 4:1 if variable-pitch sheaves are used on both shafts. These belts are commonly used on the main traction and cylinder drives of self-propelled combines. For specific details on the design of adjustable-speed drives for agricultural equipment, see Chapter 4.

IV. VARIABLE-SPEED DRIVES ON AUTOMOTIVE AND RECREATIONAL APPLICATIONS

Belt drives have been used in automotive applications for many years. Fixed-pitch V-belt drives have been used to power the accessories on the front end of automobile engines. In more recent years, variable-speed belts seem to be on the threshold of significantly new applications in the automobile industry. Belts have already been used in the main power train of a European car, and automotive companies worldwide are giving renewed interest to the use of variable-speed belts in automotive applications.

Perhaps one of the more motivating factors in recent years has been the development of the snowmobile and other recreational vehicles. In the 1960s and 1970s, snowmobiles, all-terrain vehicles, and mini-bikes became an ever-increasing market in the United States. These recreational vehicles incorporated a continuously variable transmission which was belt driven. In the early application of belts on these vehicles, the tremendous heat being generated at the drive sheaves as the power was applied to the vehicle became a significant factor in early belt failures. Service life was short, and belts had to be replaced frequently. V-belt manufacturers continued to find improved materials and better ways of manufacturing V-belts, so that they were able to give satisfactory service in these hostile environments. As belt manufacturers continued to improve the quality of their variable-speed belts, the recreational vehicle manufacturers entered a race leading to ever-increasing HP loads on continuously variable transmissions. This market became a significant testing area for new belt developments, and today, the possibility of V-belts again being used on an automotive continuously-variable transmission is very close to reality.

In addition to the continuously-variable transmission concept, developmental work has been going on worldwide in the area of controlled-speed accessory drives. The ability to drive a number of accessories from a constant speed shaft would allow for significant changes in the sizing of many automotive components, such as alternators and air-conditioning compressors.

A. Controlled-Speed Accessory Drives

Significant changes have occurred in the automotive market over the years to satisfy the demand of the buying public for an ever-increasing number of luxuries. The increased demand for electrical accessories led to larger alternators. Use of air-conditioning increased over these same years, resulting in added load from air-conditioning compressors. Power steering became almost a standard item in the automotive market. All power steering units had to be sized for adequate output at idle and, as a result, these accessories tended to have wasteful over-capacity at cruising speeds. They could easily be made smaller, lighter, and less costly, if there was some way to keep them at a "cruise speed" at idle. Significant reductions in parasitic HP loads could be achieved, and testing done by a number of companies would indicate that this could result in a significant improvement in fuel mileage.

In the design of controlled-speed accessory drives, standard shelf components do not exist. Although state-of-the-art variable-speed belts for these drives are available today, the hardware required must be custom-designed. V-belt manufacturers can provide significant engineering assistance in the design of these systems.

The typical controlled-speed accessory drive system might very well approximate those systems used on snowmobiles and other recreational vehicles today. The driveR sheave would be located on the engine crankshaft and would, in some manner, be controlled such that belt movement in the driveR sheave would be directly related to engine rpm. The driveN sheave could be spring-loaded and torque-sensitive. The torque-sensing mechanism would allow the belt drive to shift automatically to an appropriate ratio, thus accounting for varying accessory load demands. At engine idle speed, the driveR sheave would be closed and the driveN sheave would be open, resulting in a speed-up ratio. As engine speed increased, the driveR sheave would open automatically, allowing the driveN sheave to close. Speed at the driveN shaft would be reduced proportionately to an increase in engine rpm. In one of the simplest situations, the driveR sheave could be centrifugally actuated through weights which would react to engine rpm. The net effect of these ratio changes would yield a constant-speed accessory shaft from which other accessories might be driven. The torque-sensing capability in the driveN sheave would allow for speed ratio adjustments as accessory load changes, such as turning on or off the air-conditioning compressor. See Figure 11 which depicts the schematics for a constant-speed accessory drive.

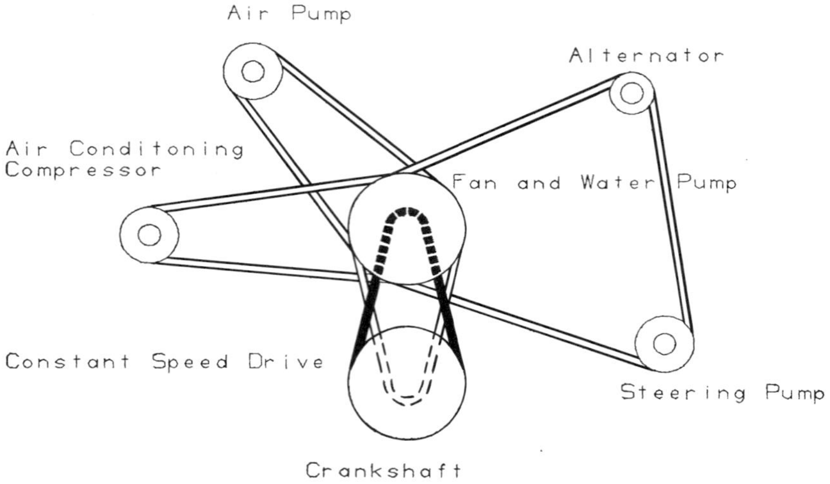

FIGURE 11 Front-end engine arrangement with a constant-speed accessory drive.

While the basic variable-speed design principles apply, there is very little published design literature available. V-belt manufacturers should be contacted for drive design assistance.

B. Continuously Variable Transmission Drives

While the large industrial variable-speed belts normally are rated at between 50–100 HP, smaller, high-capacity belts have been used in snowmobile applications up to 80 HP and over. With the trend in the automotive industry to down-sizing vehicles, engine HPs have been proportionately reduced. Developments in variable-speed belts have also been made and the likelihood of their use on the new, smaller automobiles is becoming a reality.

Much like the snowmobiles, the driveR sheave would typically be connected to a centrifugally actuated control. At idle speed, the driveR sheave halves are separated so that the belt "declutches" and no power is transmitted whatsoever. As engine speed is increased, the sheaves are brought together, engaging the belt at minimum sheave radius, which produces maximum speed reduction and maximum torque multiplication. As engine speed is increased further, centrifugal actuation continues to move the driveR sheave halves closer together while the driveN sheave halves move further apart in response to increasing belt tension. The net effect is an

FIGURE 12 Gates Power-Trac belt on a continuously variable transmission.

"upshift" increase in vehicle speed while reducing torque multiplication. As the vehicle encounters a hill or other situation such as deep snow that would increase engine load, the fall-off in engine speed combined with increase in belt tension causes the driveR sheave halves to separate, producing an automatic "downshift." The net effect of these ratio changes is similar to that produced with an automatic transmission such as a torque converter with planetary gears.

Drive ratio can be determined by engine speed alone or by a combination of engine speed and torque. The system sensitive to speed only shifts in proportion to engine speed. More complex systems include a torque-sensitive mechanism which can sense engine load and downshift even though the engine is operating at high speed.

A properly designed, continuously variable transmission incorporating V-belts could provide both optimum performance and improved fuel economy. Consequently, a small fuel-efficient engine can be made to respond with a throttle response equivalent to that of a much larger engine. The advances made in microprocessor technology allow for significant improvements in the electronic logic that could control the continuously variable V-belt transmission.

As more sophisticated hardware and control systems have emerged, variable-speed belt technology has gained acceptance throughout the automotive industry. Today, significant technological breakthroughs have been made in rubber V-belts. A significant development in this area has been the new Gates Power-Trac belt (see Figure 12). The Power-Trac belt is a unique product incorporating both state-of-the-art elastomeric technology with specially designed trusses for tensile cord support under heavy load conditions. Work is also being done to develop continuously-variable transmissions incorporating flat belts operating on pulleys with varying pitch diameters.

As with controlled speed accessory drives, the use of continuously variable belt-driven transmissions must involve the assistance of V-belt manufacturers.

7

V-Ribbed Drive Design

WALLACE D. ERICKSON *Automotive Products Technical and Development Division, The Gates Rubber Company, Denver, Colorado*

I. TYPES OF DRIVES THAT CAN USE V-RIBBED BELTS

A. Definition

As explained in Chapter 1, V-ribbed belts can generally be used on the same types of power transmission applications as V-belts. When first introduced in the 1950s, they were recommended primarily for the industrial market. A short time later, they were introduced into the agricultural market and within the last few years, they have become the dominating belt for new designs in the automotive accessory drive market.

B. Advantages and Disadvantages Compared to V-Belts

1. Tensile Stress

Like that of a V-belt, the load/life capacity of a V-ribbed belt is related to the stresses developed in the tensile section of the belt. This stress consists of the same three factors:

Tensions required to transmit the torque or power
Centrifugal tension developed as the belt travels in a circular path
Bending tension induced from bending the belt around the pulley

Like V-belts, V-ribbed belts operate in grooved pulleys but they do not have the full wedging capability of V-belts. Because of this, the normal force between the rib sidewall and the pulley face is not as high as it is with a V-belt. This must be compensated for by operating the drive with higher total belt tensions (lower tension ratios) to transmit the same power.

While V-belts are commonly designed with a tension ratio of 5 for 180° arc of contact, the V-ribbed belt drive is designed with a tension ratio of 4.

Figure 1 illustrates the different wedging characteristics between V-belts and V-ribbed belts. Note that there are two types of V-ribbed belts illustrated: one conforms to and essentially fills the bottom of

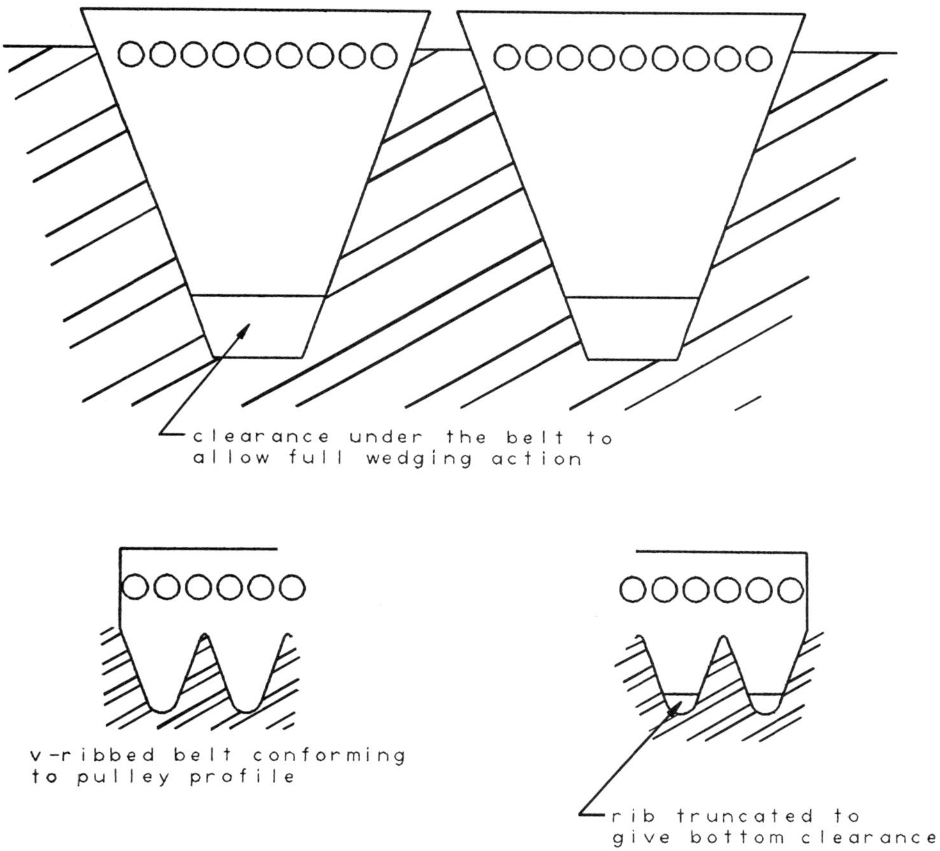

FIGURE 1 Comparison of wedging action between V-belts and V-ribbed belts.

the groove and the other is truncated to give clearance between the bottom of the belt and the pulley groove. The latter gives more wedging but is still not comparable to a V-belt because of the lack of clearance on the top of the pulley land between the grooves.

The centrifugal tension developed in a power transmission belt is directly proportional to the mass of the belt. Since the V-ribbed belt has less material in the ribs for a given width of belt, the centrifugal tension will be somewhat less for a V-ribbed belt drive than for a V-belt drive operating at the same speed.

Because of the lightweight ribs, the V-ribbed belt is also more flexible and tends to develop less bending stress in the tensile cord as it is bent around a pulley.

The lower centrifugal and bending tension in a V-ribbed belt tends to compensate for the higher tension required to transmit the torque or power of the drive. This is particularly true of drives which require small pulley diameters or are operating at high speeds.

A V-ribbed drive usually has more tensile cords to share the stress, however, This is because the tensile section is continuous across all ribs (as illustrated in Figure 1) whereas a V-belt drive has individual tensile sections for each belt.

There is no clear-cut advantage to either type belt from the standpoint of power transmission capacity. Drives requiring high-speed operation over small pulley diameters would tend to favor the V-ribbed belt whereas drives requiring the transmission of high torques over larger diameters at lower speeds would favor the V-belt.

2. Bearing and Shaft Loads

The higher tensions of the V-ribbed drive required to transmit power or torque will result in higher bearing and shaft loads than with a V-belt drive. Bearing and shaft loads resulting from belt tension should be calculated as described in Chapter 11 before deciding on the use of a V-ribbed drive. Bearings of higher capacity and/or larger shafts are often necessary.

3. Stability

To operate properly, any type of belt drive must be stable on the pulleys. That is, it must not come out of the pulley grooves nor turn over in the grooves. Instability is normally caused by the belt entering the pulley at a misaligned condition.

Misalignment can be caused either by the pulleys not rotating in a common plane as described in Chapter 12 or by the belt being displaced out of the normal plane of operation by span vibration from torque impulses or machinery vibration.

The V-ribbed belt is more sensitive to misalignment for two reasons:

The smaller, more flexible ribs on the underside of the V-ribbed belt offer less capability to guide the belt into the groove if it is entering misaligned.

The tensile cord placement above the top of the pulley grooves tends to increase the deflection stress on the ribs.

Both of the above factors can cause the V-ribbed type of belt to seat improperly if entering the pulley misaligned. It may "jump" grooves.

On the other hand, the V-ribbed belt with its wide, continuous tensile section tends to be more rigid in a lateral or transverse direction. It is similar in this regard to the joined-type V-belt (discussed in Chapter 8) which is designed specifically for drives with a tendency to induce lateral belt span vibration.

In summary, a V-ribbed type belt requires greater care on installation to assure stable operation. The required closer tolerances on alignment are often difficult to obtain on certain types of machinery—especially in adverse conditions. On drives where span vibration is a problem, however, the special joined V-belt may be required to give the same high degree of transverse or lateral rigidity obtained from a V-ribbed belt.

4. Simplicity of Inventory

The modern day V-belt has a distinct advantage over the V-ribbed belt in the simplification of inventory. V-ribbed belts of a specific cross section and length must be stocked by distributors and dealers in a wide variety of "number of ribs" to serve the wide range of power capacities required for the industrial market.

On the other hand, a V-belt of a given cross section and length can be grouped together in multiple in any required number to satisfy the power requirements of a drive. Therefore, a limited investment in belt inventory will satisfy a wide range of needs of the industrial market.

5. Reaction to Foreign Material

Because there is very little, if any, clearance between the bottom of the ribs and the bottom of the pulley grooves, foreign material coming in contact with a V-ribbed belt drive can often cause severe damage to the belt or pulley. Material caught between the belt and pulley can become tightly wedged since there usually is not sufficient clearance to allow it to pass through.

In contrast, the V-belt drive has enough clearance below the bottom of the belt to allow most foreign material to pass through without affecting the reaction between the belt and pulley.

II. BASIC CONSTRUCTION OF V-RIBBED BELTS

There are two basic types of V-ribbed belt construction. Both types use a single-unit tensile section (one row of tensile cords) which extends across the entire width of the belt. These two types differ in the method used to form the ribs on the bottom side of the belt.

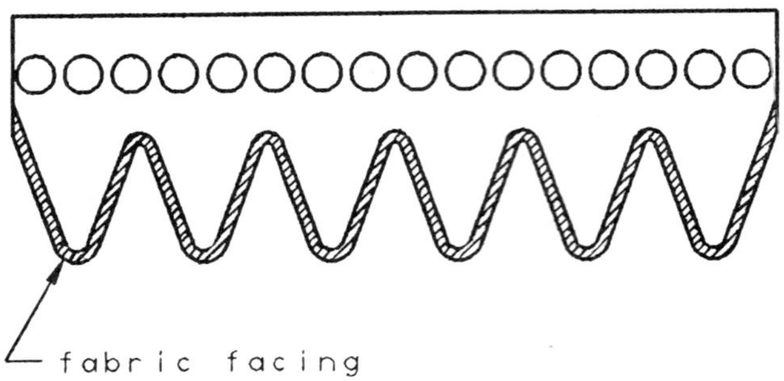

FIGURE 2 Fully molded V-ribbed belt construction.

Fully molded ribs as shown in Figure 2: these ribs are often covered with a rubber-impregnated fabric to give better wear characteristics and to reduce the noise characteristics of the drive.

Ribs formed by grinding the rib profile in a cured slab as shown in Figure 3.

Either construction will operate satisfactorily on most applications. Preference for either construction is usually predicated on the manufacturing capabilities of the individual V-ribbed belt manufacturer and user preference.

FIGURE 3 V-ribbed belt with ribs formed by grinding.

III. CROSS SECTIONS OF V-RIBBED BELTS

There are a total of five individual cross sections available for use on V-ribbed type drives. These are the H, J, K, L, and M cross section designations when working with conventional English or USA units. The cross section designations for SI metric units remain the same except that a "P" is added before the cross section designation letter (PH, PJ, PK, PL, and PM).

Table 1 and Figure 4 give the sheave or pulley groove dimensions for V-ribbed belts of these five cross sections. These sheave groove dimensions are standardized by The Rubber Manufacturers Association, The Mechanical Power Transmission and The Rubber Association of Canada in Standard IP-26.

Note that the distinctive feature of each cross section is the spacing between the ribs (Sg) rather than the groove width as it is for V-belts.

A. Nomenclature

1. V-Ribbed Belts

The standard method of specifying the size of a V-ribbed belt is by effective length, cross section designation, and the number of ribs. A description of effective length is given in Chapter 1 and the method for measuring effective length is given in the following section on industrial drive design.

An example of standard nomenclature for the English system is:

540 L 6

Belt length designation — 540; Cross section — L; Number of ribs — 6

The belt length designation is the effective length to the nearest tenth of an inch — 54.0 inches in the example.

Standard nomenclature expressed in SI metric units is:

6 PL 3672

Number of ribs — 6; Cross section designation — PL; Effective length (mm) — 3672

TABLE 1 Sheave or Pulley Groove Dimensions for V-Ribbed Belts

Cross Section	Minimum Recommended Outside Diameter	α Groove Angle ±0.25 Degrees	S_g*	r_t +0.005 −0.000	2a	r_b	h_g Minimum	d_B ±0.0005	S_e
H	0.50	40	0.063 ±0.001	0.005	0.020	0.013 +0.000 −0.005	0.041	0.0469	0.080 +0.020 −0.010
J	0.80	40	0.092 ±0.001	0.008	0.030	0.015 +0.000 −0.005	0.071	0.0625	0.125 +0.030 −0.015
K	1.50	40	0.140 ±0.002	0.010	0.038	0.020 +0.000 −0.005	0.122	0.1093	0.125 +0.050 −0.000
L	3.00	40	0.185 ±0.002	0.015	0.058	0.015 +0.000 −0.005	0.183	0.1406	0.375 +0.075 −0.030
M	7.00	40	0.370 ±0.003	0.030	0.116	0.030 +0.000 −0.010	0.377	0.2812	0.500 +0.100 −0.040

**Summation of the deviations from "S_g" for all grooves in any one sheave shall not exceed ±0.010 inches.*

Source: Courtesy of The Rubber Manufacturers Association, The Mechanical Power Transmission Association, and The Rubber Association of Canada.

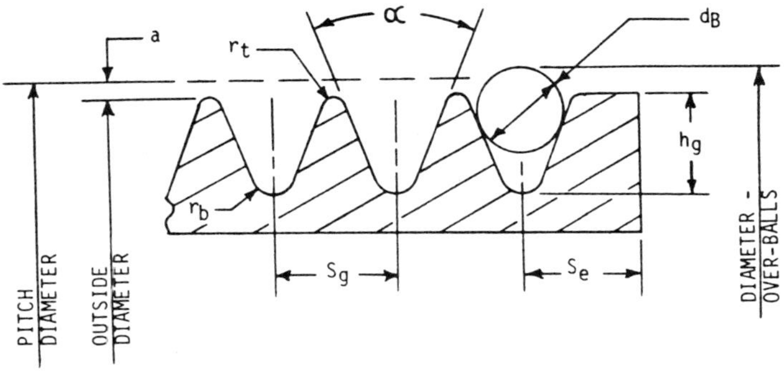

FIGURE 4 Standard groove dimensions.

2. V-Ribbed Sheaves or Pulleys

Standard nomenclature for sheaves or pulleys is not included in the industry standards. When ordering sheaves or pulleys, the following information should be included:

Cross section designation
Sheave or pulley effective (outside) diameter
Number of grooves
Bore of hub or bushing
Keyseat dimensions (if used) or any special mounting characteristics

IV. INDUSTRIAL DRIVE DESIGN USING V-RIBBED BELTS

A. Definition

Industrial drive design using V-ribbed belts is very similar in many respects to industrial drive design using V-belts. It is recommended that Chapter 3, "Industrial Drive Design Using V-Belts" be reviewed prior to continuing in this chapter. Because of the similarity, frequent reference will be made to Chapter 3 rather than repeating similar information.

Unlike the applications of V-belts to industrial-type drives, there has been no attempt to group V-ribbed belts into heavy-duty and light-duty categories. Of the five standard cross sections described earlier in this chapter, three are commonly used for industrial type drives. These are the J, L, and M cross sections.

The H section is used on special drives — usually small appliances where it is necessary to operate on very small pulley diameters. A domestic clothes dryer is a good example. The H section V-ribbed belt drives from a very small pulley on the electric motor directly to the dryer drum where it operates as a flat belt. The grooved pulley on the motor must be very small to obtain the high speed ratio necessary to drive the dryer drum slow enough to allow the clothes to tumble.

The K section is used almost exclusively for automotive accessory drives. Designing drives for this application will be discussed later in this chapter.

The intent of this (and other) drive design sections is to give the reader a general understanding of how to design drives using V-ribbed belts. Full details including up-to-date power ratings should be obtained from design manuals published by V-ribbed belt manufacturers or distributors.

B. Performance Requirements

The performance requirements for an industrial V-ribbed drive are set at a fixed life performance level the same as heavy duty industrial V-belt drives. (Refer to Chapter 4 for a more detailed explanation.)

After a drive is designed geometrically to give the desired speed of the driven unit and to fit into the available space, the capacity of the drive is obtained from the power ratings per rib for a drive of that configuration. An adequate number of ribs is chosen to give the drive the necessary power capacity.

C. Standard and Stock Components

Since V-ribbed drives are not as popular as V-belt drives, it will be more difficult to find adequate stocks of belts and pulleys for immediate delivery. The designer would be wise to check with his source of supply to determine what belt and pulley sizes are available or easily obtained before designing the drive.

1. V-Ribbed Belts

Standard V-ribbed belt effective lengths and number of ribs have been established by the belt and pulley manufacturers. These are shown in the industry standard IP-26 developed jointly by The Rubber Manufacturers Association, The Mechanical Power Transmission Association, and The Rubber Association of Canada.

Tables 2 and 3 give the standard effective lengths and the standard number of ribs per belt for J, L, and M cross sections.

2. V-Ribbed Sheaves

There are no industry standards on the effective diameter of the number of grooves of industrial V-ribbed sheaves.

Several sheave manufacturers stock certain popular sizes of sheaves but no single sheave manufacturer stocks or considers standard a wide range of diameters and number of grooves in all three industrial cross sections.

Availability of sheaves can be obtained from published catalogs from sheave manufacturers. It is recommended that the designer consult these catalogs before designing a drive to be sure of the availability of sheaves.

D. Power-Rating Formulas

1. Basic Formula

The basic power-rating formula for V-ribbed type belts is of the original type referred to in Section E of Chapter 3, "Industrial Drive Design Using V-Belts." This is of the type:

$$P_r = \left[C_1\left(\frac{S}{L}\right)^{0.09} - \frac{C_2}{d_a} - C_3 S^2\right]S \qquad (1)$$

where

P_r = HP rating per rib
S = belt speed (ft/min/1000)
L = belt length (in.)
d_a = equivalent or average pitch diameter of both sheaves (in.)
C_1, C_2, C_3 = parameters based on specific characteristics of the belt

It is suggested that the reader review the material covered in Section E of Chapter 3 for a more thorough understanding of this basic power rating formula.

The more advanced type of power rating discussed in Section E of Chapter 3 has never been developed for V-ribbed type belts. This is because this type of belt has never gained the popularity of V-belts and therefore has not warranted the extra testing and analysis required for the advanced-type formula.

2. Speed Ratio Correction

To make this basic formula easier to use, it has been converted to an rpm form with an add-on power rating for speed ratio. Again the reader is referred to Section E of Chapter 3 for a discussion on the need for a speed ratio correction.

As in Chapter 3, Equation (1) can be converted to the rpm form by using the relationship:

TABLE 2 Standard Effective Lengths for V-Ribbed Belts

J Cross Section			L Cross Section			M Cross Section		
Standard* Length Designation	Standard Effective Length	Permissible Deviation From Standard Length	Standard* Length Designation	Standard Effective Length	Permissible Deviation From Standard Length	Standard* Length Designation	Standard Effective Length	Permissible Deviation From Standard Length
180	18.0	+0.2 −0.2	500	50.0	+0.2 −0.4	900	90.0	+0.4 −0.7
190	19.0	+0.2 −0.2	540	54.0	+0.2 −0.4	940	94.0	+0.4 −0.8
200	20.0	+0.2 −0.2	560	56.0	+0.2 −0.4	990	99.0	+0.4 −0.8
220	22.0	+0.2 −0.2	615	61.5	+0.2 −0.5	1060	106.0	+0.4 −0.8
240	24.0	+0.2 −0.2	635	63.5	+0.2 −0.5	1115	111.5	+0.4 −0.9
260	26.0	+0.2 −0.2	655	65.5	+0.2 −0.5	1150	115.0	+0.4 −0.9
280	28.0	+0.2 −0.2	675	67.5	+0.3 −0.6	1185	118.5	+0.4 −0.9
300	30.0	+0.2 −0.3	695	69.5	+0.3 −0.6	1230	123.0	+0.4 −1.0
320	32.0	+0.2 −0.3	725	72.5	+0.3 −0.6	1310	131.0	+0.5 −1.1
340	34.0	+0.2 −0.3	765	76.5	+0.3 −0.6	1390	139.0	+0.5 −1.1
360	36.0	+0.2 −0.3	780	78.0	+0.3 −0.6	1470	147.0	+0.6 −1.2
380	38.0	+0.2 −0.3	795	79.5	+0.3 −0.6	1610	161.0	+0.6 −1.2
400	40.0	+0.2 −0.4	815	81.5	+0.3 −0.7	1650	165.0	+0.6 −1.3
430	43.0	+0.2 −0.4	840	84.0	+0.3 −0.7	1760	176.0	+0.7 −1.4
460	46.0	+0.2 −0.4	865	86.5	+0.3 −0.7	1830	183.0	+0.7 −1.4
490	49.0	+0.2 −0.4	915	91.5	+0.4 −0.7	1980	198.0	+0.8 −1.6
520	52.0	+0.2 −0.4	975	97.5	+0.4 −0.8	2130	213.0	+0.8 −1.6
550	55.0	+0.2 −0.4	990	99.0	+0.4 −0.8	2410	241.0	+0.9 −1.6
580	58.0	+0.2 −0.5	1065	106.5	+0.4 −0.8	2560	256.0	+1.0 −1.8
610	61.0	+0.2 −0.5	1120	112.0	+0.4 −0.9	2710	271.0	+1.1 −2.2
650	65.0	+0.2 −0.5	1150	115.0	+0.4 −0.9	3010	301.0	+1.2 −2.4
730	73.0	+0.3 −0.6	1215	121.5	+0.4 −1.0	3310	331.0	+1.3 −2.6
870	87.0	+0.3 −0.7	1230	123.0	+0.4 −1.0	3610	361.0	+1.4 −2.9
920	92.0	+0.4 −0.7	1295	129.5	+0.5 −1.0			
980	98.0	+0.4 −0.8	1310	131.0	+0.5 −1.1			
			1455	145.5	+0.6 −1.2			

**To specify belt size, use the standard length designation, followed by the letter indicating belt cross section and the number of ribs desired. For example: 865L10*

Source: Courtesy of The Rubber Manufacturers Association, The Mechanical Power Transmission Association, and The Rubber Manufacturers Association of Canada.

TABLE 3 Standard Number of Ribs for V-Ribbed Belts

Cross Section	Standard Number of Ribs
J	4,6,10,16,20
L	6,8,10,12,14,16,18,20
M	6,8,10,12,14,16,18,20

$$S = \frac{\pi}{12} dr \qquad (2)$$

where

S = belt speed (ft/min/1000)
d = pitch diameter of smaller sheave (in.)
r = rpm/1000 of faster shaft

This relationship gives a power rating formula as follows:

$$P_r = \left[K_1 \left(\frac{dr}{L}\right)^{-0.09} - K_2 \left(\frac{1}{d}\right) - K_3 (dr)^2\right] dr + \Delta P_r \qquad (3)$$

where

P_r = power rating for one rib
d = pitch diameter of smaller sheave (in.)
r = rpm/1000 of faster shaft
L = belt length (in.)
ΔP_r = add-on power rating for speed ratio
K_1, K_2, K_3 = parameters based on specific characteristics of the belt

As explained in Chapter 3, the add-on power for speed ratio is used instead of the equivalent or average diameter of both sheaves, d_a. The formula for add-on power is:

$$\Delta P_r = rK_2 \left\{1 - 2\left[\frac{1 + (R+1)/2R^{11.1111}}{2}\right]^{0.09} + 1\right\} \qquad (4)$$

where

ΔP_r = add-on power rating for speed ratio
r = rpm/1000 of faster shaft
R = speed ratio of drive
K_2 = parameter based on specific characteristics of the belt

Note that in this formula, the speed ratio of the drive must always be equal to or greater than 1. Therefore, the speed ratio is defined as the rpm of the faster shaft divided by the rpm of the slower shaft, regardless of which one is the driver or the driven.

3. Belt Length Correction

As explained in Chapter 3, it is more convenient for the designer if belt length is removed from Equation (3). Then the basic power rating can be tabulated as a function of only rpm and the smaller sheave diameter. But since belt length influences the power rating of the belt, a correction factor must be used when the length differs from the value selected as the basic length. If we put the base length into Equation (3) and correct for length by a power multiplier, we have:

$$P_r = \left\{K_1'\left[(dr)^{-0.09} - K_2\left(\frac{1}{d}\right) - K_3(dr)^2\right]dr + \Delta P_r\right\}K_L \tag{5}$$

where

P_r = power rating for one rib
d = pitch diameter of smaller sheave (in.)
r = rpm/1000 of faster shaft
ΔP_r = add-on power for speed ratio from Equation (4)
K_L = belt length correction factor
K_1', K_2, K_3 = parameters based on specific characteristics of the belt

The formula for the belt length correction factor (K_L) is:

$$K_L = 1 + \left[\left(\frac{L}{L_o}\right)^{0.09} - 1\right]2.4 \tag{6}$$

where

L_o = base belt length (in.)
L = specific belt length (in.)

4. Arc of Contact Correction

The arc of contact correction factor discussion in Chapter 3 also applies to V-ribbed type belts. However, since the V-ribbed type of belt must be operated at a higher tension, the basic formula for determining the tension ratio for a specific arc of contact on the smaller sheave will be different.

The design tension ratio for a V-belt drive with two equal diameters (arc of contact of 180°) is 5. Since the V-ribbed belt has less wedging, the design tension ratio is 4.

$$R = e^{0.441270\theta} \tag{7}$$

where

R = tension ratio, tight side tension divided by slack side tension
θ = arc of contact on the smaller sheave (rad)

From this formula, the arc correction factor (G) can be determined in the same manner as explained in Chapter 3. Since the constant in Equation (7) is 0.44127 rather than 0.51230, the equation for Factor G is:

$$\text{Factor G} = 1.3333\left(\frac{R-1}{R}\right) \tag{8}$$

where

Factor G = arc of contact correction factor

R = tension ratio = $e^{0.441270}$

where

θ = arc of contact on the small sheave diameter (rad)

Table 5 gives the arc correction factor for V-ribbed belts.

5. Summary of Power-Rating Formulas

In summary, the power-rating formulas in use today for the J, L, and M cross sections of V-ribbed type belts are of the following form:

$$P_r = dr\left[K_1' (dr)^{-0.09} - K_2\left(\frac{1}{d}\right) - K_3 (dr)^2\right] + K_2 r\left[1 - \frac{1}{K_{SR}}\right]$$

where

P_r = power rating for one belt
d = pitch diameter of smaller sheave (in.)
r = rpm/1000 of faster shaft
K_{SR} = speed ratio factor
K_1', K_2, K_3 = parameters based on specific characteristics of the belt

The simplified method of specifying the add-on power rating is $K_2 r[1 - 1/K_{SR}]$. The factor inside the brackets in Equation (4) for add-on power is only a function of speed ratio and can be expressed in a simple table which shows the specific K_{SR} value for a series of speed ratio ranges. The formula for K_{SR} derived from Equation (4) is:

$$K_{SR} = \frac{1}{2\left[\frac{1 + (R-1)/2R}{2}\right]^{0.09} - 1} \tag{9}$$

Table 4, taken from the industry standard IP-26 illustrates this point.

This power rating formula is multiplied by the two correction fac-

TABLE 4 Speed Ratio Factors (K_{SR})

Speed Ratio, D/d Range	Factor K_{SR}
1.00 To and Incl. 1.01	1.0000
Over 1.01 To and Incl. 1.04	1.0136
Over 1.04 To and Incl. 1.08	1.0276
Over 1.08 To and Incl. 1.12	1.0419
Over 1.12 To and Incl. 1.18	1.0567
Over 1.18 To and Incl. 1.24	1.0719
Over 1.24 To and Incl. 1.34	1.0875
Over 1.34 To and Incl. 1.51	1.1036
Over 1.51 To and Incl. 1.99	1.1202
Over 1.99	1.1373

Source: Courtesy of The Rubber Manufacturers Association, The Mechanical Power Transmission Association, and The Rubber Association of Canada.

tors: K_L to correct for the belt length on the drive, and Factor G to correct for the arc of contact.

E. Drive Design Procedure

As mentioned earlier , it is not the intent of this book to give drive design procedures in a step-by-step fashion along with the necessary drive design data so that a reader can design a drive. The designer should work with drive design material published by the various V-ribbed belt or sheave manufacturers. This way, he will be assured of getting the latest developments in power rating formula parameters (K_1', K_2, and K_3 values) that are compatible with the belts he will be using.

The general drive design procedure outlined in Section F of Chapter 3 can be applied to V-ribbed type belts so they will not be repeated here.

V. AGRICULTURAL APPLICATIONS FOR V-RIBBED TYPE BELTS

V-ribbed belts are not widely used on agricultural type machines. They have been used to some extent over the years on some drives for which variable speed ratios are not required, but they have never gained widespread popularity over the V-belt because of the disadvantages listed earlier in this chapter.

Most agricultural-type drives are subjected to frequent and severe shock loading. Because of the lesser wedging action of a V-ribbed type belt, it is more prone to excessive slip from shock loads than a V-belt. In the dirty and dusty environment often associated with agricultural-type drives, this excessive slip can cause abnormal belt or pulley wear.

Since agricultural drives are often subject to shock or impulse loading, the lateral or transverse rigidity of a V-ribbed belt has had significant appeal to the machinery designer. It has transverse rigidity characteristics similar to the joined type V-belt developed by The Gates Rubber Company in 1966, which is a definite advantage from a stability standpoint. (See Section I.D.3 of Chapter 3 for more information on joined belts.)

A. Geometric Drive Design

Because of its lack of popularity, the V-ribbed type of belt has not, to the present time, been standardized by the American Society of Agricultural Engineers (ASAE) as the various types of V-belts have. Whenever it is applied to agricultural-type machinery, the dimensional characteristics of the V-ribbed type drive are based on the industrial standard published jointly by the Rubber Manufacturers Association and the Mechanical Power Transmission Association. Work is currently being conducted, however, on a revised ASAE standard which will probably include V-ribbed belts.

The L and M cross sections are the ones most commonly used for agricultural-type applications. Table 1 shows the recommended industrial sheaves or pulley groove dimensions for these two cross sections.

Some of the power transmission belt manufacturers publish drive design manuals for V-ribbed type belts on agricultural machines. These manuals will provide useful information in determining the geometric characteristics of V-ribbed type drives.

It will also be useful to review Chapter 4 of this book for a review of the characteristics of power transmission drives on agricultural machinery. Many of the principles discussed in this chapter on designing V-belt drives will also apply to V-ribbed type drives.

B. Capacity Drive Design

Capacity drive design consists of selecting the proper V-ribbed cross section and the correct number of ribs to give satisfactory service life.

Again, many of the basic principles discussed in Chapter 4 on the design of V-belt drives for agricultural machines will apply to V-ribbed type drives. The tensions developed in each span of the drive are calculated in the same basic manner, keeping in mind that the V-ribbed drive must operate with higher span tensions (lower tension ratio) than a V-belt drive. Therefore, a different arc correction factor must be used than the one shown in Table 4. Table 5 gives the arc correction factors for V-ribbed belts (Factor G).

Table 5 can be used most easily with drives that have only two pulleys. Since many agricultural drives have three or more pulleys, the arc of contact can best be determined from the drive layout. Knowing the arc of contact, Factor G can be calculated from Equation (8).

The operating life of a V-ribbed belt will be influenced by the cyclic repetition of peak stress cycles the same as a V-belt. The peak stress or peak tension is comprised of the sum of the same three tensions: tightside tension necessary to transmit power or torque, the bending tension developed as the belt travels over the pulley, and the centrifugal tension which is a function of the mass of the belt and the square of the belt speed.

Fatigue rate formulas like those described in Chapter 4 have not been developed for V-ribbed belts. This is because these type formulas originated with The Gates Rubber Company and they have not, as yet, developed them for V-ribbed belts on agricultural machinery.

Some of the manufacturers of V-ribbed belts for the agricultural market do use a similar system which is based on the calculation of a *drive severity index*.

This index is based on the cyclic repetition of peak stresses or peak tensions similar to the fatigue rate system. It allows the comparison of the drive severity index for the drive being designed with a drive that is giving satisfactory service under similar conditions.

TABLE 5 Arc Correction Factors for V-Ribbed Belts (G)

$\frac{D_o - d_o}{C}$	Arc of Contact, θ on small sheave, degrees	arc correction factor G
0.00	180	1.00
0.10	174	0.98
0.20	169	0.97
0.30	163	0.95
0.40	157	0.94
0.50	151	0.92
0.60	145	0.90
0.10	139	0.88
0.80	133	0.85
0.90	127	0.83
1.00	120	0.80
1.10	113	0.77
1.20	106	0.74
1.30	99	0.71
1.40	91	0.67
1.50	83	0.63

Drive design manuals published by the manufacturers of V-ribbed belts for the agricultural market will give more detail.

VI. AUTOMOTIVE ACCESSORY APPLICATIONS FOR V-RIBBED BELTS

A. Popularity

The V-ribbed belt drive for automotive accessories is relatively new. It was first used on high-volume passenger car production in the late 1970s. It was, however, used on a limited number of heavy-duty truck and bus engines several years before that.

The advantages the V-ribbed belt has over the V-belt are ideally suited for the automotive passenger car accessory drive application. This is because most passenger car engines can operate at high speeds (often as high as 6000 or 7000 rpm) and require small pulley diameters. As discussed earlier in the chapter, the light weight of the thin V-ribbed profile develops less centrifugal tension in the belt as it operates at high belt speeds compared to a V-belt drive. The small, thin ribs also reduce the bending tension as the belt operates over very small pulley diameters such as those required on today's high-output alternators.

In addition, most automotive accessory drive applications operate in a relatively clean environment so the lack of clearance between the belt rib and the bottom of the pulley groove is not a problem. The V-ribbed belt also can be used very effectively on a serpentine-type drive to reduce the overall width of the drive package and consequently some valuable space in the engine compartment (see Sec. VI.B.2).

Because of these advantages, the V-ribbed belt is becoming increasingly popular for automotive accessory drives. Almost all new drive designs being made today in the United States utilize the V-ribbed belt. The popularity is also spreading to the European and Japanese car manufacturers.

B. Types of Automotive Drive Configurations

1. Conventional Two- or Three-Pulley Drives

The typical automotive accessory drive provides power transmission for one or possibly two accessories. The belt runs from the crankshaft as the driver pulley to the accessories in a two point or possibly a three point configuration. Figure 5 shows a typical configuration in which the belt is driving the water pump and an alternator in a three point configuration. A drive such as this is very common on most automotive engines.

When other accessories are added (such as a power steering pump or an air conditioning compressor) another belt drive – or perhaps an additional two or three belt drives – is added to provide the necessary power transmission.

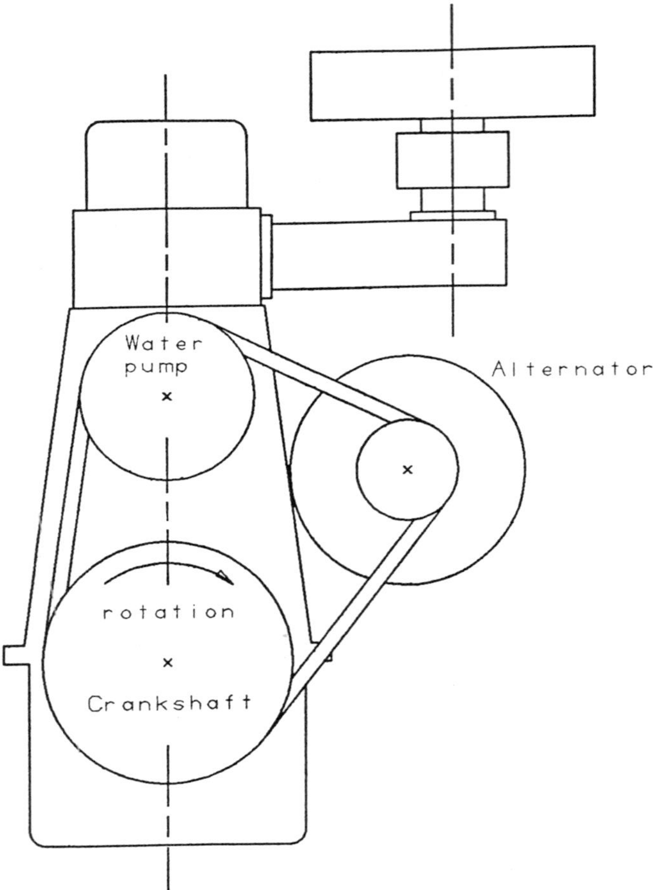

FIGURE 5 Typical 3-pulley automotive accessory drive.

Figure 6 shows the layout of accessory drives for a fully dressed engine, that is, an engine with a full array of belt driven accessories. This configuration consists of a total of four belt drives of either the two pulley or three pulley type. Each of the accessories must be mounted on a rather complex bracket which provides for movement so that the belts can be installed and tensioned.

Four belt drives on a fully dressed engine are the exception rather than the rule. Most engines will use three belt drives with the air pump – which is lightly loaded – being driven from one of the other belts or with a separate belt drive from one of the other accessories.

Typically, V-belts have been used on these types of drives. Recently V-ribbed belts have been used because of their lower centrifugal tension and, more specifically, on the alternator drive (Belt 1 in Figure 6) to give longer life on the small alternator diameter.

It is interesting to note in Figure 6 that two of the drives (Belt 1 and Belt 3) are three-pulley drives. Both drives include the water pump. When two such drives include a common accessory such as the water pump with each of the other two accessories (the air conditioning compressor and the alternator in this case) on opposite sides of the engine, the drive configuration is called a butterfly drive. This name is derived from the general appearance of the spread wings of a butterfly.

The butterfly drive configuration has been commonly used for several years. The reason for this is twofold. It is common practice to mount a cooling fan integral with the water pump. The extra power transmission capacity from two belt drives assures better performance. Such a configuration also offers an easy way to route the path of the belt to avoid interference with other parts of the engine.

2. Serpentine Drive Configuration

A serpentine drive is one with more than one driven pulley and by which power is transmitted from both the inside and outside of the belt.

This type of drive can use many different types of belts. The double V-belt (V-configuration on both inside and outside) was made especially for this type of drive and has been used for many years in industrial and agricultural applications. A regular V-belt is also used on this type of drive in the industrial and agricultural markets. Pulleys on the inside of the belt are grooved to fit the V-belt whereas those on the outside of the belt are flat. Flat belts can also be used on serpentine type drives. (See Chapter 8 for more detail on these drives.)

In the automotive market, serpentine drives using V-ribbed belts are becoming very popular. If an engine is equipped with additional accessories such as power steering and/or air conditioning, using a

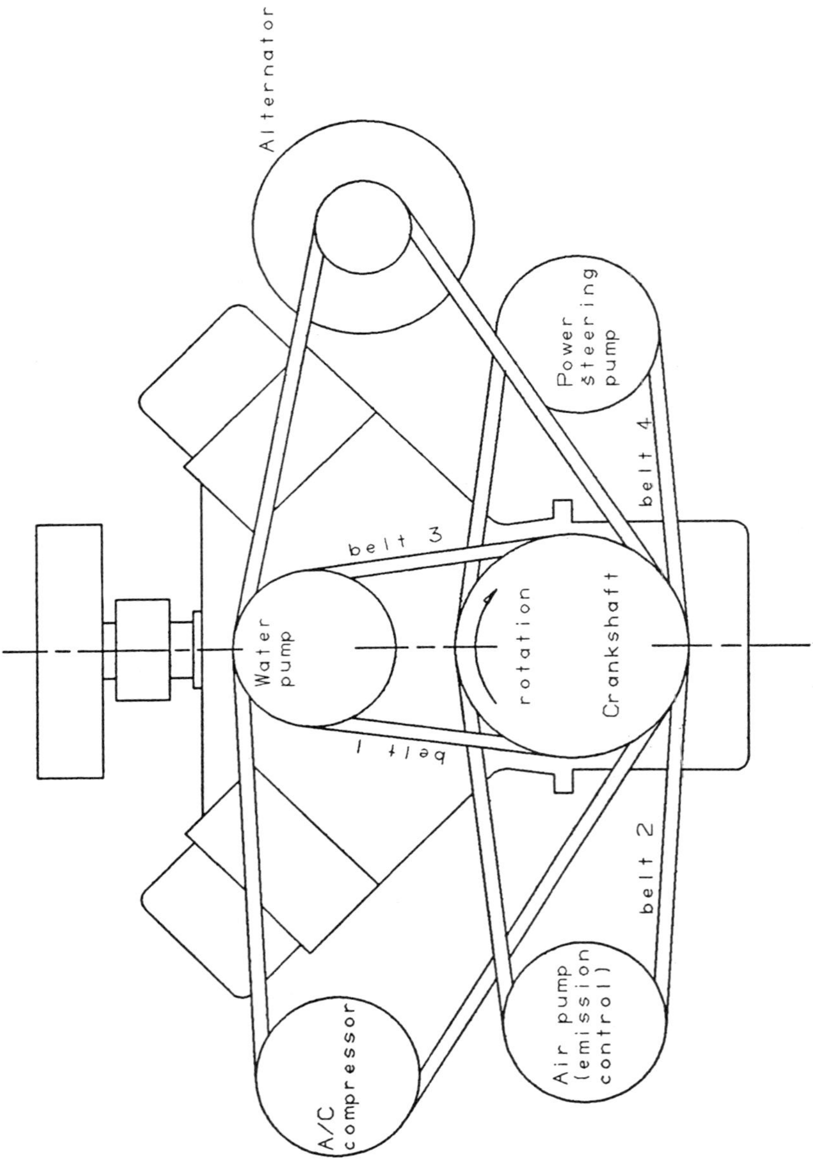

FIGURE 6 Typical accessory drive layout for a fully dressed engine.

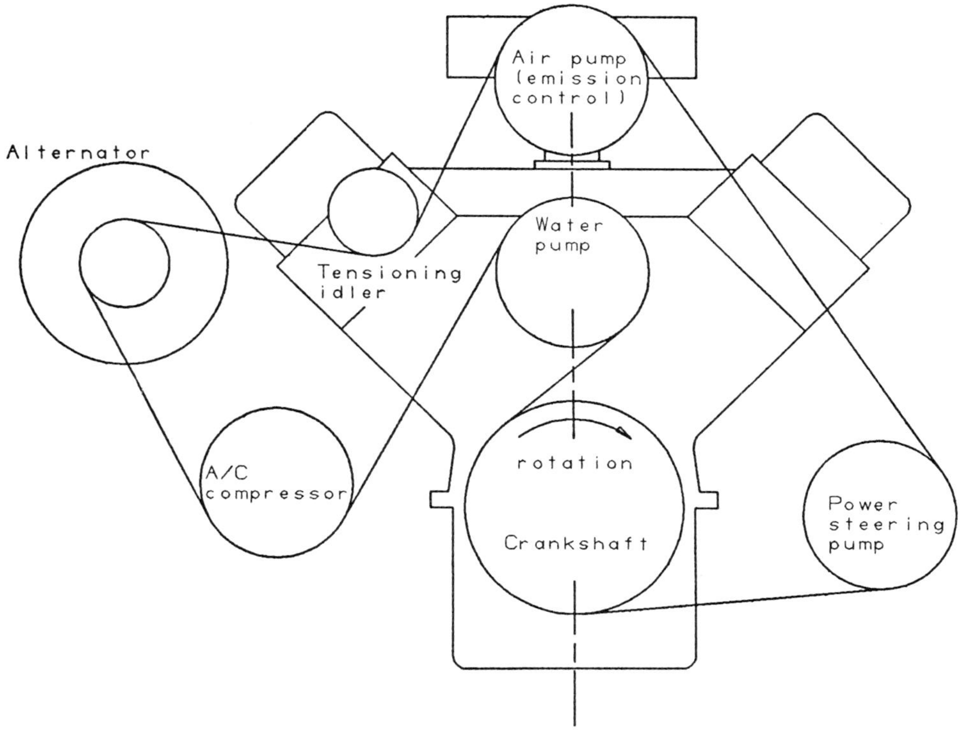

FIGURE 7 Typical serpentine accessory drive configuration for a fully dressed engine.

V-ribbed belt to drive all accessories with one drive can result in very significant space and weight savings.

Figure 7 shows a typical serpentine V-ribbed drive for a fully dressed engine. Usually a 6-rib K-section belt is used for this type of drive. A belt length range between 90 and 110 in. is common since it must include all accessories. Note in Figure 7 that the water pump is being driven from the flat backside or outside of the V-ribbed belt. Note that the water pump rotates in the opposite direction to the crankshaft which must be taken into consideration in the design of the pump and cooling fan (if used).

Care is taken in the arrangement of the accessories and the routing of the belt so that there is a large arc of contact on the flat water pump pulley. This is because a flat belt has no wedging whatsoever and must depend solely on coefficient of friction to transmit power. Since the belt must be tighter (lower tension ratio) to transmit power over a flat pulley for a given arc of contact, increasing the arc of contact will help bring the tension requirements for the flat pulley more nearly equal to those of the grooved pulleys.

The tensioning idler for the serpentine-type drive is usually a device which is designed to maintain a constant tension in the belt span it contacts. This type of tensioner usually consists of some type of spring which will exert a nearly constant force against the belt span. A dampening device of some type is also employed to minimize the idler reaction to tension fluctuations in the belt.

The advantages to the serpentine-type drive over the conventional two- or three-point drives using V-belts is the space savings in regard to drive width. Accessory drive width contributes to the overall engine length which is a very critical dimension on the transverse mounted engines commonly used on the later model front-wheel-drive cars. There is usually also a weight savings possible with the serpentine drive.

Figure 8 shows a comparison of the width of a 6-rib V-ribbed drive with a V-belt drive package consisting of 3 and 4 separate drives off the crankshaft pulley. Note that there is a very significant savings in drive width with the V-ribbed serpentine drive.

A 6-rib V-ribbed belt is commonly used on the automotive serpentine drive. The fact that the V-ribbed belt can handle the loads of the full complement of accessories with a narrower total drive width can be attributed to three of the advantages discussed earlier in this chapter: a continuous tensile member across the full width of the drive, reduced bending tension because of the thinner ribs, and less centrifugal tension when the drive is operating at high speeds.

Another important advantage to the V-ribbed serpentine drive is the ability of the drive to maintain tension over the full life of the belt when an automatic tensioning device is used. Of course, these types of tensioning devices could be (and have been) used with a V-belt drive. However, 3 or 4 individual belt drives on a fully dressed engine would require the same number of tensioning devices, which would result in a very expensive drive package.

C. Automotive Drive Design Using V-Ribbed Belts

Most of the information given in Sections II and III of Chapter 5 can be applied directly to designing drives using V-ribbed belts. Therefore, that information will not be repeated in this chapter. It is

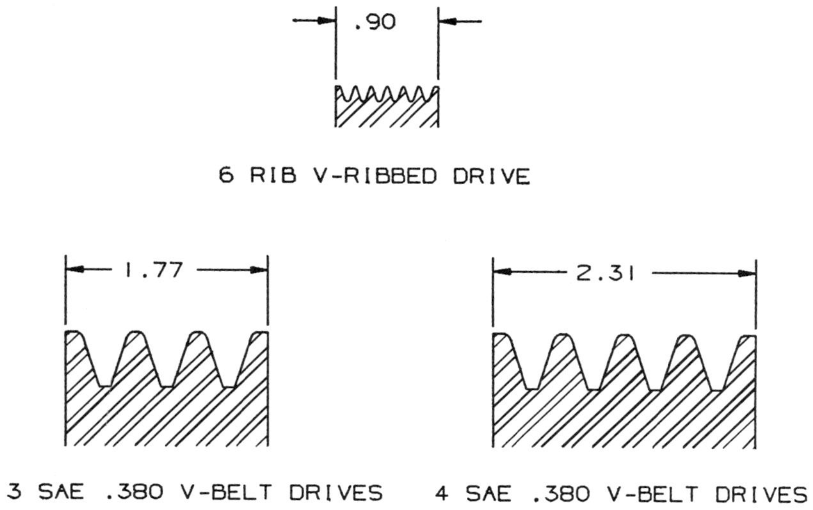

FIGURE 8 Drive width comparison between a V-ribbed serpentine drive and a V-belt package of 3 or 4 drives.

suggested that the reader study Chapter 5 before proceeding with the design of an automotive accessory drive using V-ribbed belts.

1. Standard Design Factors

The Society of Automotive Engineers (SAE) has published a standard specifying sheave groove dimensions and belt-measuring techniques and specifications based on the effective length system entitled "V-Ribbed Belts and Pulleys," SAE Standard J1459 (1984).

Table 6 gives the pulley groove dimensions in inches for the popular K-section belt as well as the L-section belt, which may find use on larger bus, truck, tractor, or stationary engines. Figure 9 defines the dimensions referred to in Table 6. Table 7 gives the groove dimension for the same two cross sections in millimeters.

The groove dimensions specified in SAE Standard J1459 are very nearly the same as those in the RMA/MPTA/RAC Standard IP-26. There is a slight difference in how the dimensions and tolerances are specified for the top of the land between the grooves since the automotive industry typically uses formed steel pulleys in contrast to the machined pulleys common in the industrial market. V-ribbed belts are interchangeable in pulleys made to either standard.

TABLE 6 Groove Dimensions (In.)

Cross section size	Minimum recommended effective diameter	S_g[a] ±0.002	F_t minimum	F_t[b] maximum	2a	r_b maximum	d_B ±0.0004	2K	S_e minimum	2x[c]
K	1.8	0.140	0.010	0.039	0.038	0.020	0.0984	0.039	0.10	0.038
L	3.0	0.185	0.016	0.069	0.061	0.016	0.1378	0.093	0.13	0.061

[a]Summation of the deviations from S_g for any 1 belt groove set in a pulley shall not exceed ±0.012 in.
[b]F_t is measured from the actual ride position of the ball or rod in the pulley.
[c]This number may vary with belt manufacturer.
Source: Courtesy of the Society of Automotive Engineers.

TABLE 7 Groove Dimensions (mm)

Cross section size	Minimum recommended effective diameter	S_g[a] ±0.05	F_t minimum	F_t[b] maximum	2a	r_b maximum	d_B ±0.010	2K	S_e minimum	2x[c]
PK	45	3.56	0.25	1.00	0.96	0.50	2.500	0.99	2.5	0.96
PL	75	4.70	0.40	1.75	1.54	0.40	3.500	2.36	3.3	1.54

[a]Summation of the deviations from S_g for any 1 belt groove set in a pulley shall not exceed ±0.30 mm.
[b]F_t is measured from the actual ride position of the ball or rod in the pulley.
[c]This number may vary with belt manufacturer.
Source: Courtesy of the Society of Automotive Engineers.

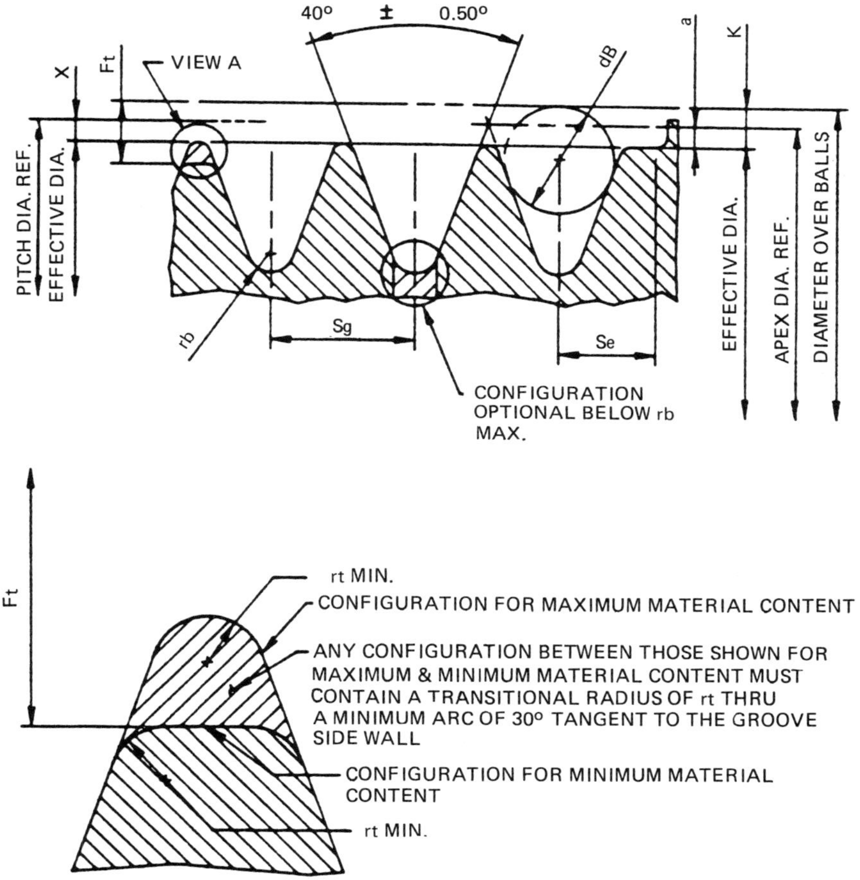

FIGURE 9 Standard groove dimensions.

The V-ribbed belt-measuring method specified in the SAE Standard J1459 is the same method as specified for other SAE and RMA/MPTA/RAC standards for V-ribbed belts and V-belts. The belt is placed on two equal-diameter pulleys of a specified effective diameter and rotated under tension. The center distance between the two pulleys is used to calculate the effective length. SAE Standard J1459 gives all of the pertinent measuring specifications as well as the allowable tolerance on center distance.

2. Design Procedures

Unlike the V-belt accessory drive which has been in use for decades, the V-ribbed accessory drive is relatively new. Therefore, accurate fatigue-rate formulas which will allow a designer to predict the operating life of a belt on a drive are not available.

Rapid advances are also being made in V-ribbed belt construction to withstand the severe environmental conditions in which an automotive accessory drive operates. The goal is to develop an elastomeric fiber-reinforced rib material which will withstand the high underhood temperatures of the modern automobile while operating over the small alternator pulleys and backside idlers.

The belts being manufactured today have a much greater resistance to high-temperature cracking than those of a few years ago. Within the near future, it is predicted that rib material will have progressed to the point where reliable predictions in stress-fatigue performance can be made based on the cyclic repetition of peak tension as explained in other chapters.

The tension in each span in a V-ribbed-type drive is calculated in the same manner as for a V-belt, which is explained in Chapter 5. The allowable tension ratio is also calculated in the same manner but it is important again to notice that the design tension ratio for a V-ribbed belt operating in V-grooved pulleys is 4 rather than 5 (that for a V-belt). This is for a pulley with 180° arc of contact. When a V-ribbed belt is operated on a flat pulley such as on a serpentine drive, the design tension ratio is 2 or 2.5.

As this discussion indicates, it is important that the machine designer get help from the application engineers which all manufacturers of V-ribbed type belts have made available.

8

Designing Special Types of Drives

CHARLES H. ARTUS *V-Belt Product Application, The Gates Rubber Company, Denver, Colorado*

I. BELTS AND SHEAVES NOT IN ONE PLANE

A. Quarter Turn

The driveR and driveN sheaves of the usual V-belt drive are in the same plane, but many V-belt drives have the sheaves purposely placed in two planes. These drives change the orientation of power transmission and perform the usual functions of V-belt drives. When properly designed and installed, they operate successfully at little, if any, increase in maintenance cost over that of usual V-belt drives.

Two of the most common two-plane V-belt drives are quarter-turn drives, which have the sheaves 90° apart (Figure 1), and eighth-turn drives, which have the sheaves at 45°. Quarter-turn drives are more widely used. An example is a vertical-shaft irrigation pump driven by a horizontal power take-off (PTO) shaft on an internal-combustion engine.

To design a quarter-turn or eighth-turn drive, first follow the standard procedure as if you were designing a regular drive to determine sheave sizes, number and type of belts, and center distance. When HP ratings cannot be read directly from tables, they must be determined from basic ratings with the various corrections applied. The arc of contact, or wrap, of V-belts on the sheave is greater on a quarter-turn or eighth-turn drive than on a one-plane drive that has the same sheave diameters and center distance. Because of this fact, and because speed ratio is usually limited to 2.5:1, Factor G, the arc-of-contact correction factor, is always taken as 1.0.

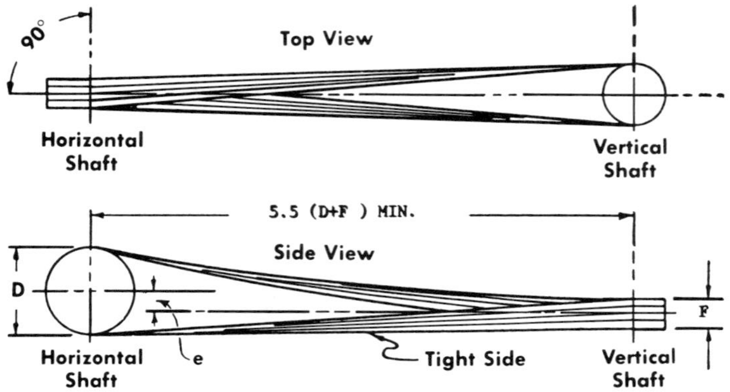

FIGURE 1 Quarter-turn drive.

If the desired speed ratio does not exceed 2.5:1, then it is only necessary to check sheave diameters and belt length so that these fulfill these minimum-center-distance requirements:

Quarter-turn, minimum center distance

$$C = 5.5\ (D + F)$$

Eighth-turn, minimum center distance

$$C = 4.0\ (D + F)$$

where

D = diameter of larger sheave (in.)
F = width of the band of belts (in.) (Table 1)

Where quarter-turn drives require a speed ratio greater than 2.5:1, a compound driver should be used. To design such a drive, the primary drive should be the quarter-turn drive (if possible) and the conventional drive should provide most of the speed ratio to optimize the drive (Figure 2). Use deep-grooved sheaves for individual V-belts and standard grooved sheaves for joined V-belts.

1. Alignment

Excessive misalignment of any V-belt drive causes increased belt-sidewall wear, and can affect belt stability and fatigue life. Alignment is a major concern for quarter-turn and eighth-turn drives.

Alignment of a V-belt entering a sheave is measured in terms of *angle of entry* or *fleeting angle*. The design and installation recommendations given below are intended to keep the angle of entry to 6° or less. *Note*: The restriction to a maximum of 6° fleeting angle applies only to belt stability. For optimum fatigue life, misalignment should be minimized.

As you look down on the drive, a line from the center of the vertical shaft should pass through the center of the face of the sheave on the horizontal shaft. The horizontal shaft should be at right angles to this line (see top view in Figure 1).

Looking at the side of the drive, the center of the horizontal shaft should be raised a distance, e, above a level line through the center of the face of the sheave on the vertical shaft (see side view in Figure 1). The tight side of the drive should be the bottom as shown and, because the slack side has more sag than the tight side, alignment is better when the vertical-shaft sheave is below the center line of the drive.

TABLE 1 Width of Band of Belts on Deep Grooved and Standard Sheaves (In.) (F)

V-belt section	Groove type	Number of belts									
		1	2	3	4	5	6	7	8	9	10
3VX	Deep groove	.38	.88	1.38	1.88	2.38	2.88	3.38	3.88	4.38	4.88
	Standard groove	.38	.79	1.19	1.60	2.00	2.41	2.82	3.22	3.63	4.03
5V/5VX	Deep groove	.62	1.43	2.24	3.06	3.87	4.68	5.50	6.30	7.12	7.93
	Standard groove	.62	1.31	2.00	2.68	3.37	4.06	4.75	5.44	6.12	6.81
8V	Deep groove	1.0	2.31	3.62	4.94	6.25	7.56	8.87	10.18	11.50	12.81
	Standard groove	1.0	2.13	3.25	4.38	5.50	6.63	7.75	8.88	10.00	11.13
A	Deep groove	.50	1.25	2.00	2.75	3.50	4.25	5.00	5.75	6.50	7.25
	Standard groove	.50	1.13	1.75	2.38	3.00	3.63	4.25	4.88	5.50	6.13
B	Deep groove	.66	1.54	2.41	3.29	4.16	5.04	5.91	6.79	7.66	8.54
	Standard groove	.66	1.41	2.16	2.91	3.66	4.41	5.16	5.91	6.66	7.41
C	Deep groove	.88	2.13	3.38	4.63	5.88	7.13	8.38	9.63	10.88	12.13
	Standard groove	.88	1.88	2.88	3.88	4.88	5.88	6.88	7.88	8.88	9.88
D	Deep groove	1.25	3.00	4.75	6.50	8.25	10.00	11.75	13.50	15.25	17.00
	Standard groove	1.25	2.69	4.13	5.56	7.00	8.44	9.88	11.32	12.75	14.19
E	Deep groove	1.50	3.56	5.62	7.69	9.75	11.81	13.87	15.93	18.00	20.06
	Standard groove	1.50	3.25	5.00	6.75	8.50	10.25	12.00	13.75	15.50	17.25

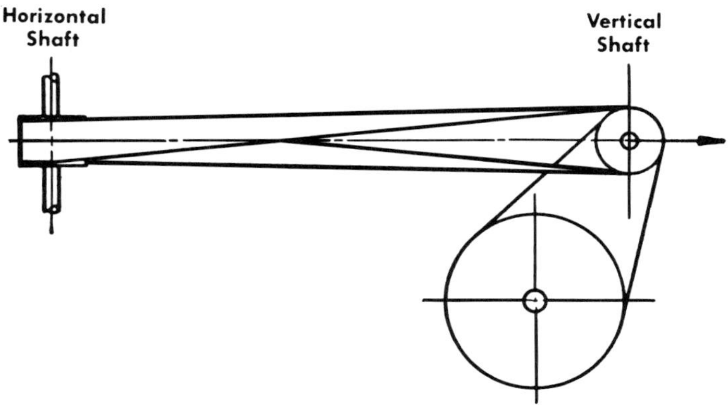

FIGURE 2 Quarter-turn drive with a conventional drive.

The amount of offset, e, of the vertical shaft (or elevating of the horizontal shaft) is sometimes called superelevation. Table 2 gives standard values of e. These can be evaluated for other belts by:

$$e = 0.15 \left(\frac{W}{T_S}\right) C^2$$

where

W = weight of belt (lb/in.)
T_S = slack-side tension (lb)
C = center distance (in.)

Remember that slack-side tensions may be different on different drives, and that belts of different stiffness can sag different amounts. Thus, consider the given value of e as a starting point. Watching the drive, particularly under starting load, will usually show any needed changes in e to obtain the best belt entry into the sheave.

Quarter-turn and eighth-turn drives, set up as stated, should not be reversed. That is, the direction of sheave rotation must not be changed. Reversing can create instability because the angles of entry change. On some vertical drives, where one sheave is above the other, or that use idlers, reversibility may be possible.

2. Maintenance

Requirements for maintaining quarter-turn and eighth-turn drives are the same as those for conventional drives, except that maintaining

TABLE 2

	Dimension e (in.)	
Center distance	A, B, C, D, E cross sections	3V, 5V, 8V cross sections
20	0.2	
30	0.2	
40	0.4	
50	0.4	
60	0.5	0.2
80	0.5	0.3
100	1.0	0.4
120	1.5	0.6
140	2.0	0.9
160	2.5	1.2
180	3.5	1.5
200	4.0	1.8
220	5.0	2.2
240	6.0	2.6

Note: For longer centers consult the manufacturer.

proper tension is somewhat more critical. Large amounts of over- or under-tensioning can create greater angles of entry and, therefore, will affect stability. There are some general precautions which should be observed:

Rotation The direction of rotation of sheaves on quarter- and eighth-turn drives must always be set up so the tight side of the drive is on the bottom.

Installation The drive should be tightened so that the belts are seated in the grooves before starting.

Idler If an idler is used on the back of the belt to reduce entry angle, or provide take-up, it should be located on the slack side, close to the driveR sheave. Consult the manufacturer or maintenance personnel for exact location of the idler.

Location of sheave on vertical shaft The center of the face of the sheave on the vertical shaft should be below the axis of the horizontal shaft by amount e which depends on the center distance (refer to Figure 1). Dimension e is given in Table 2.

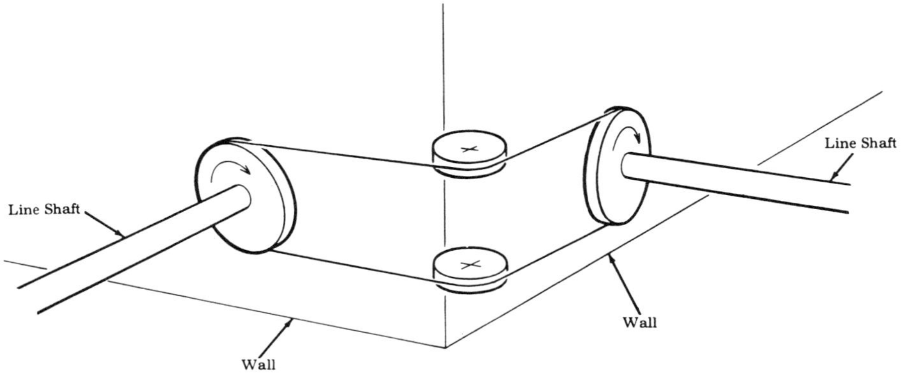

FIGURE 3 Mule pulley drive.

B. Mule Pulley

An example of a multi-plane drive is the *mule pulley drive*. Mule pulley drives are those which use a driveR pulley, a driveN pulley, and two idlers for the purpose of changing the plane of sheaves. The first of such drives was used to turn corners with factory line shafts, as illustrated in Figure 3.

There are various types of mule pulley drives. Three are shown in Figure 4. Figure 4a shows a typical mule pulley drive – one in which the driveR and driveN sheaves are the same size and are at a 90° angle to each other. Figure 4b shows one which has unequal sheave sizes. For such a drive to be properly aligned, the idlers must be cocked to lie in a plane drawn through the tangent points on the driveR and driveN sheaves. Also, the tangent points on the idlers must be in the center planes of the driveR and driveN pulleys. Figure 4c shows a mule pulley drive where the four pulleys are mounted on only two shafts --sometimes used where the center distance is too short for a regular quarter-turn drive. The idlers are free to rotate on the shafts. *Note*: In Figure 4c the driveR, driveN, idler, and sheaves can be placed in any combination to produce desired results.

1. Take-Up

If take-up on the typical mule pulley drive, Figure 4a, is made by moving either the driveR or driveN, the only special design consideration is the twist. However, when either or both of the idlers are moved for take-up, misalignment occurs as well. Misalignment is a

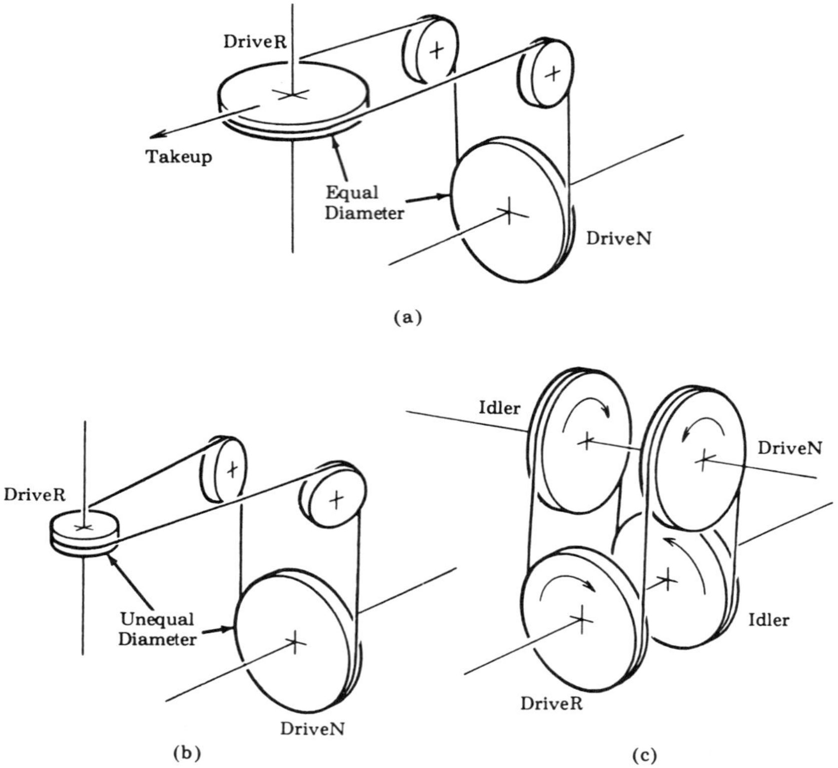

FIGURE 4 Various types of mule pulley drives.

factor in Figure 4b, regardless of takeup method; but is not a factor in Figure 4c, if the sheaves are properly spaced on the shaft and are of the same diameter on that shaft.

2. Alignment

Alignment of mule pulley drives can be investigated by laying out the drive to scale. Alignment must be checked at the two extremes of take-up idler placement. Misalignment should be kept to a minimum of 3°. It is preferable to get the best possible alignment of the tigher spans, since this generally results in better belt service.

3. Sheaves

Deep-groove sheaves are recommended for all mule pulley drives using V-belts.

C. Crossed Drive

The crossed drive is used when the driveR and driveN shafts must rotate in opposite directions.

1. Design Considerations

The drives are designed the same as conventional drives, except that the belt length is calculated as shown below:

$$L = 2C + 1.57(D + d) + \frac{(D + d)^2}{4C}$$

where

L = belt length (in.)
C = center distance (in.)
D = large sheave diameter (in.)
d = small sheave diameter (in.)

On crossed drives when the twist occurs in too short a center distance, the belts can become unstable on the drive. Follow the recommended minimum tangent lengths for crossed drives given in the following table.

Conventional belt section	Minimum tangent length (in.)
A	18
B	22
C	28
D	27

The other problem that occurs with crossed drives is belt wear due to abrasion at the crossover point where belts touch. This can be minimized by design so that this crossover point is near the center of the drive, as shown in Figure 5a. If the speed ratio is high, and the crossover point is near the small sheave, as shown in Figure 5b, both spans of belt try to occupy the same location more forcibly and wear is more severe.

An idler can be inserted in a crossed drive to prevent rubbing of the strands, as shown in Figure 6. However, this arrangement causes misalignment of the drive, and may result in reduced service. A thin, low friction plate, inserted between the strands at one point of contact, can decrease the coefficient of friction and the relative velocity, reducing wear.

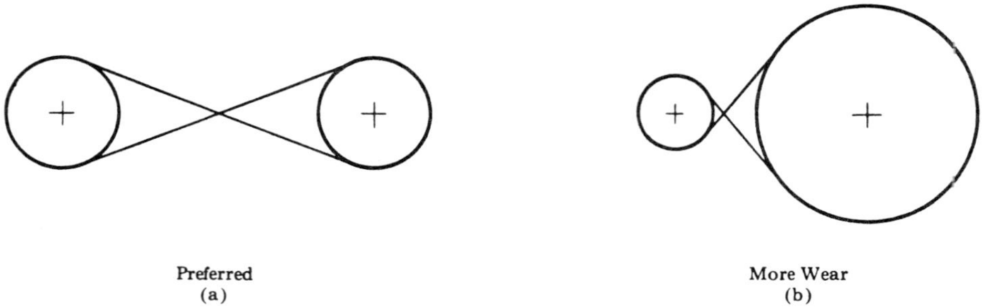

FIGURE 5 Crossed Drives.

II. SERPENTINE DRIVES

Serpentine drives are commonly defined as those having three or more sheaves or pulleys, in which the belts can drive from either inside or backside surfaces. There are many types of applications for serpentine drives, but several of the more common ones are:

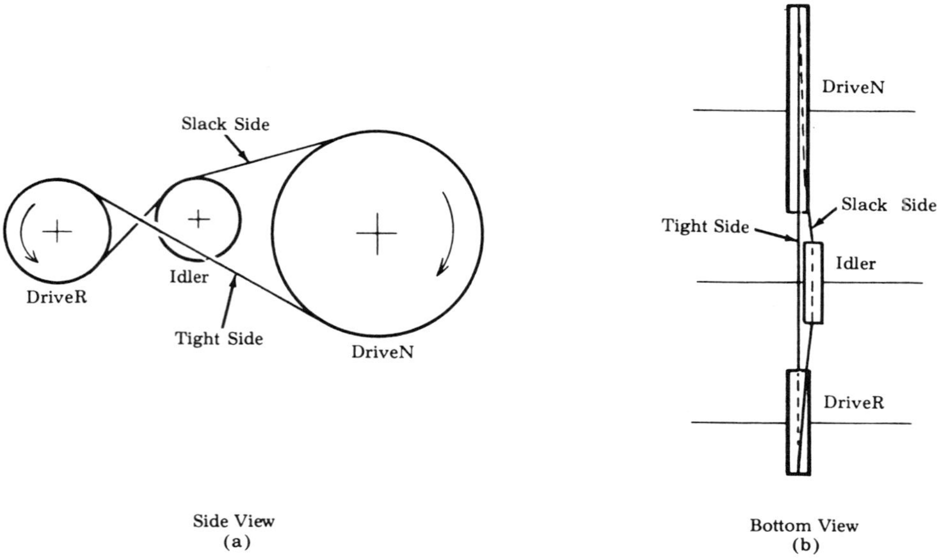

FIGURE 6 Crossed drive with idler.

Automotive front end drives: water pump, alternator, power steering, air conditioner, all driven from the crankshaft
Agricultural machines
Industrial machines such as conveyors
Machine tools
Small office machines

Many types of belts are used:

V-ribbed
Double-V
Double tooth timing
Joined, conventional, and narrow
Single, conventional, and narrow

These drives can be quite complex. Drives of this type should be referred to a supplier for recommendations and assistance with the design.

III. V-FLAT DRIVES

The typical V-flat drive is made up of a V-grooved driveR sheave and a large driveN pulley with a flat driving surface. As explained below, the most practical drive is one with a large speed reduction, which is very economical when converting a flat belt drive to V-belts, since it is not necessary to replace the large driveN pulley. These drives are also very practical for driving a large drum or cylinder where addition of V-grooves would be difficult and expensive.

In addition to the standard V-flat drives, two other combinations are used:

1. Variable speed belts using a variable pitch V-groove and flat driveN pulley.
2. V-ribbed belts, using a very small driveR with a large driveN drum as designed for clothes dryers.

1. Design Considerations

The V-flat drive generally requires less tension than a flat-belt drive. The normal operating tension ratio for a conventional drive is 5:1; for flat belts, 2:1; for V-belts operating on a V-grooved driveR, flat driveN, 2.5:1. (For 1:1 speed ratio.) As the diameter of the driveN flat pulley is increased, the arc of contact (wrap) is increased and the tension required to avoid slip is reduced. As this diameter difference

TABLE 3 Point of Optimum Geometry as Diameter Difference Increases

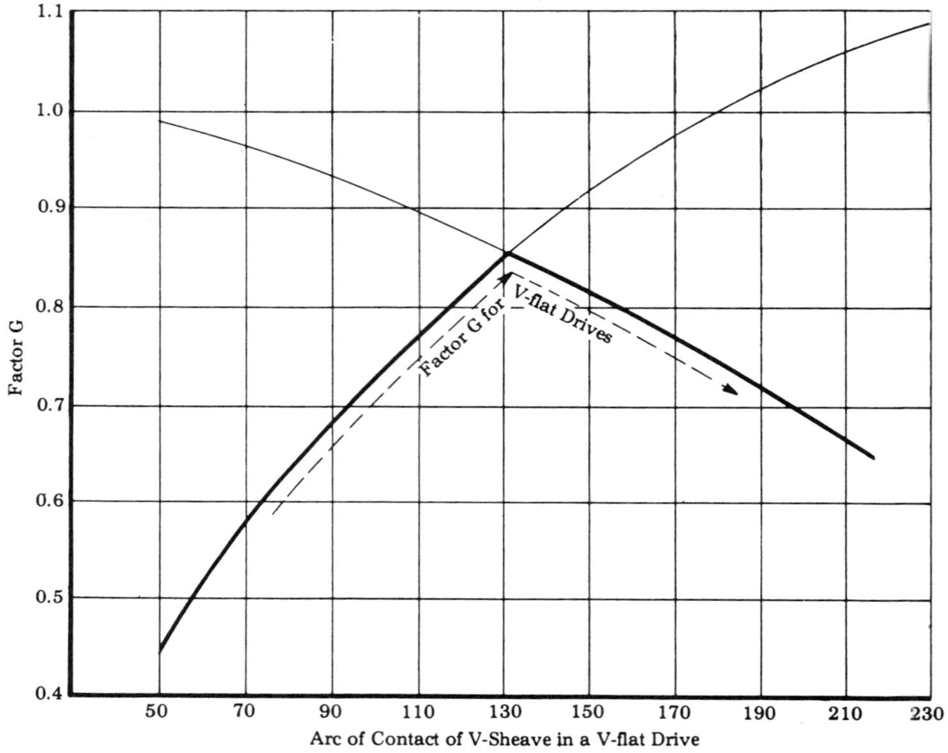

is increased, the point of optimum geometry is reached when the arc of contact is about 130° on the driveR, and 230° on the driveN (Table 3). Theoretically, the driveR and driveN would have the same pulling power before slip and would have minimum tension for a V-flat arrangement. The tension ratio at this point is 3.5:1 (see Chapter 3).

The arc of contact on the driveN sheave determines whether or not a V-flat drive is practical. The arc is proportional to the ratio:

$$\frac{D - d}{C}$$

where

D = effective outside diameter of the large, flat pulley (in.)
d = outside diameter of the small sheave (in.)
C = center distance of drive (in.)

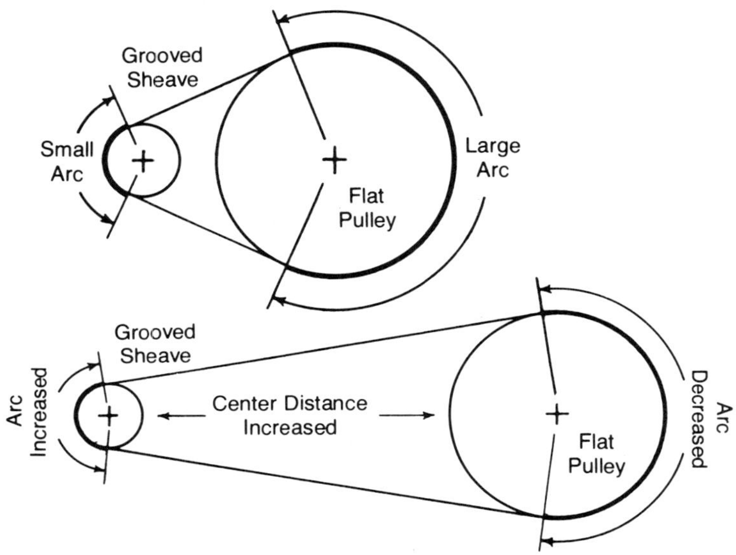

FIGURE 7 Drives with flat pulley and grooved sheave.

(*Note*: Effective outside diameter of the large, flat pulley is obtained by adding the appropriate value from information furnished by the various manufacturers.)

Whenever the ratio (D - d)/C is 0.5 or over, the large pulley or flywheel need not be grooved. The best results are obtained when this ratio is between 0.8 and 0.9.

Figure 7 shows two drives, each using the same size grooved, small sheave and large, flat pulley. In the first drive, there is very little arc of contact on the small sheave. Sheave grooves are required to give adequate power transmission capability without the need for extremely high tension to prevent slip. However, the arc of contact on the large pulley is ample. Therefore, the large pulley can have as much pulling ability as the small sheave, even though the pulley is not grooved.

In the second drive, a longer belt has been used, increasing the center distance. Note in Figure 7 that this decreases the arc of contact on the large pulley, thereby decreasing the ability of the flat pulley to transmit power without slipping. The second drive, then, requires more belt tension than the first drive to transmit the same load without belt slippage.

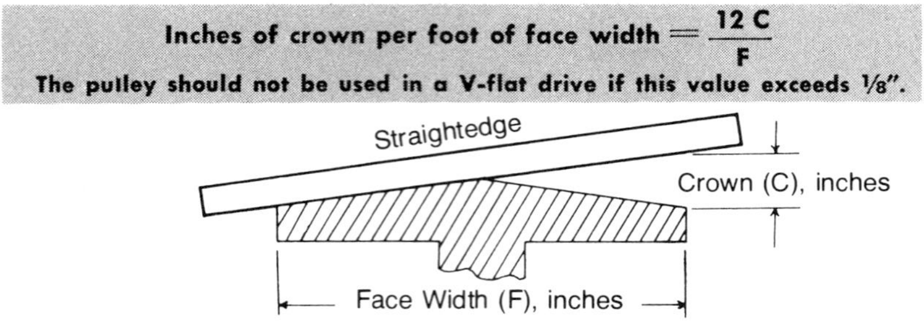

FIGURE 8 Crown on a flat pulley.

If a drive is practical, based on this ratio, then the drive should be designed using procedures found in most standard heavy-duty belt drive design manuals (also refer to Chapter 3).

The correct Factor G must be used. It is different than values for conventional drives. Check the design manual for these values.

Narrow belts, 3V, 5V, 8V, should not be used except in joined form. The bottom of this belt is quite narrow, and in individual form, could turn over on a flat pulley. Classical section belts can be used in single or joined form.

The crown on the flat pulley should not exceed 1/8 in. per foot of face width (Figure 8).

IV. SPECIAL BELTS

A. Double V-Belts (Hexagonal)

Double V-belts are used when power input or power takeoff is required on both sides of the belt. This is necessary when the driveR and driveN sheaves must rotate in opposite directions. When several sheaves are involved, the drive is termed *serpentine*.

These belts are commonly available in double A, B, C, and D sections. Standard classical A, B, C, and D sheaves are used. A design method for double V-belts is given in RMA Standard No. IP-21-1984. Points to remember in designing double V-belt drives:

Greater bending stresses occur in double V-belts because of greater thickness. On otherwise identical drives, service life may not be as great.

The pitch line is at the center of the belt. Pitch diameter is the same as sheave outside diameter for standard A, B, C, and D grooves.

Pitch length is the same as effective length for double V-belts.
These belts do not have as much sidewall contact as standard V-belts. Stability can be a problem on drives where there is misalignment or belt slip.
Considerations should be given to designing drives with conventional or joined belts and driving off the flat back side.

B. Joined V-Belts

Classical and narrow V-belts are available in the joined configuration where several belt strands are connected by a tie band across the top of the belts. The band improves lateral stability, and solves the problems of belts turning over, or coming off sheaves on severe drives. The band rides above the sheave O.D. and does not interfere with the wedging action of the individual belt strands.

This type of belt has proven very successful in eliminating belt stability problems on severe drives such as compressors, where there are large torque fluctuations, and rock-crushers or pug mills, where shock loads occur.

The drive design procedures for joined belts are identical to those for standard belts (see Chapter 3). However, certain precautions should be observed to avoid damage to the tie band:

Apply these belts only to sheaves made to RMA specifications.
Do not apply these belts to badly worn sheaves. The strands may ride down so low that the tie band will contact groove lands and be cut.
Drive should be protected from foreign material such as rocks, sawdust, sand, and chips. This material could build up in the spaces between strands and damage the tie band.
The drive should be adequately tensioned (see Chapter 11). The procedures are the same as for standard belt sections except that the extra strands increase the deflection force required. For large drives, the belt elongation method is used.

C. Spliced Belts

The standard V-belt is endless, and forms a complete loop which can be placed over the sheaves. However, there are some applications in which the endless belt cannot be used:

1. Those in which the driveR or driveN sheave is located such that related equipment prevents standard belts from being slipped over the sheave. A long line shaft is an example of this type of setup.

2. Those in which long lengths are required, such as on long conveyors or where motion, rather than power, is to be transmitted over a great distance.

For these drives, open-end belting that can be cut to the necessary length and spliced is available. The common sections available are the classical A, B, C, and D. (Some double V-belts are available in these sections.) There is also a variety of link-type belts available wherein short lengths of belting material are coupled together with metal fasteners to make up the length of belt required.

1. Belt Considerations

Open-end belting has a special construction with multiple reinforcing layers so that spliced joints will have adequate strength. Standard V-belts, when cut and then spliced together, do not give satisfactory service because they do not have the correct reinforcing material to hold the fasteners. Spliced belts can be used in multiple sets for power transmission.

The drive design procedures are similar to those for standard industrial drives, that is, on a HP-rating basis. The various manufacturers publish rating charts for reference. They also furnish the special fasteners needed. *Note*: HP ratings are lower than those for endless belts. If belts on an existing drive are being replaced, the belt section can be determined by measuring the old belt. The required length should be calculated using the following procedures:

1. Position the motor on its base so that when necessary, the center distance can be shortened by the amount shown in Table 4. (The belts can be installed over the edge of the sheaves without forcing them.)
2. After the motor has been positioned, measure the length around the outside of the sheaves. Use a steel tape so you get an accurate reading. This measurement gives the outside belt length. Subtract 0.5% from this measured outside length to allow for enough tension in the belt.

TABLE 4 Installation Allowance

Easy-splice cross-section	A-VO	V-VO	C-VO	D-VO
Center distance allowance for installation of belts	3/4 in.	1 in.	1-1/2 in.	2 in.

TABLE 5 Length Adjustment

Easy-splice cross section	A-VO	B-VO	C-VO	D-VO
Subtract these amounts:	1.10 in.	2.00 in.	2.38 in.	2.95 in.

3. Next, subtract the amount shown in Table 5. This takes care of the length of the fastener as well as the difference in length between a straight length of belting and the length of the belting around the drive.
4. This length is used for making up the belt for installation or for ordering from your supplier.
5. Table 6 shows center distance allowance for take-up.

2. Link Belting (Light Duty and ABC Sections)

These short links are made up into the desired or needed length. They are especially valuable for situations in which

Endless belts cannot be installed
Replacement on a temporary basis is necessary, since they can be made up into any desired length

There are no standard design procedures.

TABLE 6 Center Distance Allowance Necessary for Belt Take-Up

Belt length[a]	Take-up allowance	Belt length[a]	Take-up allowance
26 to 38	1 in.	195 to 240	4 in.
38 to 60	1 1/2 in.	240 to 270	4 1/2 in.
60 to 90	2 in.	270 to 330	5 in.
90 to 120	2 1/2 in.	330 to 420	6 in.
120 to 158	3 in.	420 and over	1.5% of belt length
158 to 195	3 1/2 in.		

[a]In each group the length range is up to, not including, the second length listed.

V. EXACT DRIVEN SPEEDS

On most drives, the required accuracy of the driveN sheave speed falls into three categories. These are:

1. Normal drives, in which the driveN speed is not critical, and can vary several percent
2. Those drives in which the driveN speed is held close to tolerances, but where synchronization is not required
3. Those drives in which exact synchronization is required

Normal drives will meet the noncritical requirements if design procedures shown in Chapter 3 are followed.

Where close tolerances on driveR speeds are required, V-belt creep must be considered, as well as sheave pitch diameter corrections. Creep results from the differences in tension from tight side to slack side of a sheave or pulley. Since a belt elongates slightly under tension, a greater length of belt leaves the driveN pulley than goes on it from a slack side. This makes the driveN shaft rotate at slightly less rpm than it would if there were no creep. The amount of creep may vary with the load and speed. However, the commonly used value is 0.5% loss of speed on the driveN sheave.

The numerical values that indicate the placement of the pitch line for sheaves are published in the various design manuals. These usually agree with industry standards, but these standards must essentially be an average for all belts of the various manufacturers of a given type. Therefore, they are usually not exact.

Most design manuals have a section on exact speed, and give correction factors for the exact pitch diameters. When driveN speed is critical, these correction factors should be used.

There is one other consideration when designing exact speed drives and that is to examine the driveR speeds. The speed of these prime movers is seldom constant. Internal combustion engines vary with the load imposed, and the common electric motor does not have an exact constant speed. These motors have variation due to load as well as to the electrical power characteristics. Standards for common AC electric motors allow as much as 2% variation of speed away from the rated load speed condition. This exceeds the 0.5% variation due to creep and can exceed the variation caused by slightly different pitch placement of belts in general. The only belt drive that meets the needs of the final category is the synchronous, positive drive, timing belt. Design considerations are shown in Chapter 10.

VI. DRIVES FOR WHICH BELT TAKE-UP IS NOT PROVIDED

All V-belts, over an extended operating period, will wear and stretch. Sheave grooves also wear. This causes a loss of tension that is corrected through routine maintenance by moving one of the sheaves or by adjusting the idler. If the belt is not retensioned, it may slip and burn, resulting in short service life.

On drives where belt length adjustment cannot be provided, a different type of belt should be used. Three primary types are as follows:

1. Synchronous (toothed) belts can operate for extended periods with minimal adjustment. However, these belts also require proper tension to obtain maximum life, and should not be used for nonadjustable center drives unless reduced service life is acceptable.
2. On small drives, round belts (O-ring type) or flat belts with no tensile members can be stretched and then snapped into place on special sheaves.
3. Spliced belting, link-type, can be installed and then retensioned by removing one or two of the short lengths as required. Belting with one splice can be retensioned by cutting and resplicing.

Each of these alternatives has a potential design or maintenance problem and every effort should be made to provide a proper tensioning method.

VII. IDLERS

A. Drive Design Considerations

Idlers are either grooved sheaves or flat pulleys which do not transmit power. They are used in V-belt drives to:

Provide take-up for fixed center drives
Clear obstructions
Turn corners (as in mule pulley drives)
Break up long spans where belt whip may be a problem
Maintain tension, when the idler is spring-loaded or weighted
Increase arc of contact on critically-loaded sheaves
Provide mechanism for clutching belt drives

An idler always imposes additional bending stresses on the belts, so if the above requirements can be accomplished by adjusting centers,

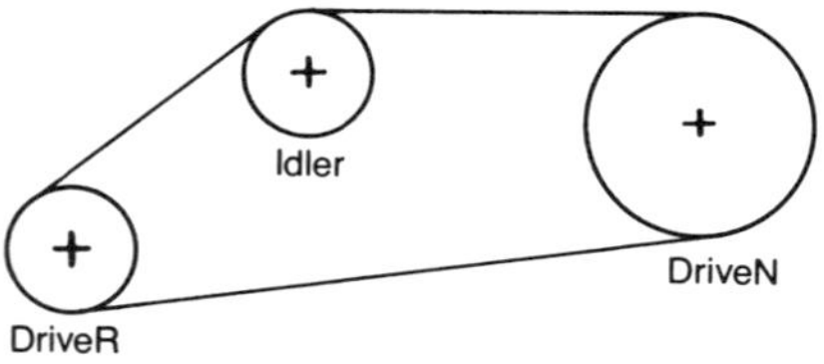

FIGURE 9 Idler on inside of drive

it is usually best to consider that method first. This is almost always more economical in the long run.

However, some marginal drives can be improved by proper design with idlers. This is especially true in the case of drives where load variations are quite large – and by proper application of spring-loaded idlers.

If idlers must be used, the important design considerations are:

Placement in drive
Flat or grooved
Diameter
Center distance, belt length, installation, and take-up
Corrections for HP rating

The basic drive design procedures are the same as those given in Chapters 3, 4, and 5. This gives the sheave sizes, number of belts, and center distance. The geometry of the idler placement must be determined by layout or special computer programs.

B. Placement

1. Inside or Outside

Idlers may be placed either inside or outside the drive, as shown in Figure 9 and Figure 10. They may be placed on the tight side or slack side of the drive as shown in Figure 11 and Figure 12.

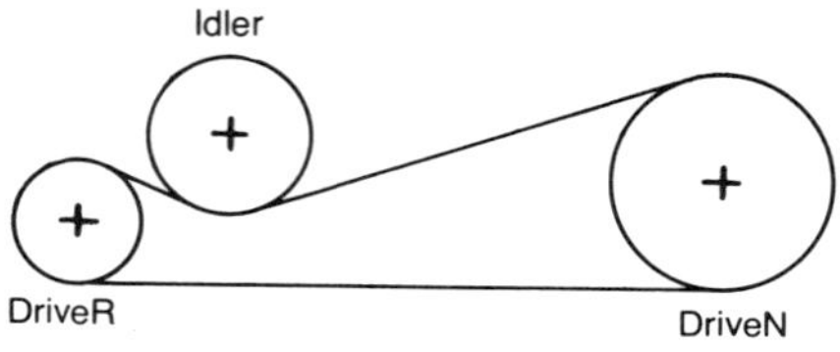

FIGURE 10 Idler on outside of drive.

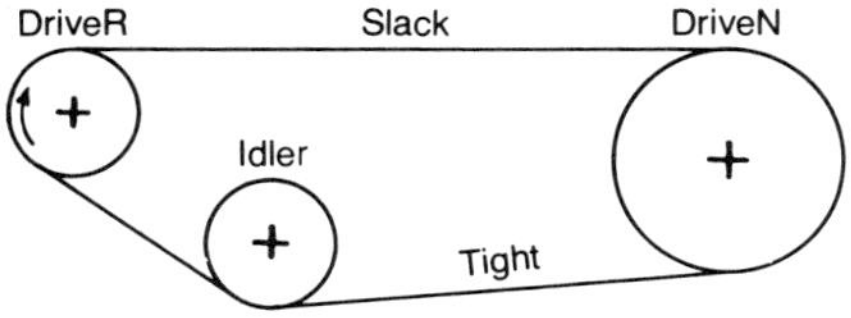

FIGURE 11 Idler on tight side of drive.

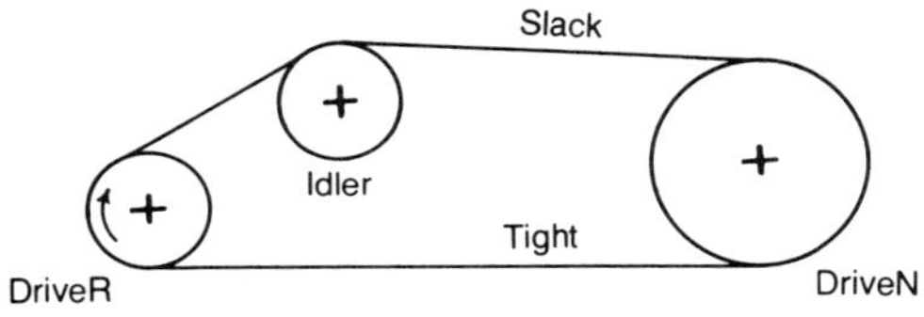

FIGURE 12 Idler on slack side of drive.

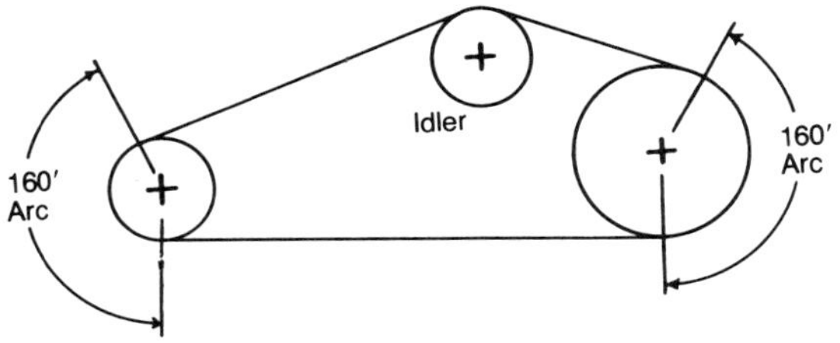

FIGURE 13 Grooved inside idler with nearly equal arcs of contact.

An inside idler decreases the arc of contact on the adjacent sheaves. An outside idler increases the arc of contact on these sheaves. Either may be used, but an outside idler must be larger, as discussed below. If you are using the idler for take-up purposes, you should remember that the amount of take-up obtained by an outside idler is limited by the belts on the opposite side of the drive. As a rule, idlers should be placed on the slack side of the drive.

Spring-loaded, or weighted idlers are discussed in Chapter 11.

A grooved inside idler may be located at any point in the span, but preferably so that it results in nearly equal arcs of contact on the two adjacent sheaves (Figure 13).

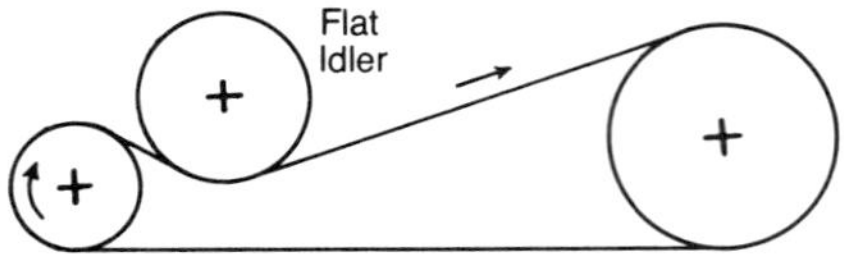

FIGURE 14 Location of a flat idler pulley.

A flat idler pulley, whether it is inside or outside, should be located as far away as is practical from the next sheave the belts are entering (in the direction the belt is traveling). V-belts move back and forth slightly on a flat pulley, and locating the idler away from the next sheave minimizes the possibility of the belts entering that sheave in a misaligned condition (Figure 14).

C. Idler Details

Flanging of idlers is recommended. When flanging is used, the inside bottom corners should not be rounded – this may cause the belts to climb off the pulley.

If the idler is to be a flat, uncrowned pulley, the minimum *face width* required should be found (between flanges, when flanged) by adding the face width of a grooved sheave (for the appropriate number of belts), in inches, to the amount given in Table 7.

Sheave face width is given in the sheaves specification tables.

TABLE 7 Additional Width for Flat Idlers

Belt cross section	Amount to add to face width of grooved sheave to find minimum uncrowned flat pulley face width (in.)
3V	0.6
5V	1.0
8V	1.3
A	0.8
B	1.0
C	1.2
D	1.5
E	1.8

1. Idler Diameters

Inside idlers should be at least as large as the smallest sheave that transmits power. Outside idlers should be at least 1/3 larger than the smallest sheave that transmits power, since the belt is being bent backwards.

2. Flat or Grooved

Outside idlers are always flat pulleys, since they contact the top of the V-belts. Inside idlers can be either grooved or flat for classical section V-belts, but are always grooved for the proper V-belt section when narrow-section V-belts or notched classical-section V-belts are used. Inside flat idlers can be used for drives using joined V-belts. Flat idlers should not be crowned.

3. Idler Support Brackets

Brackets for idlers should be of sturdy construction, designed to withstand the forces imposed by the drive. Frequently, the cause of drive problem described as belt stretch, instability, or vibration is flimsy idler bracketry.

The forces involved in the drive can be calculated from formulas given in Chapters 3 and 11.

D. Effect of Idlers on Drive Performance

If the above recommendations are followed, it is possible to design satisfactory V-belt drives using idlers. However, idlers always impose an additional bending stress point on the V-belt. This reduces the belt HP rating as is shown by the following table of approximate correction factors. The normal rating is multiplied by this factor. When establishing the normal rating, take into account the minimum arc of contact on each loaded pulley resulting from the use of idlers.

Number of idlers in drive	Rating correction factor
0	1.00
1	0.91
2	0.86
3	0.81

These factors apply only when the idler diameter is in accordance with the above recommendations. (Refer to RMA Technical Bulletin No. IP-3-6.)

9

Flat Belt Drive Design

WALTER T. NAWROCKI *V-Belt Product Application Department, The Gates Rubber Company, Denver, Colorado*

I. DEFINITION AND DESCRIPTION OF FLAT-BELT DRIVES

Flat-belt drives were some of the earliest, and at one time, one of the most widely used methods of power transmission. Their popularity began in the early 1800s and over the years, they have been used successfully in heavy duty industrial, agricultural, light machinery (appliance), and automotive accessory applications. Flat belts have practically disappeared on automotive applications, but are still used in the other areas.

In the past, industrial plants and textile mills had many flat belt drives running from line shafts suspended from the ceiling through various stages of reduction and speed change to individual machines.

The use of flat belts gradually decreased after the introduction of V-belts, but there are still applications for which flat belts are well suited. In general, a flat belt can operate more satisfactorily on a smaller pulley than a V-belt can. This makes flat belts desirable on drives where very high speed ratios require the use of a small pulley.

Flat belts should not be used on vertical shaft drives or on drives where the belt must enter the pulley misaligned. The flat belt will not stay on such drives.

In flat-belt drives, the controlling factor is the ratio between the force pulling the belt and the force pressing the belt against the pulley creating frictional force. This is referred to as the *coefficient of friction*. This frictional grip along with arc of contact of the belt on the driveN pulley is a prime factor in power transmission. The width, thickness, grade of product, number of plies, type of tensile member, and tension ratio also affect the ability of flat belts to handle the amount of HP being transmitted.

II. GENERAL DESIGN CONSIDERATIONS

Flat belt drives ordinarily do not have speed ratios exceeding 6:1 and, in the higher ratios, long center distances are required to insure sufficient arc of contact. Shorter center distances can be used if some sort of idler arrangement is used to increase arc of contact on the driveR pulley.

Flat belts lend themselves well to crossed belt drives for reversing the shaft rotation between the driveR and driveN pulleys, and for quarter-turn drive applications.

Excessive belt speed tends to throw the belt away from the pulley, causing excess slippage.

Most flat belts will have some initial stretch, so some means of take-up should be provided, including movable motor bases, special motor mount, or idler take-ups to help maintain correct belt tension. *Note*: The inherent slippage and creep in flat belt drives can be an advantage in handling shock loads or temporary drive overloads.

Pulleys should be large enough to avoid the adverse effects of severe flexing. Refer to manufacturers' drive design manuals for minimum pulley diameter recommendations.

Most flat belts require higher tension.

III. COMMON BELT TYPES

Even though many different flat belt constructions exist, there are two basic types.

A. Cordless Belts

Cordless belts include leather, woven fabric, all-elastomer, or other types of belts which use the elastic characteristics of the belt rather than a special tensile member, to resist the required tension.

B. Corded Belts

Corded belts use a load-carrying tensile cord as a working element. Sometimes a sheet of plastic is used in place of cords. The four major parts of a covered corded flat belt area are shown in Figure 1.

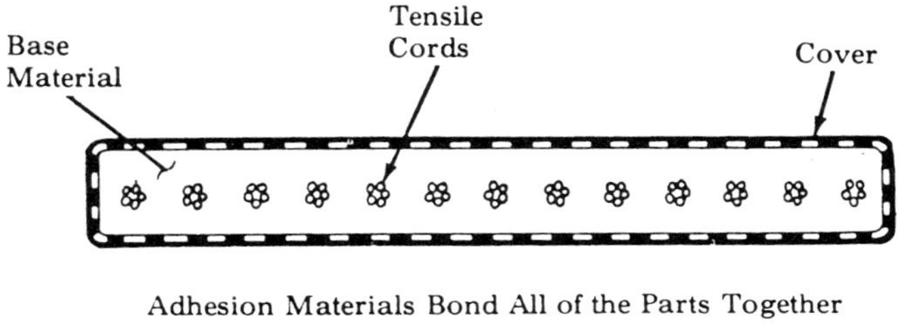

FIGURE 1 Components of a covered corded flat belt.

IV. BELT CONSTRUCTION

A. Leather Belts

Being the oldest and the first type of industrial belting, leather is the standard against which modern belt materials and constructions are compared. Leather has a high coefficient of friction and if the belt is maintained properly during the service life of the belt, the coefficient will improve.

Leather belts come in three plies or thicknesses: single, double, and triple; and in three weights: light, medium, and heavy. Combinations of these can be furnished depending on the load and power requirements, i.e., medium-single, heavy-double, and so forth.

B. Rubber Belts

Rubber belts are more widely used than any other, primarily because of the variety of types and styles available to meet varied application conditions. Rubber belts have a good coefficient of friction and are generally run at higher tensions than leather belts. They also have less initial stretch, are better tracking, and have greater lateral stiffness. They also can tolerate more severe environmental conditions.

Reinforced rubber flat belts of the cordless type construction can be made of multiple layers of rubber-impregnated canvas or cotton duck fabric. The fabric provides working strength while the rubber gives a good coefficient of friction. It also protects the fabric and bonds the layers together. This construction is economical and is easily spliced on the drive, but will stretch more than corded belts. Rubber flat belts can be raw edged (usually cut) or covered.

Corded construction rubber flat belts have better stretch resistance than the plied up construction, and can operate on smaller pulleys. These belts must be endless: they cannot be spliced because the construction allows the splice to tear out easily.

C. Plastic Belts

Plastic belts are manufactured for both high speed and general service applications. They usually employ a tensile member of sheet plastic which is often prestretched in the direction of pull to prevent excessive stretch. The tensile member is covered on both sides with other materials to provide a satisfactory wear and friction surface.

D. Fabric Belts

Fabric belts are made entirely with woven fabric, with no rubber or bonding material and in various numbers of plies. Flat fabric belts are used on special applications in the textile industry and for high speed drives, but they are rarely found on general power transmission drives.

V. DESIGNING FLAT-BELT DRIVES

Flat-belt drives can be designed by two methods: the HP rating for fixed-life method and the life-in-hours method. The American Leather Belting Association publishes HP ratings for flat leather belting. Ratings for other types of flat belts are found in individual manufacturers' design manuals. Manufacturers also publish life-in-hours design methods, which are especially useful to original equipment manufacturers. Load-life relationships for flat belts are not generally as well established as those for V-belts. The life-in-hours method of flat-belt design will not be discussed, because it is beyond the scope of this text. The HP-rating method will be discussed and illustrated with an example.

Regardless of the design method used, the objective is to design a drive which will give the desired speeds, shaft centers, and acceptable belt life with minimum maintenance. The ultimate objective of good design is to provide a drive in which the peak tension in the belt is low enough to give the desired belt life.

The use of the HP-rating method to design a flat belt drive involves a procedure in which the required sizes of the individual components are calculated in a progressive sequence. The final calculation is the width required of the belt to provide the required HP rating. The steps involved are as follows:

1. Determine the design HP, which is equal to the HP requirement of the drive multiplied by a service factor which corrects for overloads resulting from such things as the starting load or a shock load. The HP requirement of the drive is the average load that the drive must transmit. It is considered the rated HP of the prime mover unless the actual requirement of the driven machine is known.
2. Select the type of belt to use. The designer must decide whether to use an endless belt or a spliced belt and what material to use. The choice between leather and rubber, for example, is often a matter of personal preference. In other cases, the type of belt selected will be determined by the conditions to which the belt will be exposed. Those conditions may involve high temperatures, high loads, small sheaves, or any one of many other conditions. Endless belts should be used whenever possible in preference to belts using mechanical splices.

3. Select a driveR pulley diameter. The selection involves the following considerations: (1) The pulley should be equal to or larger than the minimum recommended by the National Electrical Manufacturer's Association (NEMA) if it is to be mounted on a NEMA standard motor. (2) Larger diameters reduce the required drive width and the shaft and bearing loads by reducing the required belt tensions. (3) Diameters which are too large may produce a belt speed in excess of that recommended by the manufacturer of the belt or pulleys. (4) If it is planned to buy the pulleys from standard stock in order to have immediate deliveries, the design process must be based on the selected stock sizes. (5) Some design manuals give recommendations for the minimum pulley diameters to be used on certain belts. (6) If the diameter of the smallest pulley in the drive is too small for the width of the belt finally selected, problems of excessive tracking and poor load distribution on the belt will result. Follow the manufacturer's recommendations.

For general power transmission requirements, an economical drive results if the sheave size is selected for a belt speed of about 4000 to 5000 ft/min, but this is not a fixed rule. For light-machinery drives, slower belt speeds are often more economical and result in more compact drives. For high-speed drives, belt speeds up to 15,000 ft/min are not uncommon.

4. Calculate the required driveN pulley diameter. Use the drive pulley diameter which has been selected and the desired speed ratio, which is the rpm of the faster shaft divided by the rpm of the slower shaft. If the driveN pulley is to function as a speed-down, multiply the driveR pulley diameter by the speed ratio to get the driveN diameter. If the driveN pulley is to function as a speed-up, the driveR pulley diameter is divided by the speed ratio to get the driveN diameter. In this situation, recheck to make sure driveN pulley diameter is acceptable for that belt type.

For flat belt drives the pulley pitch diameter is often assumed to be equal to the outside diameter when speed ratios and belt speed are calculated. Manufacturers' design manuals sometimes list pitch diameter-to-outside diameter (PD-to-OD) values for specific belts to be used when more accurate speed calculations are desired. In the absence of the manufacturers' manuals, a good estimate is to add one belt thickness to the pulley OD to find the pulley PD. This assumes that the pitch line is at the center of the belt.

If a pulley is to be selected from a manufacturer's stock of standard sizes, it may be necessary to round off the calculated value for the driveN pulley diameter to use the closest standard stock diameter.

5. Calculate the required belt length. The belt length formula given in Chapter 3 can be used for two pulley drives. Unless spliced belting is used, it will be necessary to select the available belt length

closest in size to the calculated length. The actual center distance for the standard belt length selected can be calculated by substituting the selected length value in the belt length formula and solving for center distance. For drives with more than two pulleys, belt length can best be found by measurement on a scale layout of the drive.

Belt drives must allow for movement of one of the shafts in order to provide for the necessary adjustments referred to as the installation allowance and the take-up allowance. Provision must be made for moving the pulley centers closer together to permit the installation of a belt which is within the length tolerance without having to pry it over the pulley. Prying the belt over the pulley can damage the belt. Also, provision must be made for adjusting the belt beyond the calculated center distance. This will permit adjustments which compensate for belt stretch and for wear of the belt and the pulleys in order to maintain the required tension throughout the life of the belt.

Manufacturers' recommendations for take-up allowance should be followed, since the requirements vary with the type of belt. In the absence of such a recommendation, provision should be made to provide take-up for at least a 3% increase in belt length.

Installation allowances must be great enough to allow the belt to slide easily over the pulley (including the flanges, if they are used). Spliced belting can be installed without any installation allowance if a clamp for stretching the belt on the drive is available. When such tools are used, the belt is cut short of the ideal length by an amount recommended by the manufacturer, usually about 2%, so that it is tensioned when spliced on the drive. By this method, it is possible to eliminate the need for take-up allowance on a drive. However, it is still preferable to provide a take-up allowance, since it minimizes the need for resplicing the belt as retensioning is required later.

6. Determine the belt width needed for the drive. First, determine the HP per inch of width for the belt selected. For leather belts, the HP ratings are usually tabulated according to belt speed and belt thickness, without regard to pulley diameter. For other types of flat belts, the rating tables usually show HP rating as a function of both pulley diameter and belt speed.

For rubber and plastic belts, the procedure for finding the required belt width is usually slightly different than the procedure used for leather belts. The basic HP rating per inch of width is corrected for arc of contact. A correction may have to be made for belt length. Rubber or plastic belts can operate at such high tensions in proportion to their weight that sag is very slight and is ignored in the design. The resultant corrected HP rating per inch of width is divided into the design HP to find the required belt width.

After the required belt width is determined, the drive system should be reviewed. If the belt is unusually wide, then larger pulleys should be investigated. This is especially important if the small pulley diameter is less than the recommended size for the width of belt chosen. If the belt is unusually narrow when compared with similar drives, then smaller pulleys should be considered.

VI. FLAT-BELT PULLEYS

The rim of a pulley used for flat belt drives is usually crowned. Crowning refers to the convex curvature given to the surface of a pulley rim. The purpose of the crown is to keep the belt in place and to permit the belt to track in a neutral position. The crown is necessary to keep the belt on the pulleys. The two types of crown are circular and cone (see Figure 2). A circular crown is preferred because it is easier on the belt, but the cone crown is easier to machine.

Crown is specified either as in. of crown per ft of pulley width, or as crown radius. The amount of crown to use varies with the pulley diameter, width, and type of belt. Excess crown is harmful to any belt. Pulleys for general service are furnished with a crown of 1/8 in. per ft of pulley face width, unless otherwise specified.

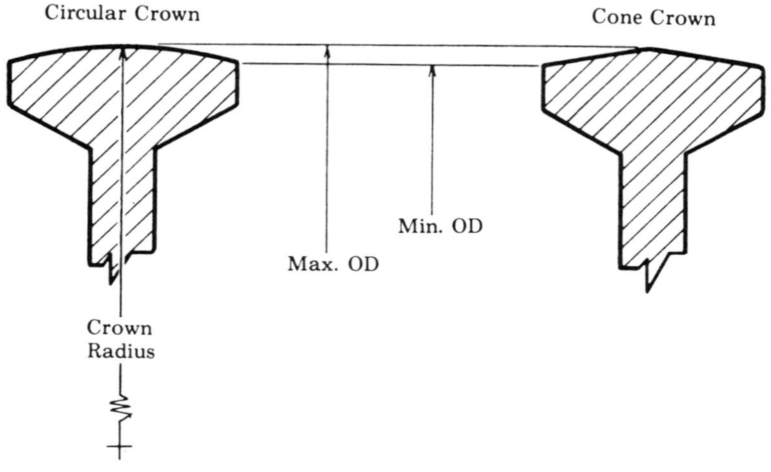

FIGURE 2 Circular and cone crowns for flat belt pulleys.

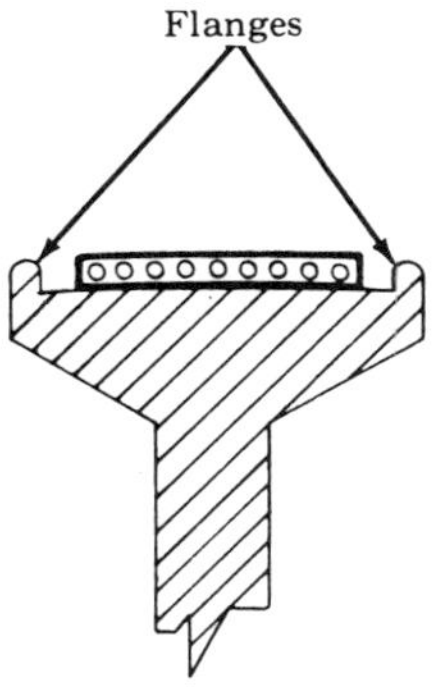

FIGURE 3 Flanged flat belt pulley.

Pulleys are sometimes flanged, particularly on small driveR pulleys on light duty or appliance drives. Flanges on the pulleys help keep the flat belt on the pulley in case of belt slip or misalignment (see Figure 3). Whether or not the pulley is flanged, it must be wider than the belt by an amount specified in the drive design information being used.

Flat belt pulleys are designated by:

Outside diameter
Face width
Bore size of hub or bushing, along with a keyway size
Crown
Flange (if any)

VII. BELT FASTENERS

Flat belts should be made endless to help insure that the belt is uniform in strength throughout, to eliminate noise, and to reduce jarring and wear on pulleys.

However, some applications make it impossible or impractical to use an endless belt. Belt fasteners are used to join the belt together. Care should be taken to select the right type and size fastener for a belt. Some things to be considered are load, belt speed, belt thickness, pulley diameters, and whether or not an idler is used.

The tyree types of fasteners are: lacing, hinges, and plates. Generally, lacing and hinges are used on light to medium loads, and plates are used on heavier loads, slower speeds, and larger pulleys.

10

Synchronous Belt Drive Design

DANIEL L. SCHWARTZ *Industrial Product Application, The Gates Rubber Company, Denver, Colorado*

I. INTRODUCTION

A. Definition

Synchronous-belt drives operate on the "tooth grip" principle. The belt may be described as resembling a flat belt with evenly spaced teeth on the inside surface. The precisely molded teeth of the belt are designed to make positive engagement with suitably shaped mating grooves on the pulley or sprocket.

Synchronous belts do not rely on friction to transmit power. As such, they should not be confused with molded-notch V-belts which transmit power by the wedging action of the V-shape. Because of the positive tooth engagement, there is little relative motion between the belt and pulley; subsequently there is no slip.

This no slip characteristic provides exact synchronization between a prime power source and a driveN unit. Thus, synchronous-belt drives are extremely useful where indexing, positioning, or a constant speed ratio is required as a machine function.

B. Terminology

Figures 1 and 2 define the belt pitch, pitch diameter, outside diameter, and pitch line differential for the conventional timing belt and newer curvilinear profiles.

It is important to note timing belt pulleys are designated as "pulleys" because of the groove design which causes the belt to actually pull on the outside of the pulley. The name pulley is carried over from flat-belt technology which is the ancestor of the timing belt. The newer curvilinear designation uses the name sprockets since they are designed to replace chain drives in some applications.

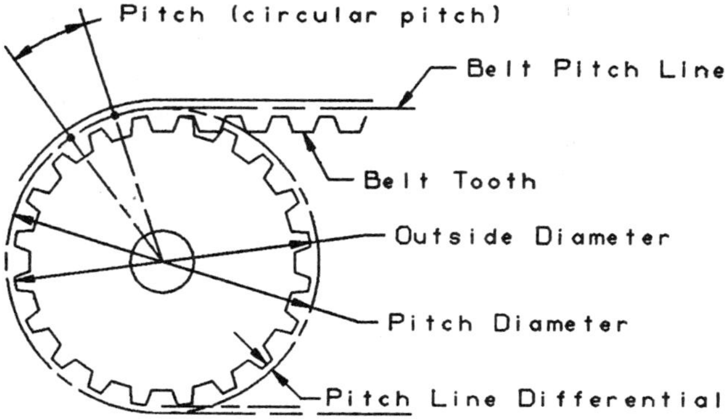

FIGURE 1 Timing belt and pulley nomenclature.

C. Applications

Synchronous belts are recommended primarily where synchronization of the driveR shaft to the driveN shaft is required. Because of the positive drive nature similar to chain or gear drives, synchronous-belt drives give 100% transfer of rotational speeds.

Synchronous belt drives are as much as 98% efficient. This high efficiency is, of course, primarily due to the positive, no-slip characteristic of synchronous belts. Reduced binding tensions also help

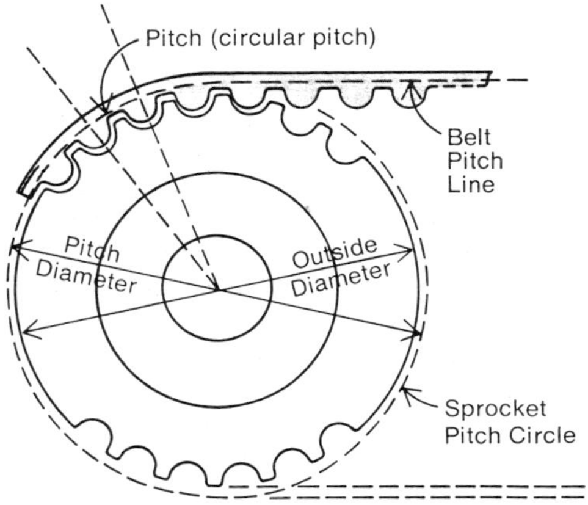

FIGURE 2 Curvilinear profile belt and sprocket nomenclature.

to increase efficiencies. Since the belt is similar to a flat belt, it operates at low flat-belt temperatures. Also, the low profile and mass reduce centrifugal tension which further increases belt efficiency.

Because of the high resistance to elongation, synchronous belts can be used where access for maintenance is limited. However, some adjustment must be provided when installing a synchronous-belt drive, as with nearly all power transmission methods, due to belt, pulley, and assembly tolerances.

II. BASIC TYPES OF INDUSTRIAL SYNCHRONOUS BELTS

A. Timing Belt

The initial development of the synchronous belt was started around 1940 by the L. H. Gilmer Company. Gilmer saw a need for a power transmission system to fill the void between roller-chain and conventional rubber V-belts. Development was interrupted by World War II, but was completed later in that same decade. The first synchronous belt product line included one trapezoidal tooth profile and pitch.

Today, a full line of various sizes of timing belts is available. They range from the mini-extra-light (MXL) pitch for subfractional HP drives up to double-extra-heavy (XXH) pitch with load capacities of over 100 HP. Figure 3 shows the six cross-sections which are in common industry use. Table 1 lists the standard belt lengths and widths which are available.

1. Belt Designation

Timing-belt sizes are indicated by a standard number. The size designation is described in the standard IP-24 published by The Rubber Manufacturers Association (RMA), Mechanical Power Transmission Association (MPTA), and The Rubber Association of Canada (RAC).

The first digits specify the belt pitch length to 0.1 in. increments. Following the belt-pitch length designation is a letter (or letters) indicating the pitch or belt section. The belt width is indicated in hundredths of an inch by the number following the section. For example, a timing belt that has a pitch length of 85 in., 0.500-in. pitch, 1 1/2-in. wide, would be designated as 850H150.

The tooth dimensions for double-sided belts are identical to those of single-sided belts. A double-sided belt that is the same size as shown in the above example would be indicated as 850DH150 per the IP-24 standard. However, most manufacturers have adopted their own designation to describe double-sided timing belts. An example is TP850H150 (TP = Twin Power^{R}).

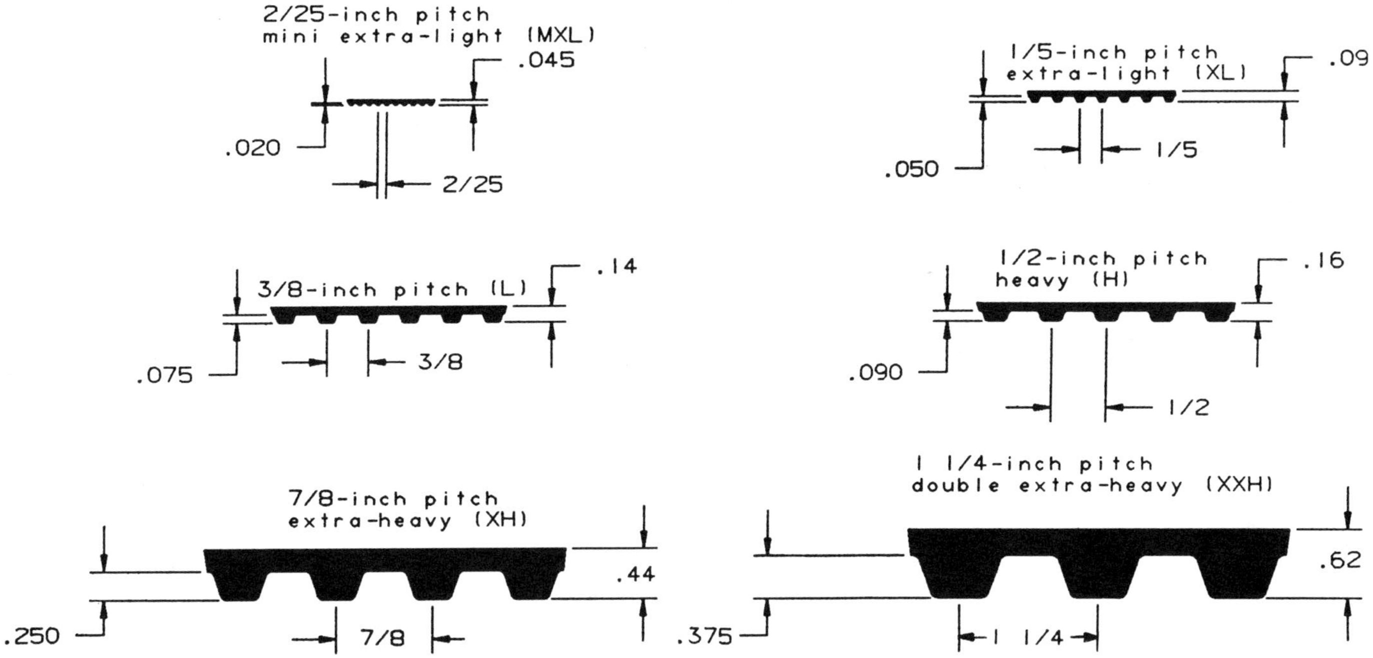

FIGURE 3 Timing belt cross section.

TABLE 1 Standard Timing Belt Lengths and Widths

.080″ pitch (MXL) — mini-pitch

length code	pitch length	no. of teeth	width code & approx. wt./grams 1/8″ 012	3/16″ 019	1/4″ 025
36MXL	3.60	45	.34	.51	.68
40MXL	4.00	50	.38	.57	.76
44MXL	4.40	55	.42	.63	.84
48MXL	4.80	60	.46	.68	.92
56MXL	5.60	70	.53	.80	1.06
64MXL	6.40	80	.61	.91	1.22
68MXL	6.80	85	.65	.97	1.30
72MXL	7.20	90	.68	1.03	1.36
76MXL	7.60	95	.72	1.08	1.44
80MXL	8.00	100	.76	1.14	1.52
88MXL	8.80	110	.84	1.25	1.68
96MXL	9.60	120	.91	1.37	1.82
104MXL	10.40	130	.99	1.48	1.98
112MXL	11.20	140	1.06	1.60	2.12
120MXL	12.00	150	1.14	1.71	2.28
132MXL	13.20	165	1.25	1.88	2.50
140MXL	14.00	175	1.33	2.00	2.66
152MXL	15.20	190	1.44	2.17	2.88
160MXL	16.00	200	1.52	2.28	3.04
168MXL	16.80	210	1.60	2.39	3.20
180MXL	18.00	225	1.71	2.56	3.42
200MXL	20.00	250	1.90	2.85	3.80
208MXL	20.80	260	1.98	2.96	3.96
236MXL	23.60	295	2.24	3.36	4.48
240MXL	24.00	300	2.28	3.42	4.56
320MXL	32.00	400	3.04	4.56	6.08

1/5″ pitch (XL) — extra light

length code	pitch length	no. of teeth	width code & approx. wt./lbs. 1/4″ 025	3/8″ 037
50XL	5.00	25	.005	.008
60XL	6.00	30	.006	.009
70XL	7.00	35	.007	.011
80XL	8.00	40	.008	.012
90XL	9.00	45	.009	.014
100XL	10.00	50	.010	.015
110XL	11.00	55	.011	.017
120XL	12.00	60	.013	.019
130XL	13.00	65	.014	.020
140XL	14.00	70	.015	.022
150XL	15.00	75	.016	.023
160XL	16.00	80	.017	.025
170XL	17.00	85	.018	.026
180XL	18.00	90	.019	.028
190XL	19.00	95	.020	.029
200XL	20.00	100	.021	.031
210XL	21.00	105	.022	.032
220XL	22.00	110	.023	.034
230XL	23.00	115	.024	.035
240XL	24.00	120	.025	.037
250XL	25.00	125	.026	.039
260XL	26.00	130	.027	.040
280XL	28.00	140	.028	.042
290XL	29.00	145	.029	.044
300XL	30.00	150	.030	.045
310XL	31.00	155	.031	.047
330XL	33.00	165	.033	.050
340XL	34.00	170	.034	.051
350XL	35.00	175	.035	.053
370XL	37.00	185	.037	.056
380XL	38.00	190	.038	.057
390XL	39.00	195	.039	.059
400XL	40.00	200	.040	.060
420XL	42.00	210	.042	.063
450XL	45.00	225	.045	.068
460XL	46.00	230	.046	.069
480XL	48.00	240	.048	.072
500XL	50.00	250	.050	.075
570XL	57.00	285	.057	.086
630XL	63.00	315	.063	.095
770XL	77.00	385	.077	.116

3/8″ pitch (L) — light duty

length code	pitch length	no. of teeth	width code & approx. wt./lbs. 1/2″ 050	3/4″ 075	1″ 100
124L	12.375	33	.02	.05	.07
135L	13.500	44	.03	.05	.07
150L	15.000	40	.03	.06	.08
165L	16.500	44	.03	.07	.08
187L	18.750	50	.04	.07	.09
195L	19.500	52	.04	.08	.10
210L	21.000	56	.04	.08	.10
225L	22.500	60	.05	.08	.11
240L	24.000	64	.06	.09	.12
255L	25.500	68	.06	.09	.13
270L	27.000	72	.06	.10	.13
285L	28.500	76	.07	.10	.14
300L	30.000	80	.07	.11	.15
315L	31.500	84	.07	.11	.15
322L	32.250	86	.08	.12	.16
345L	34.500	92	.08	.13	.17
367L	36.700	98	.09	.13	.18
390L	39.000	104	.09	.14	.19
420L	42.000	112	.10	.15	.21
450L	45.000	120	.11	.16	.22
480L	48.000	128	.12	.17	.24
510L	51.000	136	.12	.18	.25
540L	54.000	144	.13	.20	.26
600L	60.000	160	.13	.21	.27
660L	66.000	176	.14	.23	.30
817L	81.750	218	.18	.29	.37
900L	90.000	240	.20	.32	.41

1/2″ pitch (H) — heavy duty

length code	pitch length	no. of teeth	width code & approx. wt./lbs. 3/4″ 075	1″ 100	1½″ 150	2″ 200	3″ 300
210H	21.00	42	.12	.16	.24	.32	.49
220H	22.00	44	.12	.17	.25	.34	.51
230H	23.00	46	.13	.17	.26	.35	.53
240H	24.00	48	.13	.20	.28	.41	.56
270H	27.00	54	.15	.21	.31	.44	.63
300H	30.00	60	.17	.23	.35	.37	.70
320H	32.00	64	.18	.24	.37	.49	.74
330H	33.00	66	.19	.25	.38	.51	.77
340H	34.00	68	.19	.26	.39	.52	.79
350H	35.00	70	.20	.26	.40	.54	.81
360H	36.00	72	.20	.28	.42	.56	.84
370H	37.00	74	.21	.28	.43	.57	.86
390H	39.00	78	.22	.30	.45	.61	.91
400H	40.00	80	.23	.30	.46	.62	.93
410H	41.00	82	.24	.31	.47	.63	.95
420H	42.00	84	.24	.32	.49	.65	.98
450H	45.00	90	.25	.35	.52	.70	1.05
480H	48.00	96	.27	.37	.56	.75	1.12
490H	49.00	98	.28	.38	.56	.76	1.14
510H	51.00	102	.29	.39	.59	.79	1.19
540H	54.00	108	.30	.41	.63	.84	1.26
560H	56.00	112	.32	.42	.64	.86	1.30
570H	57.00	114	.32	.44	.66	.89	1.33
585H	58.50	117	.33	.44	.67	.90	1.36
600H	60.00	120	.34	.46	.70	.93	1.40
630H	63.00	126	.35	.48	.73	.98	1.47
645H	64.50	129	.36	.48	.74	.99	1.50
660H	66.00	132	.37	.51	.77	1.02	1.54
700H	70.00	140	.39	.54	.81	1.09	1.64
730H	73.00	146	.41	.55	.84	1.12	1.70
750H	75.00	150	.42	.58	.87	1.16	1.75
780H	78.00	156	.44	.59	.90	1.20	1.81
800H	80.00	160	.45	.61	.93	1.24	1.87
820H	82.00	164	.46	.62	.94	1.26	1.91
840H	84.00	168	.47	.63	.97	1.29	1.95
850H	85.00	170	.48	.65	.99	1.32	1.99
900H	90.00	180	.51	.69	1.04	1.40	2.10
960H	96.00	192	.54	.72	1.10	1.48	2.23
1000H	100.00	200	.56	.77	1.16	1.55	2.35
1100H	110.00	220	.62	.84	1.27	1.71	2.57
1140H	114.00	228	.64	.86	1.31	1.75	2.65
1250H	125.00	250	.70	.96	1.45	1.94	2.92
1400H	140.00	280	.79	1.07	1.62	2.17	3.27
1550H	155.00	310	.88	1.17	1.78	2.38	3.60
1700H	170.00	340	.96	1.30	1.95	2.63	3.97

7/8″ pitch (XH) — extra heavy

length code	pitch length	no. of teeth	width code & approx. wt./lbs. 2″ 200	3″ 300	4″ 400
507XH	50.75	58	2.11	3.16	4.22
560XH	56.00	64	2.33	3.49	4.66
630XH	63.00	72	2.62	3.93	5.24
700XH	70.00	80	2.91	4.37	5.82
770XH	77.00	88	3.20	4.81	6.41
840XH	84.00	96	3.49	5.25	6.99
980XH	98.00	112	4.08	6.12	8.15
1120XH	112.00	128	4.66	6.99	9.32
1260XH	126.00	144	5.24	7.86	10.48
1400XH	140.00	160	5.82	8.74	11.65
1540XH	154.00	176	6.41	9.61	12.81
1750XH	175.00	200	7.28	10.92	14.56

1-1/4″ pitch (XXH) double extra heavy

length code	pitch length	no. of teeth	width code & approx. wt./lbs. 2″ 200	3″ 300	4″ 400	5″ 500
700XXH	70.00	56	4.09	6.13	8.18	10.22
800XXH	80.00	64	4.67	7.01	9.34	11.68
900XXH	90.00	72	5.26	7.88	10.51	13.14
1000XXH	100.00	80	5.84	8.76	11.68	14.50
1200XXH	120.00	96	7.01	10.51	14.60	17.52
1400XXH	140.00	112	8.18	12.26	16.35	20.44
1600XXH	160.00	128	9.34	14.02	18.69	23.36
1800XXH	180.00	144	10.57	15.77	21.02	26.28

Source: Courtesy of the Gates Rubber Corp.

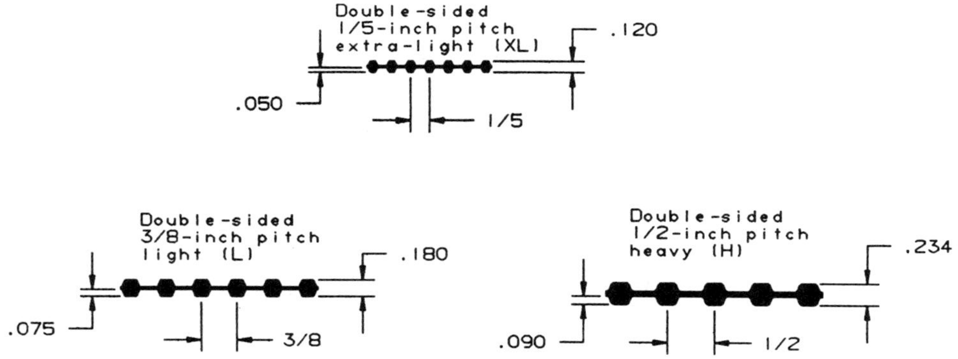

FIGURE 4 Standard double-sided timing belt cross section.

Twin-sided timing belts, having identical teeth on both sides of the belt, are useful for serpentine drives. Standard sizes available for industrial applications are shown in Figure 4. Table 2 shows standard lengths and widths.

2. Pulley Designation

Pulleys are designated by the number of grooves, the belt section, and a number indicating the appropriate width belt that may be used. The number representing the width is 100 × the nominal width. For example, a sprocket designated 30H200 has 30 pulley grooves, is used with an "H" section (0.500 in. pitch) belt and will accept a 2-in. wide belt. Table 3 shows standard timing-pulley sizes.

B. High-Torque Drive (HTD)

A more recent innovation in synchronous-belt development is a new curvilinear tooth profile called High-Torque-Drive, or, more commonly, HTD ®. This profile was originally patented under Patent #3756091 by Uniroyal, Inc. As such, all belts and/or sprockets produced today must have prior license approval from The Gates Rubber Company, Synchronous Drive Division. Several belt suppliers market and/or manufacture this product line.

The HTD has found acceptance in a fairly wide range of industrial applications. However, there has been no industry standard developed for HTD drives.

This profile has full-rounded, deeper, more closely spaced teeth than a conventional trapezoidal timing belt. Due to a greater tooth

TABLE 2 Standard Double-Sided Timing Belt Lengths and Widths

1/5-inch pitch (XL) belts

pitch length code	pitch length (in.)	no. of teeth per side
TP150XL	15.0	75
TP160XL	16.0	80
TP170XL	17.0	85
TP180XL	18.0	90
TP190XL	19.0	95
TP200XL	20.0	100
TP210XL	21.0	105
TP220XL	22.0	110
TP230XL	23.0	115
TP240XL	24.0	120
TP250XL	25.0	125
TP260XL	26.0	130
TP280XL	28.0	140
TP290XL	29.0	145
TP300XL	30.0	150
TP310XL	31.0	155
TP330XL	33.0	165
TP340XL	34.0	170

width code	width (in.)
025	1/4(.250)
037	3/8(.375)*

*Stock Belt Widths

3/8-inch pitch (L) belts

pitch length code	pitch length (in.)	no. of teeth per side
TP150L	15.0	40
TP187L	18.7	50
TP210L	21.0	56
TP225L	22.5	60
TP240L	24.0	64
TP255L	25.5	68
TP270L	27.0	72
TP285L	28.5	76
TP300L	30.0	80
TP322L	32.25	86
TP345L	34.5	92
TP367L	36.7	98
TP390L	39.0	104
TP420L	42.0	112
TP450L	45.0	120
TP480L	48.0	128
TP510L	51.0	136
TP540L	54.0	144
TP600L	60.0	160
TP660L	66.0	176

width code	width (in.)
050	1/2(.500)*
075	3/4(.750)
100	1 (1.00)*

*Stock Belt Widths

1/2-inch pitch (H) belts

pitch length code	pitch length (in.)	no. of teeth per side
TP240H	24.0	48
TP270H	27.0	54
TP300H	30.0	60
TP330H	33.0	66
TP360H	36.0	72
TP390H	39.0	78
TP420H	42.0	84
TP450H	45.0	90
TP480H	48.0	96
TP510H	51.0	102
TP540H	54.0	108
TP570H	57.0	114
TP600H	60.0	120
TP630H	63.0	126
TP660H	66.0	132
TP700H	70.0	140
TP750H	75.0	150
TP800H	80.0	160
TP850H	85.0	170
TP900H	90.0	180
TP1000H	100.0	200
TP1100H	110.0	220
TP1250H	125.0	250
TP1400H	140.0	280

width code	width (in.)
075	3/4(.750)*
100	1(1.00)*
150	1-1/2(1.50)
200	2(2.00)
300	3(3.00)

*Stock Belt Widths

Source: Courtesy of The Gates Rubber Company.

TABLE 3 Standard Timing Belt Pulleys

Number Of Grooves	BELT SECTION											
	MXL (0.080)		XL (0.200)		L (0.375)		H (0.500)		XH (0.875)		XXH (1.250)	
	Diameters		Diameters		Diameters		Diameters		Diameters		Diameters	
	Pitch	Outside	Pitch	Outside	Pitch	Outside	Pitch	Outside	Pitch	Outside	Pitch	Outside
10	0.255	0.235	0.637	0.617	1.194*	1.164						
12	0.306	0.286	0.764	0.744	1.432*	1.402						
14	0.357	0.337	0.891	0.871	1.671	1.641	2.228*	2.174				
16	0.407	0.387	1.019	0.999	1.910	1.880	2.546	2.492				
18	0.458	0.438	1.146	1.126	2.149	2.119	2.865	2.811	5.013	4.903	7.162	7.042
20	0.509	0.489	1.273	1.253	2.387	2.357	3.183	3.129	5.570	5.460	7.958	7.838
22	0.560	0.540	1.401	1.381	2.626	2.596	3.501	3.447	6.127	6.017	8.754	8.634
24	0.611	0.591	1.528	1.508	2.865	2.835	3.820	3.766	6.685	6.575	9.549	9.429
26	0.662	0.642			3.104	3.074	4.138	4.084	7.242	7.132	10.345	10.225
28	0.713	0.693	1.783	1.763	3.342	3.312	4.456	4.402	7.799	7.689		
30	0.764	0.744	1.910	1.890	3.581	3.551	4.775	4.721	8.356	8.246	11.937	11.817
32	0.815	0.795	2.037	2.017	3.820	3.790	5.093	5.039	8.913	8.803		
34	0.866	0.846									13.528	13.408
36	0.917	0.897	2.292	2.272	4.297	4.267	5.730	5.676				
40	1.019	0.999	2.546	2.526	4.775	4.745	6.366	6.312	11.141	11.031	15.915	15.795
42	1.070	1.050	2.674	2.654								
44	1.120	1.100	2.801	2.781	5.252	5.222	7.003	6.949				
48	1.222	1.202	3.056	3.036	5.730	5.700	7.639	7.585	13.369	13.259	19.099	18.979
60	1.528	1.508	3.820	3.800	7.162	7.132	9.549	9.495	16.711	16.601	23.873	23.753
72	1.833	1.813	4.584	4.564	8.594	8.564	11.459	11.405	20.054	19.944	28.648	28.528
84					10.027	9.997	13.369	13.315	23.396	23.286		
90											35.810	35.690
96							15.279	15.225	26.738	26.628		
120							19.099	19.045	33.423	33.313		

*Usually not available in all widths – consult supplier.

Source: ANSI/RMA, IP-24-1983.

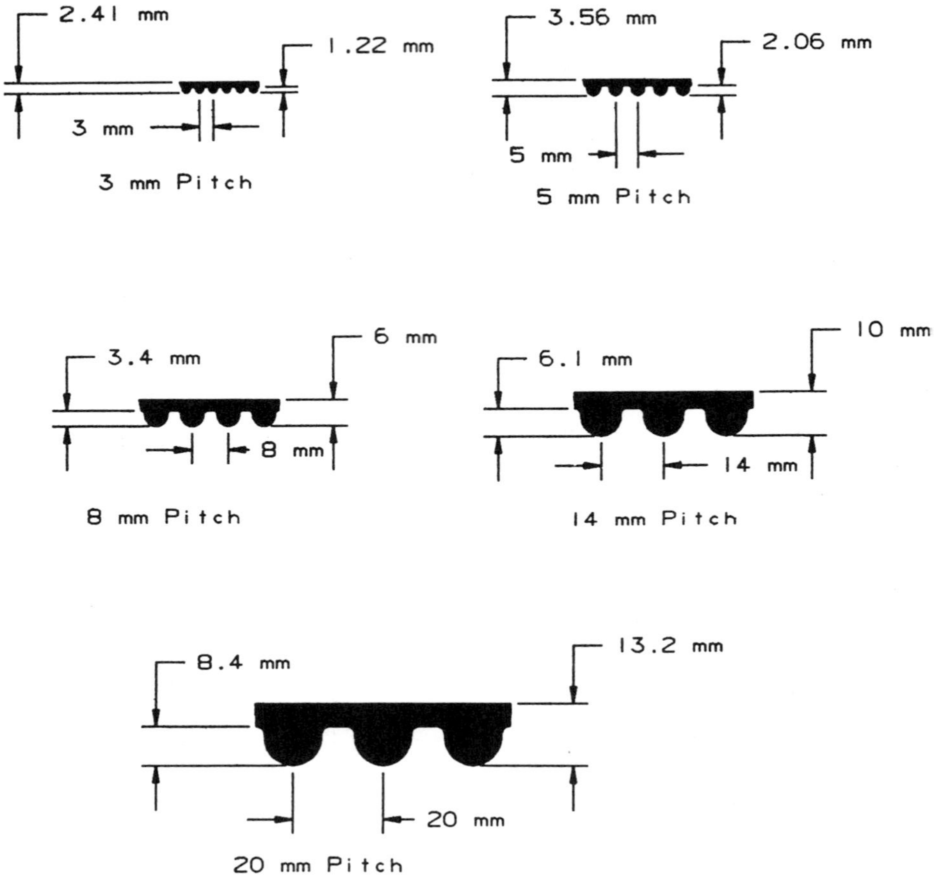

FIGURE 5 HTD belt cross sections.

shear strength and more even distribution of tooth loading to the belt tensile member, the HTD belt allows overall higher loading. Figure 5 shows the most common HTD belt sections available.

1. Belt Designation

HTD belts are identified in a similar manner to timing belts. Sizes are indicated by pitch length, pitch, and belt top width. However, HTD belt dimensions are specified in metric units. For example, a belt pitch length of 1778 mm (70 in.), 14 mm pitch (0.5512 in.), 85 mm (3.35 in.) wide is identified as 1778-14M-85. Tables 4, 5, and 6 list common belt lengths and widths for HTD belts.

2. Sprocket Designation

Sprockets are designated by number of grooves, pitch, and width. The identifying sprocket number is prefixed with the letter "P". This helps to further indicate that it fits into the HTD system. So, a 44-groove, 8-mm pitch sprocket for a 30 mm wide belt is designated P44-8M-30. Tables 7 and 8 show some common stock sizes that are available for 8-mm and 14-mm sprockets.

C. Super-Torque Positive Drive (STPD)

Another recent addition to the synchronous-belt types is STPD. STPD stands for Super-Torque Positive Drive and was developed by The Goodyear Tire and Rubber Company. It is covered under three U.S. patents: #3924481, #3969946 and #4037485. As with the HTD, there has been no industry standard developed for the STPD drives. Several belt suppliers market and/or manufacture this product line. The STPD curvilinear profile is illustrated in Figure 6. Several pitches

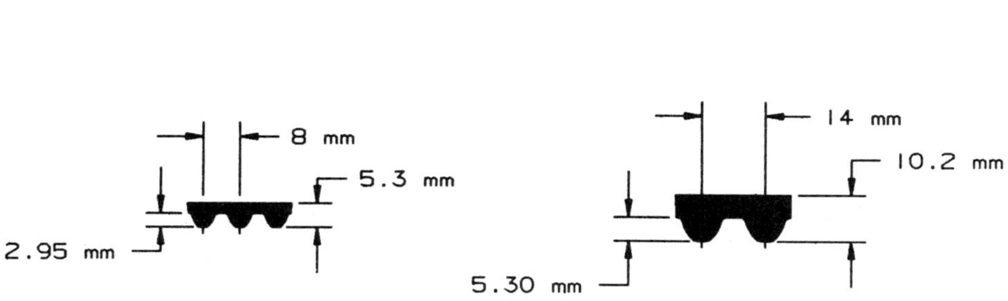

FIGURE 6 STPD belt cross sections.

TABLE 4 3-mm and 5-mm Pitch HTD Belt Lengths and Widths

3mm Pitch Belt Lengths

Belt Length & Pitch Code	Pitch Length (mm) (in.)	No. of Teeth	Belt Length & Pitch Code	Pitch Length (mm) (in.)	No. of Teeth
150-3M	150 **5.91**	50	447-3M	447 **17.60**	149
159-3M	159 **6.26**	53	474-3M	474 **18.66**	158
168-3M	168 **6.61**	56	501-3M	501 **19.72**	167
177-3M	177 **6.97**	59	513-3M	513 **20.20**	171
189-3M	189 **7.44**	63	531-3M	531 **20.91**	177
201-3M	201 **7.91**	67	564-3M	564 **22.20**	188
213-3M	213 **8.39**	71	597-3M	597 **23.50**	199
225-3M	225 **8.86**	75	633-3M	633 **24.92**	211
240-3M	240 **9.45**	80	669-3M	669 **26.34**	223
252-3M	252 **9.92**	84	711-3M	711 **27.99**	237
255-3M	255 **10.04**	85	753-3M	753 **29.65**	251
267-3M	267 **10.51**	89	795-3M	795 **31.30**	265
285-3M	285 **11.22**	95	843-3M	843 **33.19**	281
300-3M	300 **11.81**	100	882-3M	882 **34.72**	294
312-3M	312 **12.28**	104	945-3M	945 **37.20**	315
318-3M	318 **12.52**	106	1002-3M	1002 **39.45**	334
339-3M	339 **13.35**	113	1062-3M	1062 **41.81**	354
357-3M	357 **14.06**	119	1125-3M	1125 **44.29**	375
384-3M	384 **15.12**	128	1191-3M	1191 **46.89**	397
399-3M	399 **15.71**	133	1263-3M	1263 **49.72**	421
420-3M	420 **16.54**	140			

3mm Pitch Belt Widths

Belt Width Code	Belt Width (mm)	Belt Width (in.)
06	6	.236
09	9	.354
15	15	.591

TABLE 4 (Continued)

5mm Pitch Stock Belt Lengths

Belt Length & Pitch Code	Pitch Length mm in.	No. of Teeth	Width Code (mm) 15	25
			Approx. Weight (lbs)	
• 350-5M	350 13.78	70	051	086
375-5M	375 14.76	75	055	092
400-5M	400 15.75	80	059	098
425-5M	425 16.73	85	062	103
450-5M	450 17.72	90	066	110
475-5M	475 18.70	95	069	115
500-5M	500 19.69	100	073	122
535-5M	535 21.06	107	078	130
565-5M	565 22.24	113	083	138
600-5M	600 23.62	120	088	146
• 615-5M	615 24.21	123	090	150
635-5M	635 25.00	127	093	155
670-5M	670 26.38	134	098	163
710-5M	710 27.95	142	103	173
740-5M	740 29.13	148	108	180
800-5M	800 31.50	160	117	195
• 835-5M	835 32.87	167	121	203
850-5M	850 33.46	170	124	206
890-5M	890 35.04	178	130	216
• 925-5M	925 36.42	185	135	225
950-5M	950 37.40	190	138	231
1000-5M	1000 39.37	200	145	243
1050-5M	1050 41.34	210	153	255
1125-5M	1125 44.29	225	164	273
1195-5M	1195 47.05	239	170	284
1270-5M	1270 50.00	254	185	308
• 1340-5M	1340 52.76	268	195	326
1420-5M	1420 55.91	284	206	344
• 1500-5M	1500 59.06	300	217	362
1595-5M	1595 62.80	319	231	387
• 1690-5M	1690 66.54	338	240	411
1790-5M	1790 70.47	358	260	434
1895-5M	1895 74.61	379	275	459
2000-5M	2000 78.74	400	281	470

•*Check Gates for Availability

5mm Pitch Stock Belt Widths

Belt Width Code	Belt Width (mm)	Belt Width (inches)
15	15	.591
25	25	.984

Source: Courtesy of The Gates Rubber Company.

TABLE 5 8-mm and 14-mm Pitch HTD Belt Lengths and Widths

8mm Pitch Stock Belt Lengths

Belt Length & Pitch Code	Pitch Length mm in.	No. of Teeth	Width Code (mm)			
			20	30	50	85
			Approx. Weight (lbs)			
480-8M	480 18.90	60	13	20	34	57
560-8M	560 22.05	70	16	23	39	66
600-8M	600 23.62	75	17	25	42	71
640-8M	640 25.20	80	18	27	45	76
720-8M	720 28.35	90	20	30	50	85
800-8M	800 31.50	100	22	33	56	95
880-8M	880 34.65	110	25	37	61	1 04
960-8M	960 37.80	120	27	40	67	1 14
1040-8M	1040 40.94	130	29	43	74	1 23
1120-8M	1120 44.09	140	31	47	78	1 33
1200-8M	1200 47.24	150	34	50	84	1 42
1280-8M	1280 50.39	160	36	53	89	1 52
1440-8M	1440 56.69	180	40	60	1 01	1 71
1600-8M	1600 62.99	200	45	67	1 11	1 90
1760-8M	1760 69.29	220	49	73	1 23	2 08
1800-8M	1800 70.87	225	50	75	1 25	2 13
2000-8M	2000 78.74	250	56	83	1 39	2 37
2400-8M	2400 94.49	300	67	1 00	1 67	2 84
2600-8M	2600 102.36	325	73	1 09	1 81	3 09
2800-8M	2800 110.24	350	78	1 17	1 95	3 32
3048-8M	3048 120.00	381	85	1 27	2 13	3 62
3280-8M	3280 129.13	410	92	1 37	2 29	3 89
3600-8M	3600 141.73	450	1 01	1 51	2 51	4 27
4400-8M	4400 173.23	550	1 23	1 84	3 07	5 22

8mm Pitch Stock Belt Widths

Belt Width Code	Belt Width (mm)	Belt Width (inches)
20	20	.79
30	30	1.18
50	50	1.97
85	85	3.35

(continued)

TABLE 5 (Continued)

14mm Pitch Stock Belt Lengths

Belt Length & Pitch Code	Pitch Length mm in.	No. of Teeth	Width Code (mm)				
			40	55	85	115	170
			Approx. Weight (lbs)				
966-14M	966 **38.03**	69	84	1.15	1.78	2.41	3.56
1190-14M	1190 **46.85**	85	1.03	1.42	2.20	2.98	4.39
1400-14M	1400 **55.12**	100	1.21	1.67	2.58	3.50	5.16
1610-14M	1610 **63.39**	115	1.40	1.92	2.97	4.02	5.95
1778-14M	1778 **70.00**	127	1.54	2.13	3.28	4.45	6.56
1890-14M	1890 **74.41**	135	1.64	2.26	3.49	4.73	6.97
2100-14M	2100 **82.68**	150	1.82	2.51	3.88	5.25	7.75
2310-14M	2310 **90.94**	165	2.00	2.76	4.26	5.77	8.53
2450-14M	2450 **96.46**	175	2.13	2.93	4.52	6.13	9.04
2590-14M	2590 **101.97**	185	2.25	3.10	4.78	6.47	9.55
2800-14M	2800 **110.24**	200	2.43	3.34	5.17	7.00	10.33
3150-14M	3150 **124.02**	225	2.73	3.77	5.82	7.87	11.62
3360-14M	3360 **132.28**	240	2.91	4.02	6.20	8.39	12.39
3500-14M	3500 **137.80**	250	3.03	4.19	6.46	8.75	12.90
3850-14M	3850 **151.57**	275	3.33	4.60	7.10	9.62	14.20
4326-14M	4326 **170.32**	309	3.74	5.17	8.00	10.80	15.96
4578-14M	4578 **180.24**	327	3.96	5.48	8.45	11.42	16.90
4956-14M	4956 **195.12**	354	4.29	5.90	9.11	12.33	18.23
5320-14M	5320 **209.45**	380	4.61	6.33	9.28	13.24	19.57
5740-14M	5740 **225.98**	410	4.97	6.83	10.55	14.29	21.12
6160-14M	6160 **242.52**	440	5.33	7.33	11.32	15.34	22.67
6860-14M	6860 **270.08**	490	5.94	8.16	12.61	17.08	25.25

14mm Pitch Stock Belt Widths

Belt Width Code	Belt Width (mm)	Belt Width (inches)
40	40	**1.57**
55	55	**2.17**
85	85	**3.35**
115	115	**4.53**
170	170	**6.69**

Source: Courtesy of The Gates Rubber Company.

TABLE 6 20-mm Pitch HTD Belt Lengths and Widths

20mm Pitch Stock Belt Lengths

Belt Length & Pitch Code	Pitch Length mm **in.**	No. of Teeth	Width Code (mm)				
			115	170	230	290	340
			Approx. Weight (lbs)				
2000-20M	2000 **78.74**	100	6	9	12	16	19
2500-20M	2500 **98.43**	125	8	11	15	20	24
3400-20M	3400 **133.86**	170	11	16	22	27	32
4200-20M	4200 **165.35**	210	13	20	27	34	39
5000-20M	5000 **196.85**	250	16	23	32	40	47
5400-20M	5400 **212.60**	270	17	25	34	43	51
5800-20M	5800 **228.35**	290	18	27	37	46	54
6200-20M	6200 **244.09**	310	20	29	39	50	58
6600-20M	6600 **259.84**	330	21	31	42	53	62

20mm Pitch Stock Belt Widths

Belt Width Code	Belt Width (mm)	Belt Width (inches)
115	115	**4.53**
170	170	**6.69**
230	230	**9.06**
290	290	**11.42**
340	340	**13.39**

[a]Dimensions are given in inches and millimeters. Inches are shown in color in bold face type. Millimeters are shown in black in light face type.
Source: Courtesy of The Gates Rubber Company.

are shown, but there is no established stock line of STPD belts or sprockets. Table 11 lists some 8-mm belt sizes and widths which are available on a made-to-order basis.

1. Belt Designation

Belts are identified using the same convention as other types of synchronous belts except the order is reversed. They are identified by belt top width × 10, pitch, and pitch length, with all units in mm. However, the letter "S" has been added as a prefix to the belt pitch in order to identify the belt as the STPD profile. For example, a 20-mm-wide, 8-mm pitch, 2000-mm pitch length belt is designated as 200-S8M-2000.

2. Pulley Designation

Pulleys are specified in a similar manner to belts. They are identified by a number of grooves and by "S" which indicates STPD system, pitch, and width. For example, a 24-groove, 8-mm pitch sprocket for a 30-mm-wide belt is designated 24-S8M-30.

D. Metric and Special-Pitch Belts

1. Metric Timing Belts

In addition to the six standard timing-belt profiles, two metric sizes have been produced by European manufacturers. These belts are most commonly made of polyurethane with steel cable used as the tensile member. Figure 7 shows the overall dimensions of these profiles.

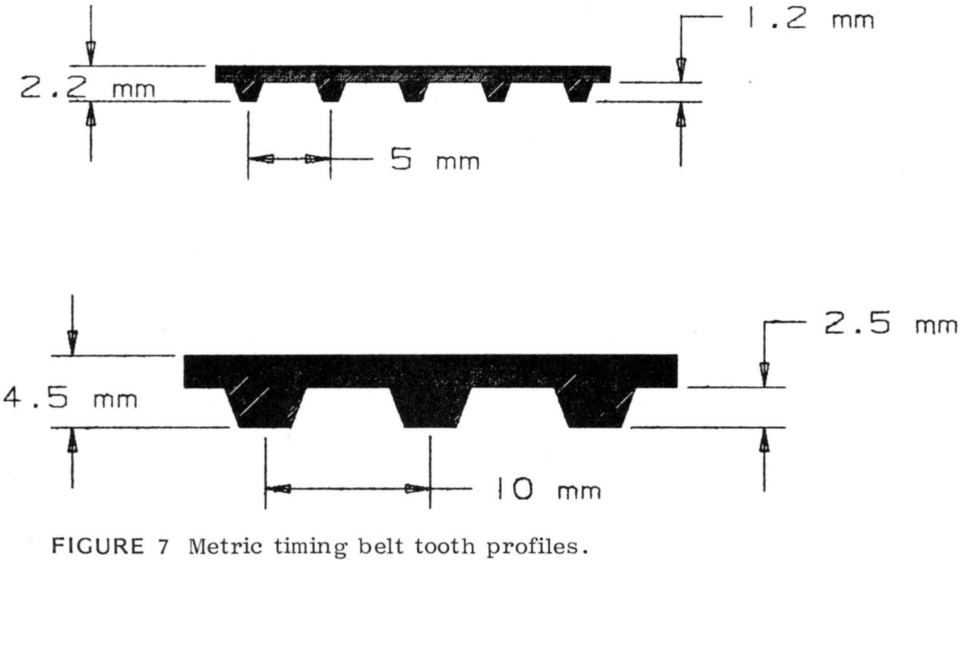

FIGURE 7 Metric timing belt tooth profiles.

TABLE 7 HTD Stock Sprockets

SPROCKET TYPES

The HTD sprockets listed in the tables below are all stock sizes. All dimensions given are for the sprocket with the QD bushing in place and are in inches only.

Type A | Type B | Type C | Type D | Type E | Type 6

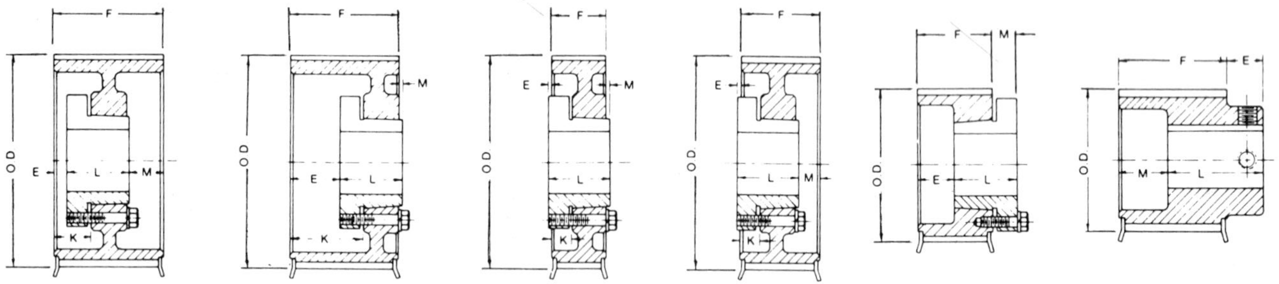

Type AF | Type BF | Type CF | Type DF | Type EF | **Type 6F** Bored-To-Suit

The figure following the sketch reference letter in the "Type" column indicates the construction of the sprocket (1 = Solid, 2 = Web, and 3 = Arms), and the letter "F" indicates that the sprocket has flanges.

SPROCKET NUMBER	NO. OF GRVS.	DIAMETERS			TYPE	DIMENSIONS (INCHES)					BORE RANGE		APP. □ WT. (Lbs.)
		P.D.	O.D.	Flange		E	F	K	L	M	Min.	Max.	
FOR BELTS 20mm (.79 in.) WIDE • 8mm PITCH (8M-20)													
P22-8M-20	22	2.206	2.160	2.559	6F	5/8	1 1/8	. . .	1 3/4	0	* 1/2	1 3/16	1.6
P24-8M-20-JA	24	2.406	2.359	2.756	E1F	9/16	1 1/8	. . .	1	7/16	1/2	1 1/4	1.5
P26-8M-20-JA	26	2.607	2.559	2.953	E1F	9/16	1 1/8	. . .	1	7/16	1/2	1 1/4	1.8
P28-8M-20-JA	28	2.807	2.759	3.150	B1F	1/8	1 1/8	9/16	1	0	1/2	1 1/4	1.5
P30-8M-20-JA	30	3.008	2.958	3.346	B1F	1/8	1 1/8	9/16	1	0	1/2	1 1/4	2.0
P32-8M-20-JA	32	3.208	3.156	3.543	B1F	1/8	1 1/8	9/16	1	0	1/2	1 1/4	2.1
P34-8M-20-SH	34	3.409	3.355	3.819	D1F	3/16	1 1/8	5/16	1 1/4	1/16	1/2	1 11/16	2.2
P36-8M-20-SH	36	3.609	3.555	3.937	D1F	3/16	1 1/8	5/16	1 1/4	1/16	1/2	1 11/16	2.5
P38-8M-20-SH	38	3.810	3.756	4.134	D1F	3/16	1 1/8	5/16	1 1/4	1/16	1/2	1 11/16	2.8
P40-8M-20-SH	40	4.010	3.956	4.331	D1F	3/16	1 1/8	5/16	1 1/4	1/16	1/2	1 11/16	2.3
P44-8M-20-SDS	44	4.411	4.357	4.764	C1F	3/16	1 1/8	3/8	1 5/16	0	1/2	2	2.5
P48-8M-20-SDS	48	4.812	4.758	5.157	C1F	3/16	1 1/8	3/8	1 5/16	0	1/2	2	3.2
P56-8M-20-SDS	56	5.614	5.560	5.945	C1F	3/16	1 1/8	3/8	1 5/16	0	1/2	2	4.5
P64-8M-20-SDS	64	6.416	6.362	6.772	C1F	3/16	1 1/8	3/8	1 5/16	0	1/2	2	6.0
P72-8M-20-SDS	72	7.218	7.164	7.598	C2F	3/16	1 1/8	3/8	1 5/16	0	1/2	2	5.7
P80-8M-20-SDS	80	8.020	7.966	8.386	C2F	3/16	1 1/8	3/8	1 5/16	0	1/2	2	6.5
P90-8M-20-SDS	90	9.023	8.969	. . .	C3	3/16	1 1/8	3/8	1 5/16	0	1/2	2	7.0

□ Wt. does not include bushing.
* MPB with no keyseat or setscrews.

(continued)

TABLE 7 (Continued)

SPROCKET NUMBER	NO. OF GRVS	DIAMETERS			TYPE	DIMENSIONS (INCHES)					BORE RANGE		APP. □ WT. (Lbs.)
		P.D.	O.D.	Flange		E	F	K	L	M	Min.	Max.	
FOR BELTS 30mm (1.18 in.) WIDE • 8mm PITCH (8M-30)													
P22-8M-30	22	2.206	2.160	2.559	6F	5/8	1 1/2	. . .	2 1/8	0	*1/2	1 3/16	2.0
P24-8M-30	24	2.406	2.359	2.756	6F	5/8	1 1/2	. . .	2 1/8	0	*1/2	1 1/4	2.5
P26-8M-30	26	2.607	2.559	2.953	6F	3/4	1 1/2	. . .	2 1/4	0	*1/2	1 1/4	3.2
P28-8M-30-JA	28	2.807	2.759	3.150	B1F	1/2	1 1/2	15/16	1	0	1/2	1 1/4	1.8
P30-8M-30-JA	30	3.008	2.958	3.346	B1F	1/2	1 1/2	15/16	1	0	1/2	1 1/4	2.0
P32-8M-30-JA	32	3.208	3.156	3.543	B1F	1/2	1 1/2	15/16	1	0	1/2	1 1/4	2.2
P34-8M-30-SH	34	3.409	3.355	3.819	A1F	3/16	1 1/2	11/16	1 1/4	1/16	1/2	1 11/16	2.4
P36-8M-30-SH	36	3.609	3.555	3.937	A1F	3/16	1 1/2	11/16	1 1/4	1/16	1/2	1 11/16	2.8
P38-8M-30-SH	38	3.810	3.756	4.134	A1F	3/16	1 1/2	11/16	1 1/4	1/16	1/2	1 11/16	3.2
P40-8M-30-SH	40	4.010	3.956	4.331	A1F	3/16	1 1/2	11/16	1 1/4	1/16	1/2	1 11/16	2.7
P44-8M-30-SDS	44	4.411	4.357	4.764	B1F	3/16	1 1/2	3/4	1 5/16	0	1/2	2	3.0
P48-8M-30-SDS	48	4.812	4.758	5.157	B1F	3/16	1 1/2	3/4	1 5/16	0	1/2	2	3.7
P56-8M-30-SDS	56	5.614	5.560	5.945	B1F	3/16	1 1/2	3/4	1 5/16	0	1/2	2	5.0
P64-8M-30-SK	64	6.416	6.362	6.772	C1F	3/8	1 1/2	1/4	1 7/8	0	1/2	2 5/8	8.5
P72-8M-30-SK	72	7.218	7.164	7.598	C2F	3/8	1 1/2	1/4	1 7/8	0	1/2	2 5/8	7.8
P80-8M-30-SK	80	8.020	7.966	8.386	C2F	3/8	1 1/2	1/4	1 7/8	0	1/2	2 5/8	9.8
P90-8M-30-SK	90	9.023	8.969	. . .	C2	3/8	1 1/2	1/4	1 7/8	0	1/2	2 5/8	11.5
P112-8M-30-SK	112	11.229	11.175	. . .	C3	3/8	1 1/2	1/4	1 7/8	0	1/2	2 5/8	13.4

SPROCKET NUMBER	NO. OF GRVS	DIAMETERS			TYPE	DIMENSIONS (INCHES)					BORE RANGE		APP. □ WT. (Lbs.)
		P.D.	O.D.	Flange		E	F	K	L	M	Min.	Max.	
FOR BELTS 50mm (1.97 in.) WIDE • 8mm PITCH (8M-50)													
P28-8M-50-JA	28	2.807	2.759	3.150	A1F	3/8	2 3/8	13/16	1	1	1/2	1 1/4	2.4
P30-8M-50-JA	30	3.008	2.958	3.346	A1F	3/8	2 3/8	13/16	1	1	1/2	1 1/4	2.7
P32-8M-50-JA	32	3.208	3.156	3.543	A1F	3/8	2 3/8	13/16	1	1	1/2	1 1/4	3.0
P34-8M-50-SH	34	3.409	3.355	3.819	A1F	0	2 3/8	1/2	1 1/4	1 1/8	1/2	1 11/16	3.0
P36-8M-50-SH	36	3.609	3.555	3.937	A1F	0	2 3/8	1/2	1 1/4	1 1/8	1/2	1 11/16	3.8
P38-8M-50-SH	38	3.810	3.756	4.134	A1F	0	2 3/8	1/2	1 1/4	1 1/8	1/2	1 11/16	4.2
P40-8M-50-SH	40	4.010	3.956	4.331	A1F	0	2 3/8	1/2	1 1/4	1 1/8	1/2	1 11/16	3.7
P44-8M-50-SD	44	4.411	4.357	4.764	A1F	0	2 3/8	9/16	1 13/16	9/16	1/2	2	4.7
P48-8M-50-SD	48	4.812	4.758	5.157	A1F	0	2 3/8	9/16	1 13/16	9/16	1/2	2	6.0
P56-8M-50-SK	56	5.614	5.560	5.945	D1F	1/16	2 3/8	9/16	1 7/8	9/16	1/2	2 5/8	7.6
P64-8M-50-SK	64	6.416	6.362	6.772	D1F	1/16	2 3/8	9/16	1 7/8	9/16	1/2	2 5/8	10.3
P72-8M-50-SK	72	7.218	7.164	7.598	D1F	1/16	2 3/8	9/16	1 7/8	9/16	1/2	2 5/8	13.3
P80-8M-50-SF	80	8.020	7.966	8.386	D2F	1/16	2 3/8	9/16	2	7/16	1/2	2 15/16	12.7
P90-8M-50-SF	90	9.023	8.969	. . .	D2	1/16	2 3/8	9/16	2	7/16	1/2	2 15/16	16
P112-8M-50-SF	112	11.229	11.175	. . .	D2	1/16	2 3/8	9/16	2	7/16	1/2	2 15/16	21
P144-8M-50-E	144	14.437	14.383	. . .	D2	1/2	2 3/8	3/8	2 5/8	1/4	7/8	3 1/2	35
P192-8M-50-E	192	19.249	19.195	. . .	D3	1/2	2 3/8	3/8	2 5/8	1/4	7/8	3 1/2	45

(continued)

TABLE 7 (Continued)

SPROCKET NUMBER	NO. OF GRVS	DIAMETERS			TYPE	DIMENSIONS (INCHES)					BORE RANGE		APP. □ WT. (Lbs.)
		P.D.	O.D.	Flange		E	F	K	L	M	Min.	Max.	
FOR BELTS 85mm (3.35 in.) WIDE • 8mm PITCH (8M-85)													
P34-8M-85-SH	34	3.409	3.355	3.819	A1F	1	3¾	1½	1¼	1½	½	1 11/16	4.0
P36-8M-85-SH	36	3.609	3.555	3.937	A1F	1	3¾	1½	1¼	1½	½	1 11/16	5.2
P38-8M-85-SH	38	3.810	3.756	4.134	A1F	1	3¾	1½	1¼	1½	½	1 11/16	5.8
P40-8M-85-SD	40	4.010	3.956	4.331	A1F	11/16	3¾	1¼*	1 13/16	1¼	½	2	4.6
P44-8M-85-SD	44	4.411	4.357	4.764	A1F	11/16	3¾	1¼	1 13/16	1¼	½	2	6.2
P48-8M-85-SD	48	4.812	4.758	5.157	A1F	11/16	3¾	1¼	1 13/16	1¼	½	2	7.8
P56-8M-85-SK	56	5.614	5.560	5.945	A1F	⅝	3¾	1¼	1⅞	1¼	½	2⅝	9.8
P64-8M-85-SK	64	6.416	6.362	6.772	A1F	⅝	3¾	1¼	1⅞	1¼	½	2⅝	13.0
P72-8M-85-SF	72	7.218	7.164	7.598	A1F	⅝	3¾	1¼	2	1⅛	½	2 15/16	16
P80-8M-85-SF	80	8.020	7.966	8.386	A2F	⅝	3¾	1¼	2	1⅛	½	2 15/16	17
P90-8M-85-SF	90	9.023	8.969	. . .	A2	⅝	3¾	1¼	2	1⅛	½	2 15/16	20
P112-8M-85-SF	112	11.229	11.175	. . .	A2	⅝	3¾	1¼	2	1⅛	½	2 15/16	28
P144-8M-85-E	144	14.437	14.383	. . .	A2	3/16	3¾	1 1/16	2⅝	15/16	⅞	3½	46
P192-8M-85-E	192	19.249	19.195	. . .	A3	3/16	3¾	1 1/16	2⅝	15/16	⅞	3½	61

□ Wt. does not include bushing
* MPB with no keyseat or setscrews

Source: Courtesy of The Gates Rubber Company.

2. 40DP Timing Belt

Prior to the adoption of the standard MXL (0.080 in.) pitch by RMA, a mini-pitch belt was designed to run in a 40 diametral-pitch spur gear. The belt pitch for the 40DP is 0.0816 in. It is very close to the standard MXL 0.080 in. pitch, making it very difficult to visually tell the difference between the two belts.

Because the 40DP pitch and profile are very close to the MXL timing belt, accurate measuring tools must be used to distinguish the difference between the two. The two systems are not compatible, i.e., a 40DP belt will not run in a MXL sprocket and vice versa. Figure 8 shows a 40DP cross section belt.

3. Other Unique Pitches

In addition to the belt pitches mentioned above, there are many other nonstandard-pitch synchronous belts in use. For example, there are at least 8 different pitches between XL (0.200 in.) and L (0.375 in.), such as 0.233 in., 0.246 in., 0.250 in., etc., being used in industry. Some are common to a special industry segment, such as cotton-cleaner belts. These are 1-in. pitch and have a trapezoidal-shaped tooth, very similar to a timing belt.

III. AUTOMOTIVE SYNCHRONOUS BELTS

A. STA, AL, and ZA

These designations are for a belt section most commonly used in the automotive industry. It is derived from the industrial L-section timing belt. The "A" is used to distinguish it from the industrial L-section. The main difference between the industrial and automotive

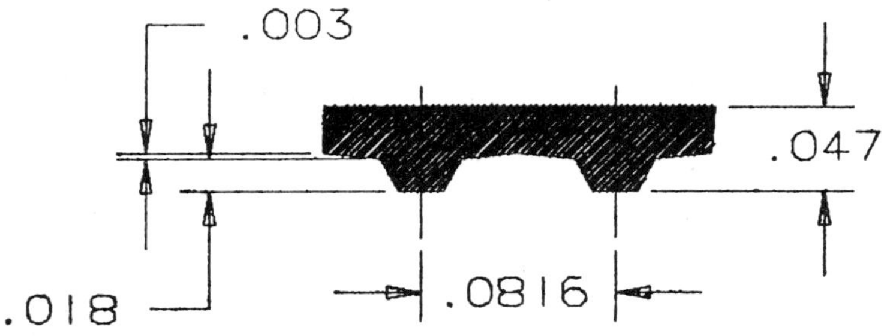

FIGURE 8 40DP tooth profile.

TABLE 8 HTD Stock Sprockets

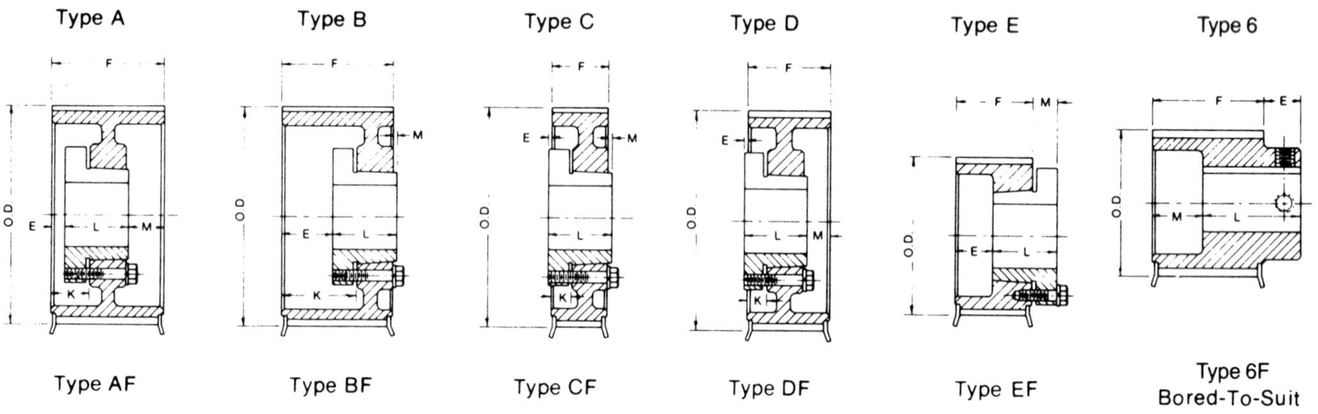

The figure following the sketch reference letter in the "Type" column indicates the construction of the sprocket. (1 = Solid, 2 = Web, and 3 = Arms), and the letter F indicates the sprocket has flanges.

SPROCKET NUMBER	NO. OF GRVS	DIAMETERS			TYPE	DIMENSIONS (INCHES)					BORE RANGE		APP. □ WT. (Lbs.)
		P.D.	O.D.	Flange		E	F	K	L	M	Min.	Max.	
FOR BELTS 40mm (1.57 in.) WIDE • 14mm PITCH (14M-40)													
P28-14M-40-SK	28	4.912	4.808	5.56	E1F	7/8	2 1/8	. . .	1 7/8	5/8	1/2	2 5/8	5.5
P29-14M-40-SK	29	5.088	4.983	5.56	E1F	7/8	2 1/8	. . .	1 7/8	5/8	1/2	2 5/8	6.5
P30-14M-40-SK	30	5.263	5.157	6.13	D1F	3/16	2 1/8	7/16	1 7/8	7/16	1/2	2 5/8	6.0
P32-14M-40-SK	32	5.614	5.507	6.13	D1F	3/16	2 1/8	7/16	1 7/8	7/16	1/2	2 5/8	8.0
P34-14M-40-SK	34	5.965	5.858	6.50	D1F	3/16	2 1/8	7/16	1 7/8	7/16	1/2	2 5/8	8.5
P36-14M-40-SF	36	6.316	6.208	6.81	D1F	3/16	2 1/8	7/16	2	5/16	1/2	2 15/16	9.5
P38-14M-40-SF	38	6.667	6.559	7.16	D1F	3/16	2 1/8	7/16	2	5/16	1/2	2 15/16	11.5
P40-14M-40-SF	40	7.018	6.909	7.50	D1F	3/16	2 1/8	7/16	2	5/16	1/2	2 15/16	13.0
P44-14M-40-E	44	7.720	7.610	8.22	D1F	5/8	2 1/8	1/4	2 5/8	1/8	7/8	3 1/2	16.5
P48-14M-40-E	48	8.421	8.311	8.94	D1F	5/8	2 1/8	1/4	2 5/8	1/8	7/8	3 1/2	20.0
P56-14M-40-E	56	9.825	9.715	10.38	D1F	5/8	2 1/8	1/4	2 5/8	1/8	7/8	3 1/2	28.0
P64-14M-40-E	64	11.229	11.119	11.75	D2F	5/8	2 1/8	1/4	2 5/8	1/8	7/8	3 1/2	29.0
P72-14M-40-E	72	12.632	12.522	13.19	D2F	5/8	2 1/8	1/4	2 5/8	1/8	7/8	3 1/2	33.0
P80-14M-40-E	80	14.036	13.926	14.63	D2F	5/8	2 1/8	1/4	2 5/8	1/8	7/8	3 1/2	38.0
P90-14M-40-E	90	15.790	15.680	. . .	D3	5/8	2 1/8	1/4	2 5/8	1/8	7/8	3 1/2	39.0
P112-14M-40-E	112	19.650	19.540	. . .	D3	5/8	2 1/8	1/4	2 5/8	1/8	7/8	3 1/2	51.0
P144-14M-40-E	144	25.264	25.154	. . .	D3	5/8	2 1/8	1/4	2 5/8	1/8	7/8	3 1/2	80.0

(continued)

TABLE 8 (Continued)

SPROCKET NUMBER	NO. OF GRVS	DIAMETERS P.D.	DIAMETERS O.D.	DIAMETERS Flange	TYPE	DIMENSIONS (INCHES) E	DIMENSIONS (INCHES) F	DIMENSIONS (INCHES) K	DIMENSIONS (INCHES) L	DIMENSIONS (INCHES) M	BORE RANGE Min.	BORE RANGE Max.	APP. □ WT. (Lbs.)
FOR BELTS 55mm (2.17 in.) WIDE • 14mm PITCH (14M-55)													
P28-14M-55-SK	28	4.912	4.808	5.56	E1F	1½	2¾	. . .	1⅞	⅝	½	2⅝	7.0
P29-14M-55-SK	29	5.088	4.983	5.56	E1F	1½	2¾	. . .	1⅞	⅝	½	2⅝	8.0
P30-14M-55-SK	30	5.263	5.157	6.13	A1F	⅛	2¾	¾	1⅞	¾	½	2⅝	7.5
P32-14M-55-SK	32	5.614	5.507	6.13	A1F	⅛	2¾	¾	1⅞	¾	½	2⅝	9.0
P34-14M-55-SK	34	5.965	5.858	6.50	A1F	⅛	2¾	¾	1⅞	¾	½	2⅝	10.0
P36-14M-55-SF	36	6.316	6.208	6.81	A1F	⅛	2¾	¾	2	⅝	½	2 15/16	11.0
P38-14M-55-SF	38	6.667	6.559	7.16	A1F	⅛	2¾	¾	2	⅝	½	2 15/16	13.0
P40-14M-55-SF	40	7.018	6.909	7.50	A1F	⅛	2¾	¾	2	⅝	½	2 15/16	15.0
P44-14M-55-E	44	7.720	7.610	8.22	D1F	5/16	2¾	9/16	2⅝	7/16	⅞	3½	19.0
P48-14M-55-E	48	8.421	8.311	8.94	D1F	5/16	2¾	9/16	2⅝	7/16	⅞	3½	23.0
P56-14M-55-E	56	9.825	9.715	10.38	D1F	5/16	2¾	9/16	2⅝	7/16	⅞	3½	32.0
P64-14M-55-F	64	11.229	11.119	11.75	C1F	⅞	2¾	⅛	3⅝	0	1	4	53.0
P72-14M-55-F	72	12.632	12.522	13.19	C2F	⅞	2¾	⅛	3⅝	0	1	4	49.0
P80-14M-55-F	80	14.036	13.926	14.63	C2F	⅞	2¾	⅛	3⅝	0	1	4	54.0
P90-14M-55-F	90	15.790	15.680	. . .	C3	⅞	2¾	⅛	3⅝	0	1	4	55.0
P112-14M-55-F	112	19.650	19.540	. . .	C3	⅞	2¾	⅛	3⅝	0	1	4	71.0
P144-14M-55-F	144	25.264	25.154	. . .	C3	⅞	2¾	⅛	3⅝	0	1	4	106.0
P168-14M-55-F	168	29.475	29.365	. . .	C3	⅞	2¾	⅛	3⅝	0	1	4	124.0
P192-14M-55-F	192	33.686	33.576	. . .	C3	⅞	2¾	⅛	3⅝	0	1	4	146.0

□ Weight does not include bushing.

SPROCKET NUMBER	NO. OF GRVS	DIAMETERS			TYPE	DIMENSIONS (INCHES)					BORE RANGE		APP. □ WT. (Lbs.)
		P.D.	O.D.	Flange		E	F	K	L	M	Min.	Max.	
FOR BELTS 85mm (3.35 in.) WIDE • 14mm PITCH (14M-85)													
P28-14M-85-MPB	28	4.912	4.808	5.56	6F	1	4	. . .	4	1	* 1¼	2¹¹⁄₁₆	17
P29-14M-85-MPB	29	5.088	4.983	5.56	6F	1	4	. . .	4	1	* 1¼	2¹¹⁄₁₆	18
P30-14M-85-SK	30	5.263	5.157	6.13	A1F	¾	4	1⅜	1⅞	1⅜	½	2⅝	10
P32-14M-85-SK	32	5.614	5.507	6.13	A1F	¾	4	1⅜	1⅞	1⅜	½	2⅝	13
P34-14M-85-SK	34	5.965	5.858	6.50	A1F	¾	4	1⅜	1⅞	1⅜	½	2⅝	14
P36-14M-85-SF	36	6.316	6.208	6.81	A1F	¾	4	1⅜	2	1¼	½	2¹⁵⁄₁₆	15
P38-14M-85-SF	38	6.667	6.559	7.16	A1F	¾	4	1⅜	2	1¼	½	2¹⁵⁄₁₆	18
P40-14M-85-SF	40	7.018	6.909	7.50	A1F	¾	4	1⅜	2	1¼	½	2¹⁵⁄₁₆	20
P44-14M-85-E	44	7.720	7.610	8.22	A1F	⁵⁄₁₆	4	1³⁄₁₆	2⅝	1¹⁄₁₆	⅞	3½	25
P48-14M-85-E	48	8.421	8.311	8.94	A1F	⁵⁄₁₆	4	1³⁄₁₆	2⅝	1¹⁄₁₆	⅞	3½	29
P56-14M-85-F	56	9.825	9.715	10.38	D1F	¼	4	¾	3⅝	⅝	⅞	3½	46
P64-14M-85-F	64	11.229	11.119	11.75	D1F	¼	4	¾	3⅝	⅝	1	4	62
P72-14M-85-F	72	12.632	12.522	13.19	D2F	¼	4	¾	3⅝	⅝	1	4	60
P80-14M-85-F	80	14.036	13.926	14.63	D2F	¼	4	¾	3⅝	⅝	1	4	66
P90-14M-85-F	90	15.790	15.680	. . .	D3	¼	4	¾	3⅝	⅝	1	4	69
P112-14M-85-F	112	19.650	19.540	. . .	D3	¼	4	¾	3⅝	⅝	1	4	89
P144-14M-85-F	144	25.264	25.154	. . .	D3	¼	4	¾	3⅝	⅝	1	4	127
P168-14M-85-F	168	29.475	29.365	. . .	D3	¼	4	¾	3⅝	⅝	1	4	148
P192-14M-85-F	192	33.686	33.576	. . .	D3	¼	4	¾	3⅝	⅝	1	4	177
P216-14M-85-F	216	37.896	37.786	. . .	D3	¼	4	¾	3⅝	⅝	1	4	251

TABLE 8 (Continued)

SPROCKET NUMBER	NO. OF GRVS	DIAMETERS			TYPE	DIMENSIONS (INCHES)					BORE RANGE		APP. □ WT. (Lbs.)
		P.D.	O.D.	Flange		E	F	K	L	M	Min.	Max.	
FOR BELTS 115mm (4.53 in.) WIDE • 14mm PITCH (14M-115)													
P32-14M-115SK	32	5.614	5.507	6.13	A1F	1 3/8	5 1/4	2	1 7/8	2	1/2	2 5/8	16
P34-14M-115-SK	34	5.965	5.858	6.50	A1F	1 3/8	5 1/4	2	1 7/8	2	1/2	2 5/8	17
P36-14M-115-SF	36	6.316	6.208	6.81	A1F	1 3/8	5 1/4	2	2	1 7/8	1/2	2 15/16	18
P38-14M-115-SF	38	6.667	6.559	7.16	A1F	1 3/8	5 1/4	2	2	1 7/8	1/2	2 15/16	22
P40-14M-115-SF	40	7.018	6.909	7.50	A1F	1 3/8	5 1/4	2	2	1 7/8	1/2	2 15/16	25
P44-14M-115-E	44	7.720	7.610	8.22	A1F	15/16	5 1/4	1 13/16	2 5/8	1 11/16	7/8	3 1/2	30
P48-14M-115-E	48	8.421	8.311	8.94	A1F	15/16	5 1/4	1 13/16	2 5/8	1 11/16	7/8	3 1/2	35
P56-14M-115F	56	9.825	9.715	10.38	A1F	3/8	5 1/4	1 3/8	3 5/8	1 1/4	1	4	53
P64-14M-115-J	64	11.229	11.119	11.75	D1F	3/16	5 1/4	1	4 1/2	15/16	1 7/16	4 1/2	76
P72-14M-115-J	72	12.632	12.522	13.19	D1F	3/16	5 1/4	1	4 1/2	15/16	1 7/16	4 1/2	99
P80-14M-115-J	80	14.036	13.926	14.63	D2F	3/16	5 1/4	1	4 1/2	15/16	1 7/16	4 1/2	87
P90-14M-115-J	90	15.790	15.680	. . .	D2	3/16	5 1/4	1	4 1/2	15/16	1 7/16	4 1/2	95
P112-14M-115-J	112	19.650	19.540	. . .	D3	3/16	5 1/4	1	4 1/2	15/16	1 7/16	4 1/2	114
P144-14M-115-J	144	25.264	25.154	. . .	D3	3/16	5 1/4	1	4 1/2	15/16	1 7/16	4 1/2	166
P168-14M-115-J	168	29.475	29.365	. . .	D3	3/16	5 1/4	1	4 1/2	15/16	1 7/16	4 1/2	198
P192-14M-115-J	192	33.686	33.576	. . .	D3	3/16	5 1/4	1	4 1/2	15/16	1 7/16	4 1/2	232
P216-14M-115-J	216	37.896	37.786	. . .	D3	3/16	5 1/4	1	4 1/2	15/16	1 7/16	4 1/2	307

SPROCKET NUMBER	NO. OF GRVS	DIAMETERS			TYPE	DIMENSIONS (INCHES)					BORE RANGE		APP. □ WT. (Lbs.)
		P.D.	O.D.	Flange		E	F	K	L	M	Min.	Max.	
FOR BELTS 170mm (6.69 in.) WIDE • 14mm PITCH (14M-170)													
P38-14M-170-MPB	38	6.667	6.559	7.16	6F	2 7/16	7 3/8	3 1/16	2	2 15/16	* 1/2	2 15/16	28
P40-14M-170-MPB	40	7.018	6.909	7.50	6F	2 7/16	7 3/8	3 1/16	2	2 15/16	* 1/2	2 15/16	33
P44-14M-170-E	44	7.720	7.610	8.22	A1F	2	7 3/8	2 7/8	2 5/8	2 3/4	7/8	3 1/2	38
P48-14M-170-E	48	8.421	8.311	8.94	A1F	2	7 3/8	2 7/8	2 5/8	2 3/4	7/8	3 1/2	45
P56-14M-170-F	56	9.825	9.715	10.38	A1F	1 7/16	7 3/8	2 7/16	3 5/8	2 5/16	1	4	65
P64-14M-170-J	64	11.229	11.119	11.75	A1F	15/16	7 3/8	2 1/8	4 1/2	1 15/16	1 7/16	4 1/2	91
P72-14M-170-J	72	12.632	12.522	13.19	A1F	15/16	7 3/8	2 1/8	4 1/2	1 15/16	1 7/16	4 1/2	115
P80-14M-170-J	80	14.036	13.926	14.63	A2F	15/16	7 3/8	2 1/8	4 1/2	1 15/16	1 7/16	4 1/2	107
P90-14M-170-J	90	15.790	15.680	. . .	A2	15/16	7 3/8	2 1/8	4 1/2	1 15/16	1 7/16	4 1/2	116
P112-14M-170-M	112	19.650	19.540	. . .	A3	0	7 3/8	1 7/16	6 3/4	5/8	2	5 1/2	175
P144-14M-170-M	144	25.264	25.154	. . .	A3	0	7 3/8	1 7/16	6 3/4	5/8	2	5 1/2	240
P168-14M-170-M	168	29.475	29.365	. . .	A3	0	7 3/8	1 7/16	6 3/4	5/8	2	5 1/2	278
P192-14M-170-M	192	33.686	33.576	. . .	A3	0	7 3/8	1 7/16	6 3/4	5/8	2	5 1/2	322
P216-14M-170-M	216	37.896	37.786	. . .	A3	0	7 3/8	1 7/16	6 3/4	5/8	2	5 1/2	399

□ Weight shown is for pulley without bushing. * MPB furnished less keyseat or setscrews.

Source: Courtesy of The Gates Rubber Company.

sections is the pitch line differential (PLD), shown as value "a" in Figure 9. As such, the automotive and industrial belts are not interchangeable.

B. ZB and RH

Industrie Pirelli, spa, originally developed this belt in the RH-section. It is basically an H-section industrial timing belt. The ZB is the same tooth configuration but its "a" value differs from the RH belt as shown in Figure 9. The "a" value difference between ZB and RH profiles does not permit interchangeability.

C. Automotive Super-Torque Positive Drive

The 9.525-mm pitch was added to be equivalent in length and pitch to the STA profile.

D. LR

The LR developed by Uniroyal is an 8-mm pitch industrial HTD belt with the teeth stretched to a 9.525-mm pitch. The LR was derived from the L-pitch (0.375 in.) and the R for the round tooth design.

IV. SYNCHRONOUS-BELT CONSTRUCTIONS

A. Basic Parts

Figure 10 shows the basic component parts of a synchronous belt. All four component parts are integrally molded with specially compounded adhesion gums to ensure that all parts work together as a unit.

1. Tooth and Body

The teeth engage the grooves on the pulley or sprocket. They are precisely spaced so that their root lies substantially at the pitch line. As such, their spacing is not changed by the forces imposed on them. The shear strength of the teeth is greater than the strength of the tensile member when at least six belt teeth are in mesh with the pulley or sprocket grooves.

As previously pointed out, the tooth configuration has great influence on how the load is distributed to the tensile member. The trapezoidal-shaped timing-belt profile has a fairly high stress concentration near the root of the tooth. The newer curvilinear profiles have a more even and uniform strain distribution across the belt tooth thickness. Since the curvilinear design improves stress distribution, the overall loading on many drives can be higher in comparison to the conventional timing-belt profiles.

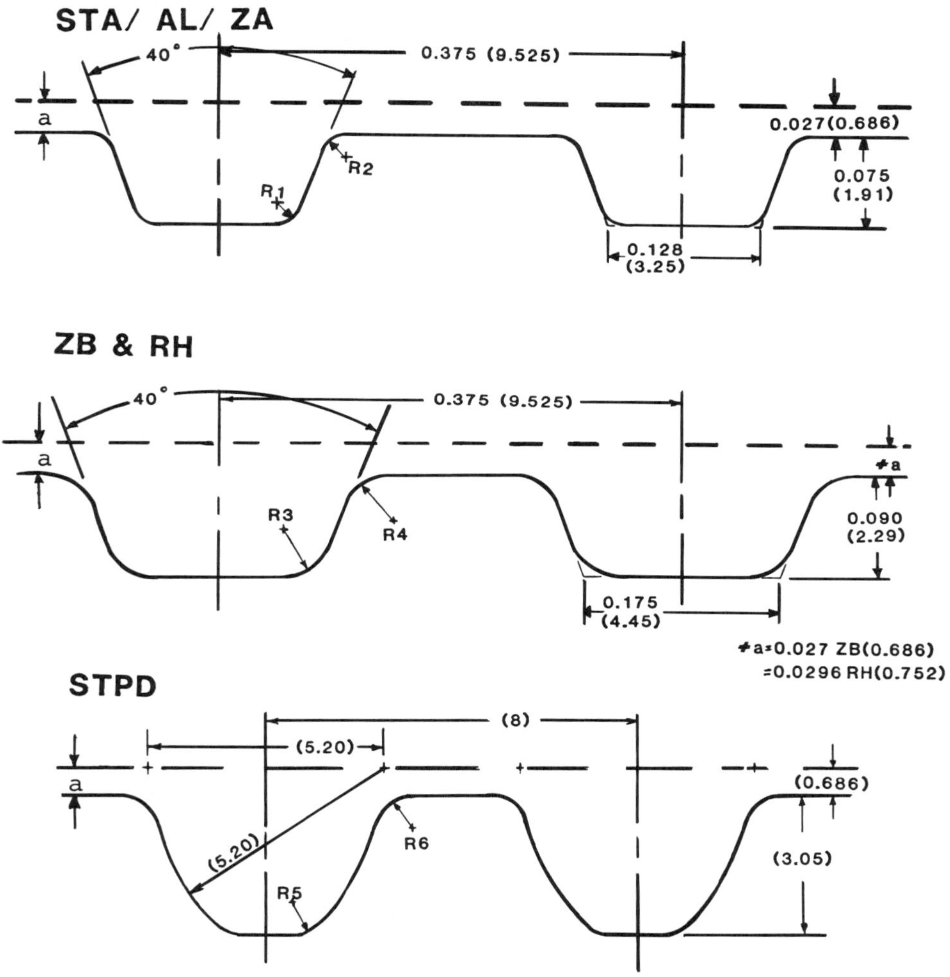

FIGURE 9 Automotive synchronous belt-tooth profiles.

Most synchronous belts are made of a premium oil- and heat-resistant elastomer stock. Special constructions are available to provide high oil resistance, static resistance, static conductance, and nonmarking characteristics. A belt manufacturer should be consulted for further details.

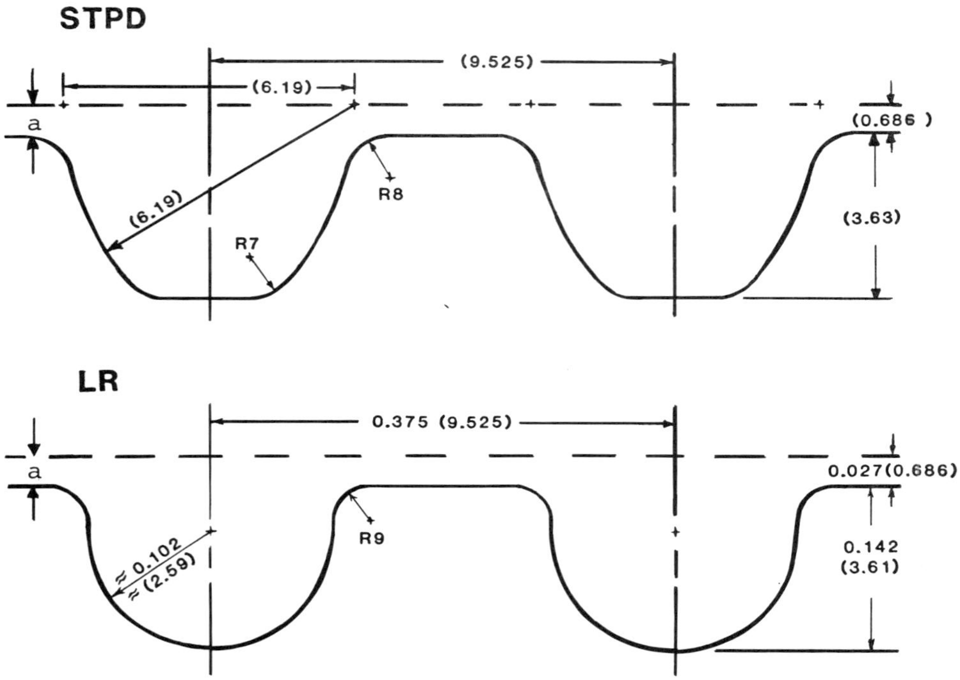

NOTE
Dimensions in ( ) are millimeters
Dashed lines indicate belt pitch line location

A more recent development made by The Gates Rubber Company in synchronous-belt performance is called Poly Chain belt. It is a special combination of polyurethane, DuPont Kevlar tensile member, and nylon fabric tooth facing. The material improvements made permit a much higher overall belt-loading in comparison to any other synchronous belt. Specific details about Poly Chain belts may be obtained by contacting The Gates Rubber Company.

2. Fabric Tooth Facing

Most synchronous belts have a wear-resistant fabric for a tooth facing material. This is usually a high wear-resistant material with a low coefficient of friction such as a nylon duck. It covers and protects the tooth surfaces and usually outlasts the other components of belt.

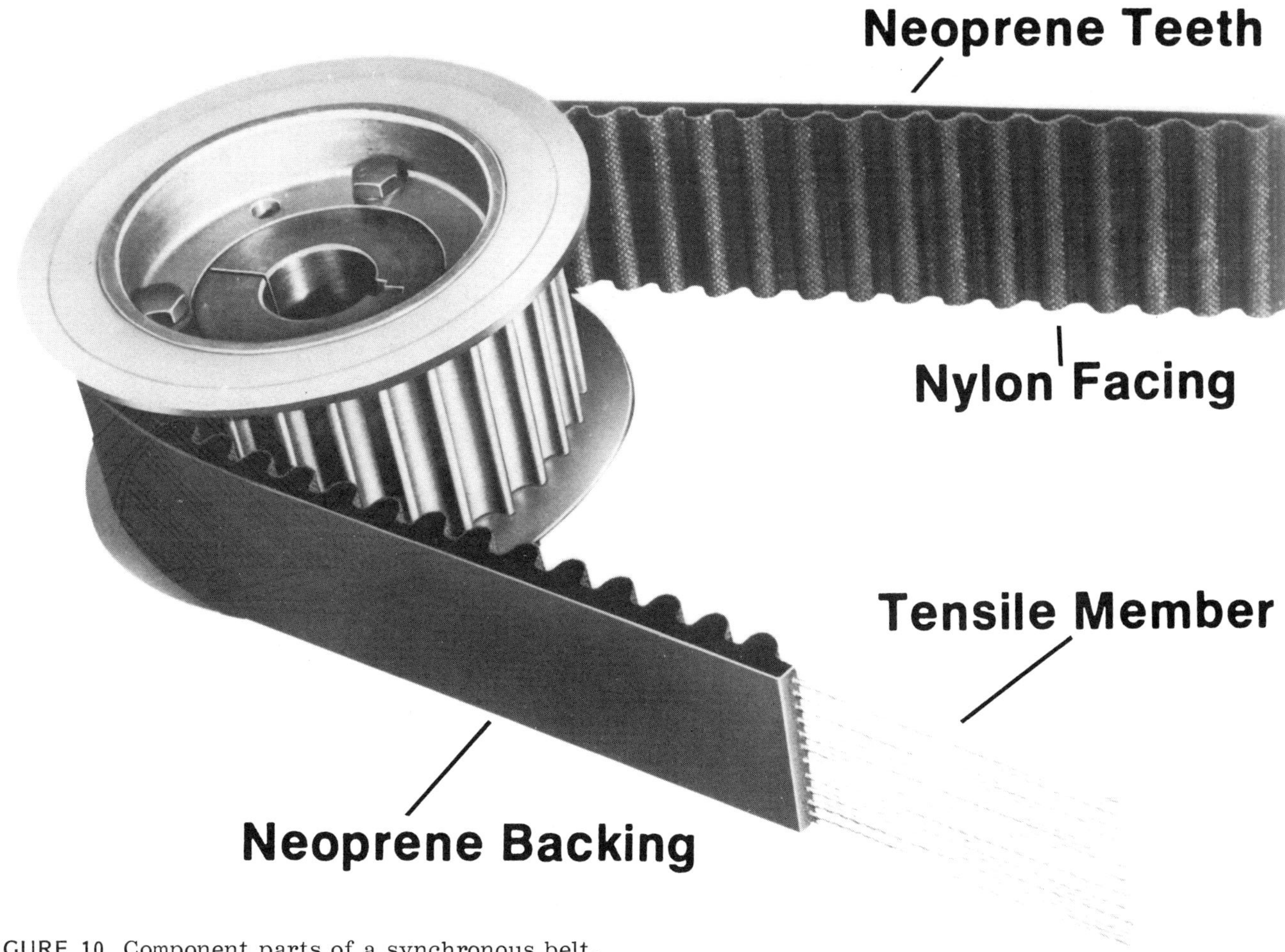

FIGURE 10 Component parts of a synchronous belt.

TABLE 9 8-mm Pitch STPD Belt Lengths and Widths

Pitch Lengths		No.	Widths*	
in mm	in Inches	of Teeth	mm	Inches*
560	22.047	70	15	0.59
632	24.882	79	25	0.98
712	28.031	89	40	1.57
800	31.496	100	60	2.36
896	35.276	112		
1000	39.370	125		
1120	44.094	140		
1248	49.134	156		
1400	55.118	175		
1600	62.992	200		
1800	70.866	255		
2000	78.740	250		
2240	88.189	280		
2496	98.268	312		
2800	110.236	350		

*Other widths available on request.

Source: Courtesy of the Goodyear Tire and Rubber Company.

3. Backing or Overcord

The overcord is bonded to the tensile member and provides protection against grime, oil, and moisture. It also protects from frictional wear if a backside idler is used or if power is transmitted from the backside of the belt. Special overcords can be made for specialty belts such as a thick backing. Backings may also be specially compounded for applications where nonmarking, such as on printing equipment is critical. Low durometer or soft gum backings are also available for use on certain applications.

It is also possible to incorporate special features in the belt backing or overcord such as lugs which are used in conveying small parts. Another example of a special feature in present use is a timing belt with micro-V-ribs ground in the overcord. It is used on a serpentine front-engine automotive application.

In some cases, when a special belt thickness tolerance is required, the belt overcord is ground. Presently there is no industry standard establishing thickness tolerance for various classes of ground back belts. However, special classes of grinds are in wide use even though there is no formal standard. An example of grind class thickness tolerances is shown in Table 10.

4. Tensile Member

The tensile cord is the muscle of the belt. Fiberglass is the standard and most commonly used tensile cord in synchronous belts. Its high-

TABLE 10 Grind Class Thickness Tolerances

Belt pitch	Thickness	Class 1	Class 2	Class 3
0.080 (MXL)	0.045	±0.005	±0.010	N/A
0.200 (XL)	0.090	±0.005	±0.010	±0.025[a]
0.375 (L) IND	0.140	±0.005	±0.010	±0.025[a]
0.375 (L) Auto	0.160	±0.005	±0.010	±0.015[a]
0.500 (H)	0.170	±0.005	±0.010	±0.025[a]
0.875 (XH)	0.440	N/A	±0.010	±0.025[b]
1.250 (XXH)	0.620	N/A	±0.010	±0.025[b]

[a]Standard manufacturing tolerance but not ground.
[b]All XH and XXH belts are ground to ±0.025 in standard product.
Note: The above-noted grind classes also apply to other tooth designs including LR, STPD, RH, STA plus the many nonstandard pitch trapezoidal constructions.
Source: Courtesy of The Gates Rubber Company.

modulus, low-stretch characteristics make it excellent for a wide range of general applications.

Care must be exercised in handling, storing, and installing synchronous belts with fiberglass tensile cords. Tightly crimping or bending the belt may cause invisible damage to the cord, substantially reducing its useful service life. Since fiberglass tensile cords have very little or no compression strength, it may not be suitable for applications where high-shock or reversing loads are likely.

Alternate cord materials are DuPont Kevlar, steel, and polyester. Kevlar and steel generally have slightly higher ultimate strengths than fiberglass. However, this does not directly translate into increased belt capacity. A belt's overall performance is optimized by a balance of factors such as fatigue life, environmental stability, strength, and the characteristics such as shock loading and ambient temperature of the application. There is not a direct relationship between any two of these factors. So, a stronger belt will not necessarily have a greater fatigue life.

Kevlar cord is an excellent replacement for fiberglass or steel cord on some problem drives. Kevlar has excellent shock resistance compared to other high-tensile modulus cords used in timing belts. It is replacing steel, particularly where low weight, high fatigue life, corrosion resistance, and flexibility are important drive factors.

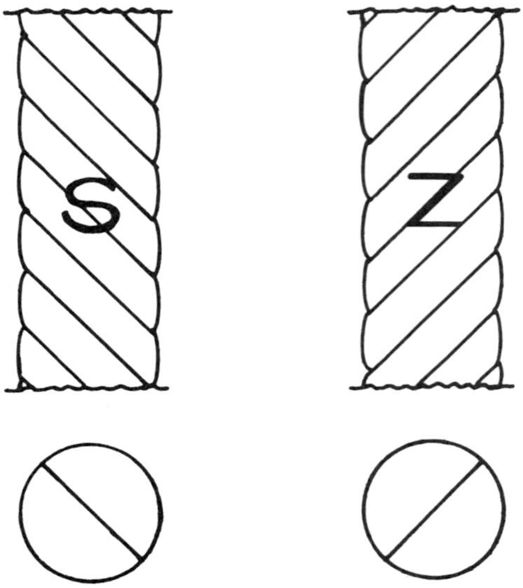

FIGURE 11 Tensile cord twist.

Polyester cord is used as an alternate tensile member only in some mini-pitch synchronous belts. These belts are used on drives that provide motion transfer or on which loads are extremely light. Producers claim the belts will operate smoothly over small-diameter sprockets or pulleys, and that the more elastic polyester cord will absorb shock and dampen vibration.

Tensile cords are typically made up of several filaments that are twisted around each other similar to a wound rope. The cord or cords are then helically wound at precise intervals on a mold mandrel as the belt is built. The helix of the cord-lay on the mandrel and the cord-twist both influence the belt tracking. Tracking is a term used to describe the side force or thrust exerted by a belt in the direction of the motor shaft axis. Figure 11 shows cords designated by the direction of the spiral as "S" and "Z" twist.

Standard synchronous belts are built with both S and Z cords wound simultaneously. On drives where belt tracking is a problem, special belts made with S or Z cord only may be appropriate. A belt produced with an S cord will track in an opposite direction to a belt with a Z cord.

Technically, a belt made with S and Z cords will behave as a belt made with only an S or Z cord due to the helix or cord-lay. Of course, this theoretical behavior is dependent on the cord being laid right to left or vice versa. However, in actual practice, the helix of the wind on the mandrel is so slight that its influence on belt tracking is not measurable.

V. OPERATING CHARACTERISTICS

A. Operating Tensions

1. Bearing Loads

As previously mentioned, synchronous-belt drives differ from other belt drives in that friction is not involved in the transmission of power. Belt slip is prevented by the mechanical locking of the teeth in the pulley grooves. However, laboratory testing has shown that synchronous belts operate at tension ratios similar to V-belts. (For additional information on belt tension, see Chapter 3.)

A common misconception is that a synchronous belt drive will have lower bearing loads in comparison to a V-belt drive. Since operating tensions are very similar for both synchronous- and V-belt drives, the subsequent bearing loads are the same for a continuous load condition. (For further discussion on bearing loads, see Chapter 11.)

On drive applications where the load requirements fluctuate, belt tensions for a V-belt will not always be equal to the tension in the synchronous belt. In this case, synchronous belts will have a lower average bearing load. This is due to the self-generated tension characteristic of synchronous belt drives.

2. Self-Generated Tension

The illustration in Figure 12 shows how the pulley teeth act as an inclined plane with the belt exerting a normal force "a" to the pulley groove. Most of the normal force is used to transmit the power of the required load. However, a small portion causes the belt tooth to slide up, moving away from the pulley. This is known as the *ramp effect* and the source of the self-generated tension.

As the load is increased, belt tension is increased because the belt moves further out of the pulley groove. Therefore, a synchronous belt generates tension in proportion to the load being transmitted. This concept is also true for the HTD and STPD tooth profiles, but the ramp effect is somewhat reduced due to the more perpendicular tooth configuration of the curvilinear profiles.

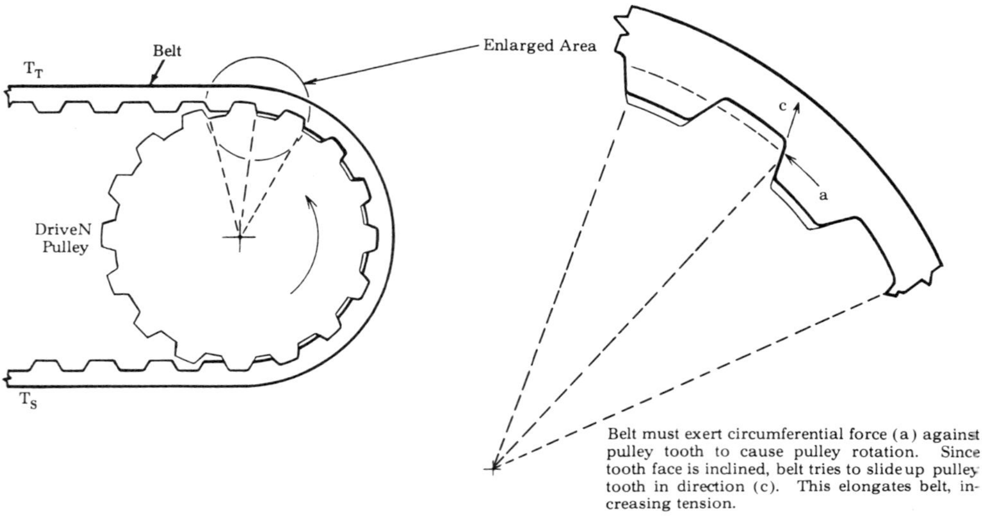

FIGURE 12 Self-generated tension.

3. Static Tension

Static or installation tension is generally lower for synchronous belts than for V-belts. A V-belt's performance and operating tension depends on the static tension imposed for a particular application. However, synchronous belts, due to their self-generated tension characteristic, do not require a specific static tension in order to operate at 5:1 tension ratio.

It is very important to note that although synchronous belts do have some degree of self-tensioning, performance problems will result if the belt is improperly tensioned on installation. For example, an undertensioned drive may be noisy due to the belt teeth not mating properly with the pulley or sprocket grooves. Premature belt wear may also be a result of under- or overtensioning the belt on installation.

Belt slip on a synchronous-belt drive, commonly referred to as ratcheting, may also occur on an undertensioned belt. When ratcheting occurs, damage to the belt may not be visible, but it is likely that the belt's service life will be dramatically reduced.

Belt ratcheting, allowed by undertensioning, may also cause damage to shafts, bearing, and support hardware. The high forces generated during belt ratcheting will be transmitted directly to these components.

FIGURE 13 Undertensioned drive.

As discussed above, one effect of an undertensioned synchronous-belt drive is wear. This is due mainly to the belt pitch not properly matching the sprocket pitch. Figure 13 illustrates a running, undertensioned drive. It clearly shows the mismatch of the belt pitch to the sprocket pitch. In fact, on the slack side of the drive, the photo shows the tip of the belt tooth riding on the tip (OD) of the sprocket tooth.

In addition to extreme belt-tooth wear on an undertensioned drive, tooth cracking may also be a problem. This is due to increased belt tooth loading. Figure 14 illustrates tooth engagement for both a proper and an undertensioned drive. In a properly tensioned drive, the load is transmitted along the belt-tooth apex near the tensile cord line. By summing the moments about point "0", it becomes apparent that tooth loading is greatly increased in an undertensioned drive. In effect, the load-carrying capacity of the belt is reduced, shortening the service life.

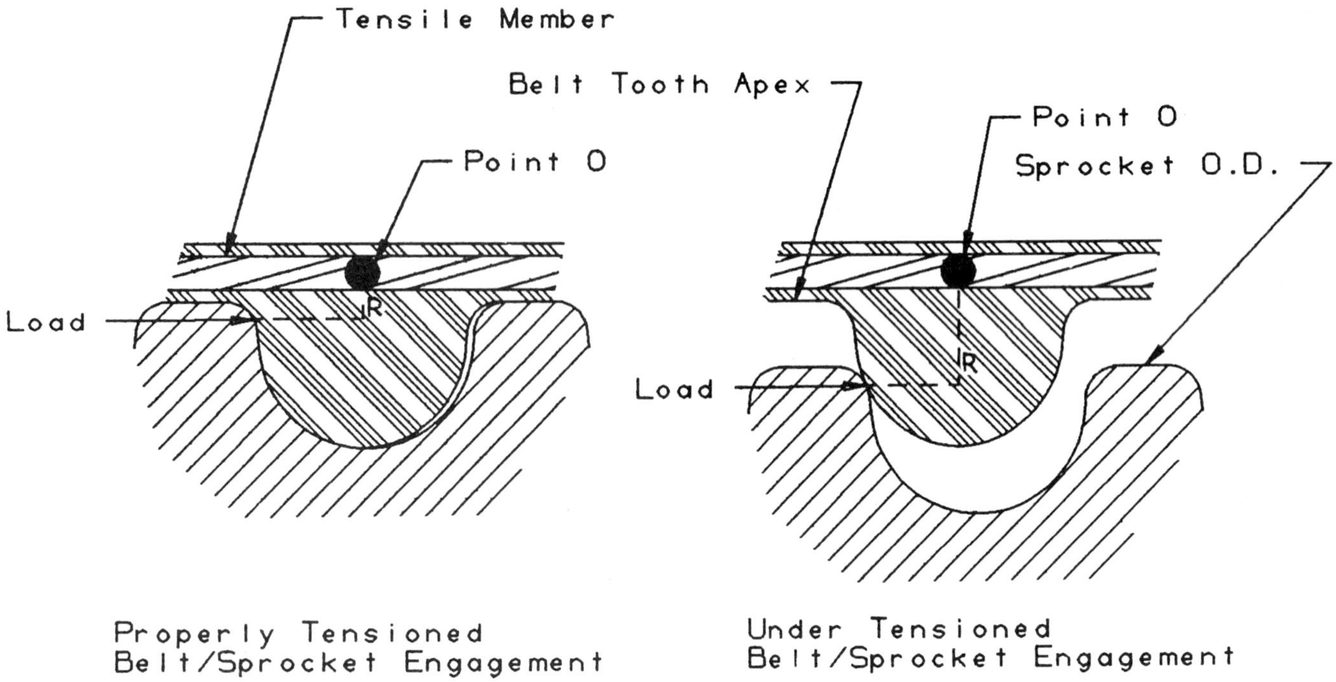

FIGURE 14 Belt-tooth loading.

4. Spring-Loaded Idlers

In most cases, the use of a spring-loaded idler is not recommended on synchronous-belt drives unless locked in place prior to running the drive. This is due to the self-generated tension principle previously discussed.

A synchronous belt will normally generate enough tension to overcome a spring-loaded idler. The belt will ratchet because the effectiveness of the idler has been eliminated.

Any spring force sufficient to prevent belt tension from overcoming the spring-loaded idler will be excessive and will significantly reduce belt life. A spring-loaded idler may be used on a drive that is used for motion transfer only.

5. Fixed-Center Drives

It is not advisable to apply a synchronous-belt drive where two shafts are fixed (not movable) unless an idler is used to properly tension the belt. An exception is drives where very little or no power is being transmitted.

Fixed centers imply exact tolerances on pulleys, belts, and the center distance of the equipment. Although length tolerances for synchronous belts are less than for other belts, no belt can be manufactured to exact specifications. So, the combination of tolerances for belt length, pulley diameters, and drive geometry are such that drive performance is reduced. Some applications do use fixed centers, but not without some difficulties.

B. Teeth-In-Mesh

Horsepower and torque ratings for synchronous belts are based on a minimum of six teeth-in-mesh between the belt and pulley or sprocket. The ratings must be corrected for excessive belt-tooth loading if there are less than six teeth-in-mesh.

The number of teeth-in-mesh may be calculated by using this formula:

$$\text{teeth-in-mesh} = \left[0.5 - \left(\frac{D-d}{6C}\right)\right] Ng$$

where

D = pitch diameter, large pulley (in.)
d = pitch diameter, small pulley (in.)
C = center distance between shafts (in.)
Ng = number of grooves in small pulley

TABLE 11 Teeth-In-Mesh Factor

Teeth-in-mesh	Factor K_{tm}
6 or more	1.00
5	0.80
4	0.60
3	0.40
2	0.20

Source: Courtesy of The Gates Rubber Company.

In cases where fewer than six teeth are in full contact, 20% of the HP or torque rating must be subtracted for each tooth less than six not in full contact. After computing the teeth-in-mesh, the belt rating should be multiplied by the appropriate K_{tm} factors shown in Table 11.

C. Drive Installation and Alignment

1. Drive Installation

When installing a synchronous belt, one should never force it over a flange or pulley. This will cause unseen damage to the belt's tensile member. The drive center distance or idler position should be changed so that the belt can be easily slipped on the drive. Table 12 shows typical installation allowances for drives using both unflanged and flanged pulley or sprockets.

Table 12 also shows the tensioning (take-up) allowance required over the belt life. The very small allowances shown reflect the high resistance to elongation of synchronous belts.

2. Alignment

Synchronous belts are sensitive to misalignment and should not be used on drives where misalignment is inherent to the drive operation. Misalignment leads to inconsistent belt wear. Premature tensile failure may result from misalignment due to uneven load distribution across the belt width. Tracking problems may also result from drive misalignment.

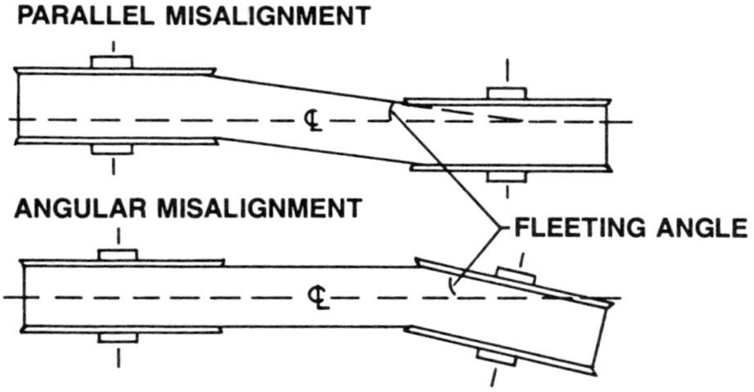

FIGURE 15 Two types of drive misalignment.

The two types of misalignment, parallel and angular, are shown in Figure 15.

Parallel misalignment occurs when the driveR and driveN shaft are parallel, but the two pulleys or sprockets lie in different planes. When the two shafts are not parallel, the drive is angularly misaligned.

A fleeting angle is the angle at which the belt enters and exits the pulley or sprocket. It equals the sum of the parallel and angular misalignments.

Any degree of pulley or sprocket misalignment will result in some reduction of belt life which is not accounted for in the normal design procedure. Misalignment of all synchronous belt drives should be less than 1/4°.

D. Pulley or Sprocket Guide Flange

Most stock pulleys or sprockets are supplied such that the standard practices are followed. For drives where special pulleys or sprockets must be made, the flanges should be produced in conjunction with RMA standard IP-24 for synchronous belts. Figure 16 shows the RMA recommended flange dimensions for timing-belt pulleys.

The following conditions should be considered when specifying guide flanges for synchronous-belt drives:

1. On all two-point drives, the minimum flanging requirement are two flanges on one pulley or sprocket or one flange on each pulley or sprocket on opposite sides.
2. On drives where the center distance is more than eight times the diameter of the small pulley or sprocket, both pulleys or sprockets should be flanged on both sides.

TABLE 12 Synchronous-Belt Installation and Tension Allowances

(a) Timing belts. (Center distance allowance for installation and tensioning.)

Belt length (in.)	Standard Installation allowance (flanged pulleys removed for installation (in.)	Tensioning allowance (any drive) (in.)
3.6 to 5.0	0.02	0.02
Over 5.0 to 10.0	0.03	0.03
Over 10.0 to 20.0	0.04	0.03
Over 20.0 to 40.0	0.05	0.04
Over 40.0 to 60.0	0.07	0.05
Over 60.0 to 180.0	0.12	0.08

Additional center distance allowance for installation over flanged pulleys[a] (add to standard installation allowance above)

Pitch (in.)	One pulley flanged (in.)	Both pulleys flanged (in.)
2/25 (MXL)	.33	.49
1/5 (XL)	.46	.71
3/8 (L)	.64	.85
1/2 (H)	.64	.96
7/8 (XH)	1.14	1.92
1-1/4 (XXH)	1.53	2.65

(continued)

3. On vertical shaft drives, one pulley or sprocket should be flanged on both sides. All other pulleys or sprockets in the system should be flanged on the bottom only.
4. On drives with more than two pulleys or sprockets, the minimum flanging requirements are two flanges on every other pulley or sprocket or one flange on every pulley or sprocket, alternating sides around the system.

TABLE 12 (Continued)

(b) HTD belts. (Center distance allowance for installation and tensioning.)

Belt length (in.)	Standard installation allowance (flanged sprockets removed for installation) (in.)	Tensioning allowance (any drive) (in.)
20 and under	0.04	0.03
Over 20.0 to 40.0	0.05	0.03
Over 40.0 to 60.0	0.07	0.04
Over 60.0 to 90.0	0.09	0.05
Over 90.0 to 120.0	0.11	0.05
Over 120.0 to 160.0	0.14	0.05
Over 160.0 to 190.0	0.17	0.05
Over 190.0 to 260.0	0.21	0.05

Additional center distance allowance for installation over flanged sprockets[b] (add to standard installation allowance in above table)

Pitch (in.)	One sprocket flanged (in.)	Both sprockets flanged (in.)
3 mm	0.34	0.48
5 mm	0.53	0.75
8 mm	0.85	1.29
14 mm	1.40	2.29
20 mm	1.85	3.05

[a]For drives that require installation of the belt over one pulley at a time, use the value for both pulleys flanged – even if only one pulley is flanged.
[b]For drives that require installation of the belt over one sprocket at a time, use the value for both sprockets flanged – even if only one sprocket is flanged.
Source: Courtesy of The Gates Rubber Company.

Belt Section	Minimum Flange Thickness	Minimum Flange Height*
MXL (0.080)	0.023	0.020
XL (0.200)	0.029	0.040
L (0.375)	0.050	0.065
H (0.500)	0.050	0.080
XH (0.875)	0.098	0.190
XXH (1.250)	0.127	0.245

*Flange outside diameter equals pulley outside diameter plus twice flange height.

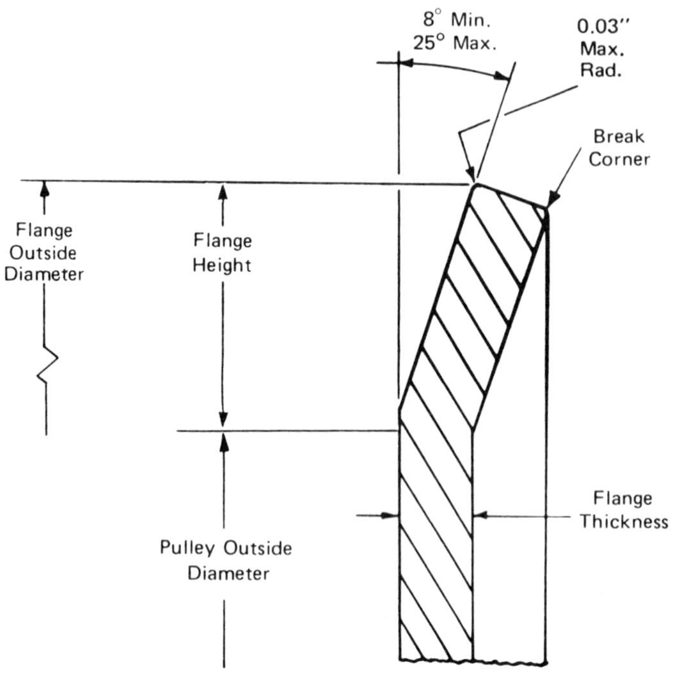

FIGURE 16 Guide flange dimensions for timing belt pulleys.

E. Idlers

Idlers on synchronous-belt drives may be used for take-up on fixed center drives. They may also be used to avoid obstructions in the drive area. A backside idler may be used to increase belt wrap (teeth-in-mesh) on the small pulley or sprocket of a high speed ratio drive.

Inside idlers should be sized such that they are no smaller than the smallest loaded pulley or sprocket in the drive. Backside idlers should follow the convention for V-belts, i.e., they should be at least 30% larger than the smallest loaded pulley diameter.

F. Debris

Synchronous belts should not be used in environments where excessive debris is present. Debris can be more damaging to the synchronous-belt drive than to the V-belt drive, which has a tendency to remove debris from the sheave grooves through drive operation.

Entrapment of debris in synchronous drives is a major concern. Debris can be packed into sprocket grooves causing improper belt-tooth engagement, reducing belt life. Care must be exercised to provide adequate shielding to drives in environments where debris is likely to be present.

Completely enclosing a synchronous-belt drive may be acceptable. Since synchronous belts generate little heat during drive operation, air circulation is not critical except when extremely high temperatures are already present.

VI. SYNCHRONOUS-BELT DRIVE SELECTION PROCEDURE

A. Data Required

In order to design a synchronous-belt drive, all of the following five items are needed:

1. DriveN horsepower requirement
2. Rpm of the driveR shaft
3. Rpm of the driveN shaft
4. Drive geometry (center distance; size restrictions that apply, if any)
5. Hours per day operation

B. Seven-Step Procedure

Design of synchronous-belt drives are usually done by the HP-rating method as shown in the following steps.

1. Find the Designed Horsepower

As in other belt drive design, the actual HP requirement is multiplied by a service factor to obtain design HP. It is important to note that synchronous-belt service factors are typically higher compared to other types of belt drives. Table 12 shows typical service factors to be used in determining design HP for timing and HTD drives.

2. Determine the Proper Belt Pitch

The belt pitch is obtained from a selection guide by using the design HP and rpm of the faster pulley or sprocket. Figures 17 and 18 are pitch selection guides used for determining the correct belt pitch.

3. Select the Pulley Combination, Belt Length, and Center Distance

Most manufacturers provide drive selection tables for this purpose. Table 14 is an example from a drive selection table.

4. Find the Corrected Horsepower Rating

A drive selection table is typically coded to correct for teeth-in-mesh as shown at the bottom of Table 15. The HP correction factor should be multiplied by the appropriate basic HP rating as determined from a rating table. Table 16 shows the horsepower rating for an L-pitch timing belt.

5. Calculate the Belt Width

The corrected HP rating and the design HP are used to determine the belt width factor. Divide the design HP by the corrected HP rating. This gives the belt width factor. The belt width factor as shown on the bottom of Table 16 can be used to determine the required belt width necessary for the drive. (Table 16, for a L-section, 3/8" pitch belt.)

6. Installation and Take-Up Allowances

Because of their high resistance to elongation, synchronous-belt drives sometimes need minimal take-up. However, some adjustment must be provided when installing synchronous belts, as with nearly all power transmission methods, because of manufacturing tolerances, wear of pressure surfaces, and tensioning requirements. Some typical allowances are shown in Table 12.

TABLE 13 Synchronous-Belt Service Factors

DriveN Machine	DriveR					
The driveN machines listed below are representative samples only. Select a driveN machine whose load characteristics most closely approximate those of the machine being considered.	**AC Motors:** Normal Torque, Squirrel Cage, Synchronous, Split Phase. **DC Motors:** Shunt Wound. **Engines:** Multiple Cylinder Internal Combustion.			**AC Motors:** High Torque, High Slip, Repulsion-Induction, Single Phase, Series Wound, Slip Ring. **DC Motors:** Series Wound, Compound Wound. **Engines:** Single Cylinder Internal Combustion. **Line shafts** **Clutches**		
	Intermittent Service	**Normal Service**	**Continuous Service**	**Intermittent Service**	**Normal Service**	**Continuous Service**
	3-5 Hours Daily or Seasonal	**8-10 Hours Daily**	**16-24 Hours Daily**	**3-5 Hours Daily or Seasonal**	**8-10 Hours Daily**	**16-24 Hours Daily**
Display Equipment Dispensing Equipment Instrumentation Projection Equipment Measuring Devices Medical Equipment	1.0	1.2	1.4	1.2	1.4	1.6
Appliances, Sweepers, Sewing Machines Office Equipment	1.2	1.4	1.6	1.4	1.6	1.8
Conveyors: Belt, Light Package Screens, Oven Screens, Drum, Conical Wood Lathes, Band Saws	1.4	1.5	1.7	1.5	1.7	1.9
Agitators for Liquids Dough Mixers Drill Press, Lathes, Screw Machines, Jointers Circular Saws, Planers Laundry Machinery Paper Machinery (except Pulpers) Printing Machinery	1.4	1.6	1.8	1.6	1.8	2.0
Agitators for semi-liquids Conveyor Belt: Ore, Coal, Sand Line Shafts Machine Tools: Grinder, Shaper, Boring Mill, Milling Machines Pumps: Centrifugal, Gear, Rotary Pipeline Screens: Vibrating Cam Type Textile: Warpers, Reels Compressor: Centrifugal	1.5	1.7	1.9	1.7	1.9	2.1
Brick Machinery (except Pug Mills) Conveyor: Apron, Pan, Bucket, Elevator Extractors, Washers Fans, Blowers: Centrifugal, Induced Draft Exhausters Generators & Exciters Hoists, Elevators Rubber Calender, Mills, Extruders Saw Mill Machinery Textile Machinery: Looms, Spinning Frames, Twisters	1.6	1.8	2.0	1.8	2.0	2.2
Centrifuges Conveyors: Flight, Screw Hammer Mills Paper Pulpers	1.7	1.9	2.1	1.9	2.1	2.3
Brick & Clay Pug Mills Fans, Blowers, Propellor Mine Fans, Positive Blowers	1.8	2.0	2.2	2.0	2.2	2.4
Reciprocating Compressors Mills: Ball, Rod, Pebble, etc. Pumps: Reciprocating	1.9	2.1	2.3	2.1	2.3	2.5

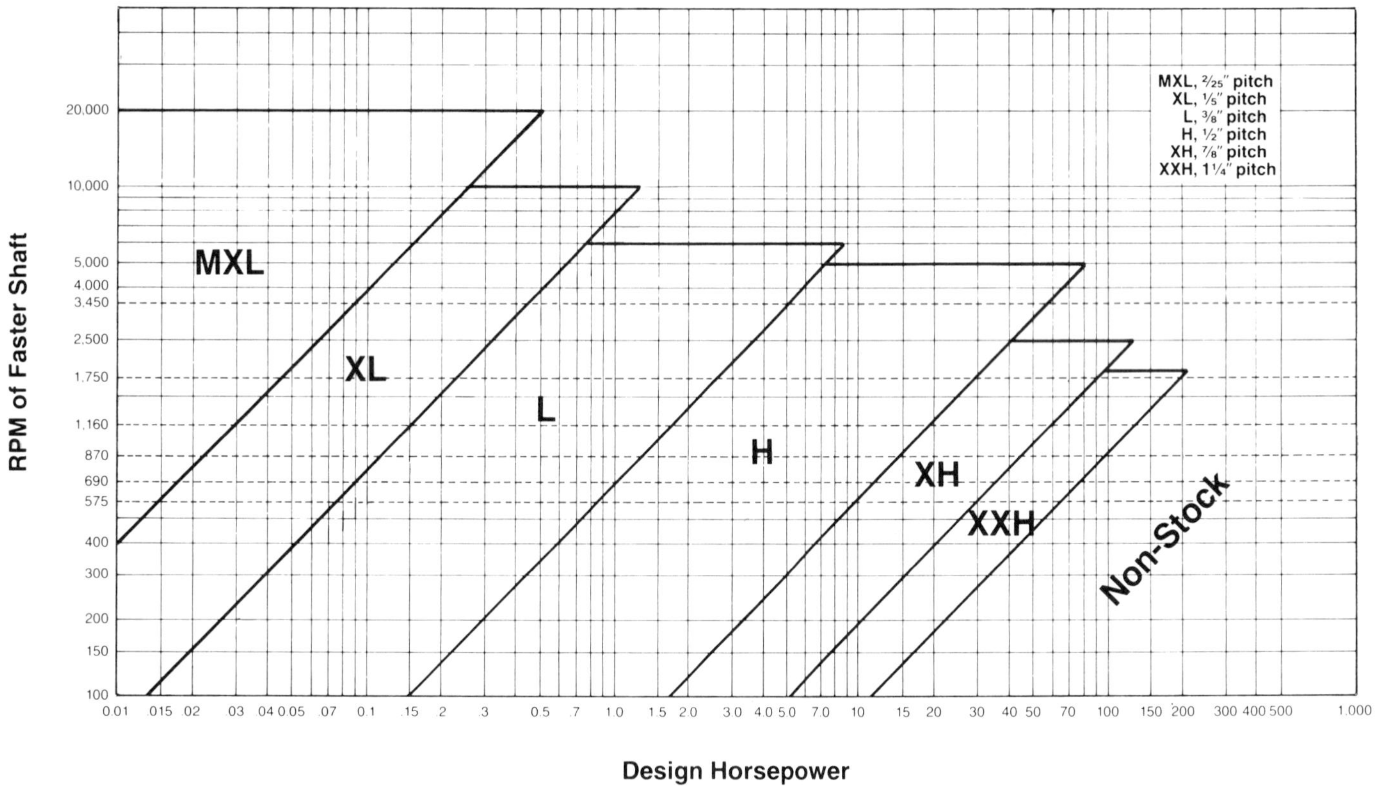

FIGURE 17 Timing belt pitch selection guide.

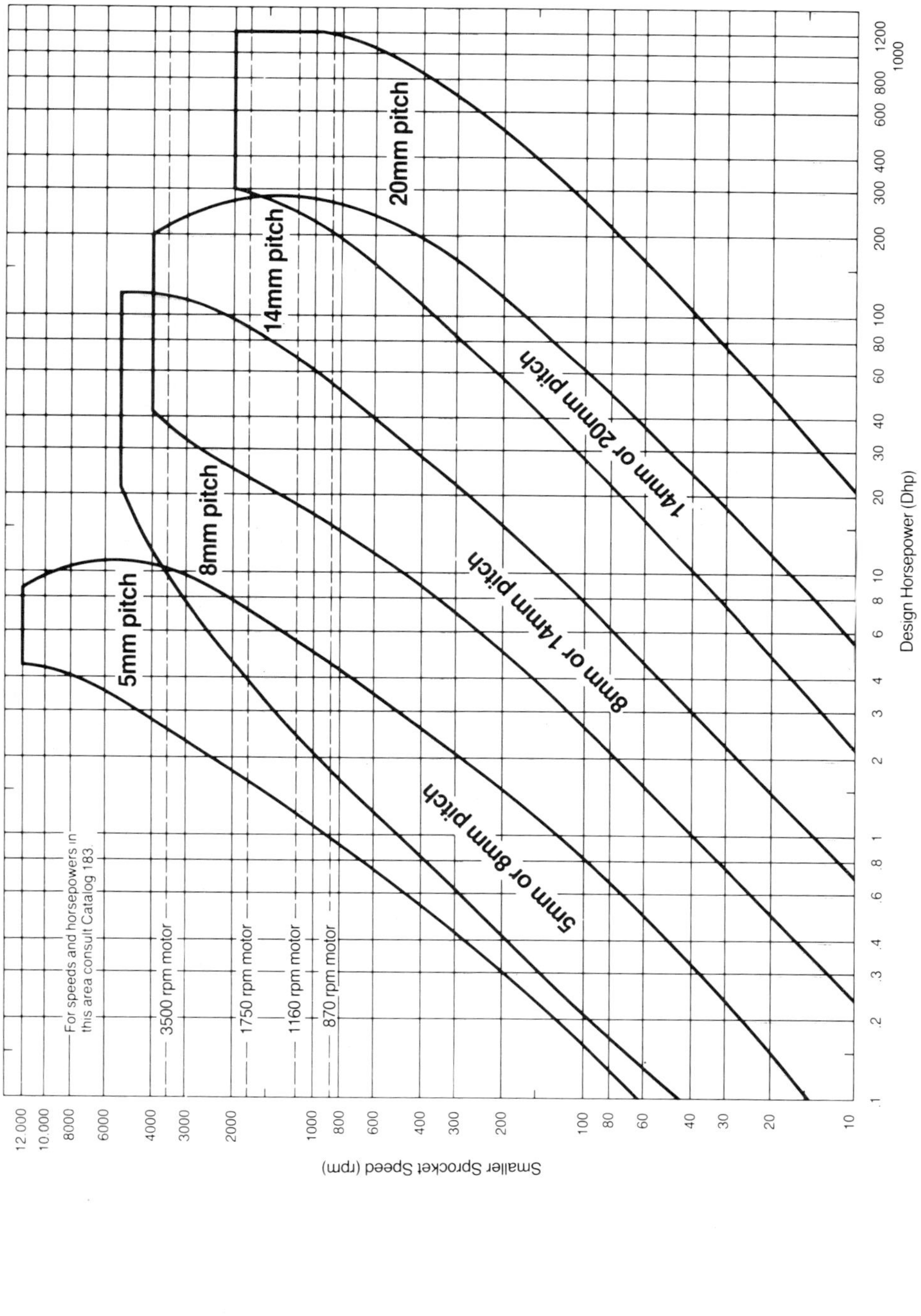

FIGURE 18 HTD belt pitch selection guide.

TABLE 14 Timing Belt Drive Selection Table (XXH, 1 1/4 in. pitch belts)

DriveN Speed			Pulley Combination					Center Distance, Inches								Horsepower Rating for 5″ wide belt		
For motor speed of			DriveR		DriveN											For motor speed of		
1750 rpm	1160 rpm	870 rpm	No. of grooves	Pitch diam., inches	No. of grooves	Pitch diam., inches	Speed Ratio	700 XXH P.L. 70.00 56 teeth	800 XXH P.L. 80.00 64 teeth	900 XXH P.L. 90.00 72 teeth	1000 XXH P.L. 100.00 80 teeth	1200 XXH P.L. 120.00 96 teeth	1400 XXH P.L. 140.00 112 teeth	1600 XXH P.L. 160.00 128 teeth	1800 XXH P.L. 180.00 144 teeth	1750 rpm	1160 rpm	870 rpm
1750	1160	870	40 XXH	15.915	40 XXH	15.915	1.00			20.00	25.00	35.00	45.00	55.00	65.00	148.78■	161.32	137.11
1750	1160	870	34 XXH	13.528	34 XXH	13.528	1.00		18.75	23.75	28.75	38.75	48.75	58.75	68.75	166.26	148.71	121.44
1750	1160	870	30 XXH	11.937	30 XXH	11.937	1.00	16.25	21.25	26.25	31.25	41.25	51.25	61.25	71.25	166.95	137.12	109.64
1750	1160	870	26 XXH	10.345	26 XXH	10.345	1.00	18.75	23.75	28.75	33.75	43.75	53.75	63.75	73.75	160.04	123.31	96.91
1750	1160	870	24 XXH	9.549	24 XXH	9.549	1.00	20.00	25.00	30.00	35.00	45.00	55.00	65.00	75.00	154.06■	115.66	90.23
1750	1160	870	22 XXH	8.754	22 XHH	8.754	1.00	21.25	26.25	31.25	36.25	46.25	56.25	66.25	76.25	146.56■	107.58■	83.37
1750	1160	870	20 XXH	7.958	20 XXH	7.958	1.00	22.50	27.50	32.50	37.50	47.50	57.50	67.50	77.50	137.67■	99.09■	76.34■
1750	1160	870	18 XXH	7.162	18 XXH	7.162	1.00	23.75	28.75	33.75	38.75	48.75	58.75	68.75	78.75		90.23■	69.14■
1615	1071	803	24 XXH	9.549	26 XXH	10.345	1.08	19.37	24.37	29.37	34.37	44.37	54.37	64.37	74.37	154.06■	115.66	90.23
1604	1063	798	22 XXH	8.754	24 XXH	9.549	1.09	20.62	25.62	30.62	35.62	45.62	55.62	65.62	75.62	146.56■	107.58■	83.37
1591	1055	791	20 XXH	7.958	22 XXH	8.754	1.10	21.87	26.87	31.87	36.87	46.87	56.87	66.87	76.87	137.67■	99.09■	76.34■
1575	1044	783	18 XXH	7.162	20 XXH	7.958	1.11	23.12	28.12	33.12	38.12	48.12	58.12	68.12	78.12		90.23■	69.14■
1544	1024	768	30 XXH	11.937	34 XXH	13.528	1.13	14.98	19.98	24.99	29.99	39.99	49.99	59.99	70.00	166.95	137.12	109.64
1517	1005	754	26 XXH	10.345	30 XXH	11.937	1.15	17.48	22.49	27.49	32.49	42.49	52.49	62.49	72.50	160.04	123.31	96.91
1488	986	740	34 XXH	13.528	40 XXH	15.915	1.18		16.83	21.84	26.85	36.86	46.86	56.86	66.86	166.26	148.71	121.44
1481	982	736	22 XXH	8.754	26 XXH	10.345	1.18	19.98	24.99	29.99	34.99	44.99	54.99	65.00	75.00	146.56■	107.58■	83.37
1458	967	725	40 XXH	15.915	48 XXH	19.099	1.20				22.44	32.46	42.47	52.48	62.48	148.78■	161.32	137.11
1458	967	725	20 XXH	7.958	24 XXH	9.549	1.20	21.24	26.24	31.24	36.24	46.24	56.24	66.25	76.25	137.67■	99.09■	76.34■
1432	949	712	18 XXH	7.162	22 XXH	8.754	1.22	22.49	27.49	32.49	37.49	47.49	57.49	67.50	77.50		90.23■	69.14■
1400	928	696	24 XXH	9.549	30 XXH	11.937	1.25	18.09	23.09	28.10	33.10	43.11	53.11	63.11	73.12	154.06■	115.66	90.23

DriveN Speed			Pulley Combination					Center Distance, Inches								Horsepower Rating for 5″ wide belt		
For motor speed of			DriveR		DriveN											For motor speed of		
1750 rpm	1160 rpm	870 rpm	No. of grooves	Pitch diam., inches	No. of grooves	Pitch diam., inches	Speed Ratio	700 XXH P.L. 70.00 56 teeth	800 XXH P.L. 80.00 64 teeth	900 XXH P.L. 90.00 72 teeth	1000 XXH P.L. 100.00 80 teeth	1200 XXH P.L. 120.00 96 teeth	1400 XXH P.L. 140.00 112 teeth	1600 XXH P.L. 160.00 128 teeth	1800 XXH P.L. 180.00 144 teeth	1750 rpm	1160 rpm	870 rpm
1346	892	669	20 XXH	7.958	26 XXH	10.345	1.30	20.59	25.60	30.60	35.60	45.61	55.61	65.61	75.62	137.67■	99.09■	76.34■
1338	887	665	26 XXH	10.345	34 XXH	13.528	1.31	16.17	21.19	26.20	31.21	41.22	51.23	61.23	71.23	160.04	123.31	96.91
1312	870	652	30 XXH	11.937	40 XXH	15.915	1.33		18.02	23.04	28.05	38.07	48.08	58.09	68.10	166.95	137.12	109.64
1313	870	653	18 XXH	7.162	24 XXH	9.549	1.33	21.84	26.85	31.85	36.86	46.86	56.86	66.86	76.87		90.23■	69.14■
1283	851	638	22 XXH	8.754	30 XXH	11.937	1.36	18.68	23.70	28.71	33.71	43.72	53.73	63.73	73.73	146.56■	107.58■	83.37
1240	822	616	34 XXH	13.528	48 XXH	19.099	1.41			19.17	24.21	34.26	44.29	54.30	64.31	166.26	148.71	121.44
1235	819	614	24 XXH	9.549	34 XXH	13.528	1.42	16.76	21.78	26.80	31.81	41.83	51.84	61.84	71.85	154.06■	115.66	90.23
1212	803	602	18 XXH	7.162	26 XXH	10.345	1.44	21.19	26.20	31.21	36.22	46.22	56.23	66.23	76.23		90.23	69.14■
1167	773	580	40 XXH	15.915	60 XXH	23.873	1.50					28.47	38.54	48.59	58.61	148.78■	161.32	137.11
1167	773	580	20 XXH	7.958	30 XXH	11.937	1.50	19.27	24.29	29.31	34.32	44.33	54.34	64.34	74.35	137.67■	99.09■	76.34■
1138	754	566	26 XXH	10.345	40 XXH	15.915	1.54		19.17	24.21	29.24	39.28	49.30	59.31	69.32	160.04	123.31	96.91
1132	751	563	22 XXH	8.754	34 XXH	13.528	1.55	17.34	22.37	27.40	32.41	42.43	52.45	62.45	72.46	146.56■	107.58■	83.37
1094	725	544	30 XXH	11.937	48 XXH	19.099	1.60			20.31	25.37	35.44	45.48	55.51	65.53	166.95	137.12	109.64
1050	696	522	24 XXH	9.549	40 XXH	15.915	1.67	14.65	19.74	24.80	29.83	39.87	49.90	59.92	69.93	154.06■	115.66	90.23
1050	696	522	18 XXH	7.162	30 XXH	11.937	1.67	19.86	24.89	29.90	34.92	44.94	54.95	64.96	74.96		90.23■	69.14■

Key to Horsepower Correction Factor □ = **1.0**

■ Use this pulley and rpm only if required to obtain speed ratio or to meet diameter limitations. See Page 58.

For XXH 1¼″ Pitch Drives

Belt Width Factor	0-0.35	0.36-0.56	0.57-0.78	0.79-1.00
Stock Belt Width	2″	3″	4″	5″

Source: Courtesy of The Gates Rubber Company.

TABLE 15 Timing Belt Drive Selection Table (L, 3/8 in. pitch belts)

DriveN Speed			Pulley Combination					Center Distance, Inches									
For motor speed of			DriveR		DriveN												
3450 rpm	1750 rpm	1160 rpm	No. of grooves	Pitch diam., inches	No. of grooves	Pitch diam., inches	Speed Ratio	**124 L** P.L. 12.375 33 teeth	**150 L** P.L. 15.00 40 teeth	**187 L** P.L. 18.75 50 teeth	**210 L** P.L. 21.00 56 teeth	**225 L** P.L. 22.50 60 teeth	**240 L** P.L. 24.00 64 teeth	**255 L** P.L. 25.50 68 teeth	**270 L** P.L. 27.00 72 teeth	**285 L** P.L. 28.50 76 teeth	**300 L** P.L. 30.00 80 teeth
1150	583	387	28 L	3.342	84 L	10.027	3.00										
1150	583	387	24 L	2.865	72 L	8.594	3.00										
1150	583	387	20 L	2.387	60 L	7.162	3.00								5.47	6.29	7.09
1150	583	387	16 L	1.910	48 L	5.730	3.00					4.87	5.68	6.47	7.25	8.02	8.79
1150	583	387	12 L	1.432	36 L	4.297	3.00			4.65	5.82	6.59	7.36	8.12	8.88	9.64	10.40
1150	583	387	10 L	1.194	30 L	3.581	3.00		3.55	5.49	6.64	7.40	8.16	8.92	9.68	10.43	11.19
1098	557	369	14 L	1.671	44 L	5.252	3.14				4.72	5.52	6.31	7.08	7.86	8.63	9.39
1078	547	363	10 L	1.194	32 L	3.820	3.20		3.30	5.27	6.43	7.19	7.95	8.71	9.47	10.23	10.98
1068	542	359	26 L	3.104	84 L	10.027	3.23										
1054	535	354	22 L	2.626	72 L	8.594	3.27										
1035	525	348	18 L	2.149	60 L	7.162	3.33								5.62	6.44	7.25
1035	525	348	12 L	1.432	40 L	4.775	3.33			4.16	5.36	6.15	6.92	7.69	8.46	9.22	9.98
1006	510	338	14 L	1.671	48 L	5.730	3.43					5.02	5.83	6.62	7.41	8.18	8.96
986	500	331	24 L	2.865	84 L	10.027	3.50										
958	486	322	20 L	2.387	72 L	8.594	3.60										
958	486	322	10 L	1.194	36 L	4.297	3.60			4.81	5.99	6.76	7.53	8.29	9.05	9.81	10.57
941	477	316	12 L	1.432	44 L	5.252	3.67			3.61	4.87	5.68	6.47	7.25	8.02	8.79	9.56
920	467	309	16 L	1.910	60 L	7.162	3.75							4.90	5.77	6.59	7.40
904	458	304	22 L	2.626	84 L	10.027	3.82										
863	438	290	18 L	2.149	72 L	8.594	4.00										

DriveN Speed: For motor speed of 3450 rpm	1750 rpm	1160 rpm	Pulley Combination: DriveR No. of grooves	DriveR Pitch diam., inches	DriveN No. of grooves	DriveN Pitch diam., inches	Speed Ratio	Center Distance, Inches: **124 L** P.L. 12.375 33 teeth	**150 L** P.L. 15.00 40 teeth	**187 L** P.L. 18.75 50 teeth	**210 L** P.L. 21.00 56 teeth	**225 L** P.L. 22.50 60 teeth	**240 L** P.L. 24.00 64 teeth	**255 L** P.L. 25.50 68 teeth	**270 L** P.L. 27.00 72 teeth	**285 L** P.L. 28.50 76 teeth	**300 L** P.L. 30.00 80 teeth
863	438	290	12 L	1.432	48 L	5.730	4.00				4.33	5.17	5.99	6.78	7.57	8.35	9.12
863	438	290	10 L	1.194	40 L	4.775	4.00			4.31	5.52	6.31	7.09	7.86	8.63	9.39	10.15
821	417	276	20 L	2.387	84 L	10.027	4.20										
805	408	271	14 L	1.671	60 L	7.162	4.29							5.05	5.91	6.75	7.56
784	398	264	10 L	1.194	44 L	5.252	4.40		3.29	3.75	5.02	5.83	6.62	7.41	8.18	8.96	9.72
767	389	258	16 L	1.910	72 L	8.594	4.50										5.75
739	375	249	18 L	2.149	84 L	10.027	4.67										
719	365	242	10 L	1.194	48 L	5.730	4.80				4.47	5.32	6.14	6.94	7.73	8.51	9.28
690	350	232	12 L	1.432	60 L	7.162	5.00							5.19	6.06	6.90	7.71
671	340	226	14 L	1.671	72 L	8.594	5.14										5.89
657	333	221	16 L	1.910	84 L	10.027	5.25										
575	292	193	14 L	1.671	84 L	10.027	6.00										
575	292	193	12 L	1.432	72 L	8.594	6.00										6.03
575	292	193	10 L	1.194	60 L	7.162	6.00							5.33	6.21	7.05	7.86
493	250	166	12 L	1.432	84 L	10.027	7.00										
479	243	161	10 L	1.194	72 L	8.594	7.20										6.16
411	208	138	10 L	1.194	84 L	10.027	8.40										

Key to Horsepower Correction Factor ☐ = **1.0** ⬚ = **0.8** ▢ = **0.6** ▢ = **0.4** ■ = **0.2**

▪ Use this pulley and rpm only if required to obtain speed ratio or to meet diameter limitations.

	For L ⅜″ Pitch Drives		
Belt Width Factor	0-0.45	0.46-0.72	0.73-1.00
Stock Belt Width	½″	¾″	1″

Source: Courtesy of The Gates Rubber Company.

TABLE 16 Timing Belt Horsepower Rating Table (L, 3/8 in. Pitch Belts)

RPM of Faster Shaft	Rated Horsepower for Small Pulley (Number of Grooves and Pitch Diameter, Inches)																		
	10L 1.194	12L 1.432	14L 1.671	16L 1.910	17L 2.029	18L 2.149	19L 2.268	20L 2.387	21L 2.507	22L 2.626	24L 2.865	26L 3.104	28L 3.342	30L 3.581	32L 3.820	36L 4.297	40L 4.775	44L 5.252	48L 5.730
725	0.38	0.45	0.53	0.60	0.64	0.68	0.72	0.75	0.79	0.83	0.90	0.98	1.05	1.13	1.20	1.35	1.50	1.64	1.79
870	0.45	0.54	0.63	0.72	0.77	0.81	0.86	0.90	0.95	0.99	1.08	1.17	1.26	1.35	1.44	1.61	1.79	1.96	2.14
950	0.49	0.59	0.69	0.79	0.84	0.89	0.94	0.99	1.03	1.08	1.18	1.28	1.37	1.47	1.57	1.76	1.95	2.14	2.32
1160	0.60	0.72	0.84	0.96	1.02	1.08	1.14	1.20	1.26	1.32	1.44	1.56	1.67	1.79	1.91	2.14	2.36	2.59	2.81
1425	0.74	0.89	1.03	1.18	1.25	1.33	1.40	1.47	1.54	1.62	1.76	1.90	2.04	2.18	2.32	2.60	2.87	3.14	3.40
1750	0.91	1.09	1.27	1.45	1.53	1.62	1.71	1.80	1.89	1.97	2.15	2.32	2.49	2.66	2.82	3.15	3.47	3.77	4.07
2850		1.76	2.04	2.32	2.46	2.60	2.74	2.87	3.01	3.14	3.40	3.65	3.89	4.13	4.36	4.79	5.17	5.52	5.81
3450			2.46	2.79	2.95	3.11	3.27	3.42	3.58	3.73	4.02	4.30	4.57	4.82	5.06	5.48	5.84	6.11	6.29
100	0.052	0.062	0.073	0.083	0.088	0.094	0.099	0.10	0.11	0.11	0.12	0.14	0.15	0.16	0.17	0.19	0.21	0.23	0.25
200	0.10	0.12	0.15	0.17	0.18	0.19	0.20	0.21	0.22	0.23	0.25	0.27	0.29	0.31	0.33	0.37	0.42	0.46	0.50
300	0.16	0.19	0.22	0.25	0.27	0.28	0.30	0.31	0.33	0.34	0.37	0.41	0.44	0.47	0.50	0.56	0.62	0.69	0.75
400	0.21	0.25	0.29	0.33	0.35	0.37	0.40	0.42	0.44	0.46	0.50	0.54	0.58	0.62	0.67	0.75	0.83	0.91	1.00
500	0.26	0.31	0.36	0.42	0.44	0.47	0.49	0.52	0.55	0.57	0.62	0.68	0.73	0.78	0.83	0.93	1.04	1.14	1.24
600	0.31	0.37	0.44	0.50	0.53	0.56	0.59	0.62	0.65	0.69	0.75	0.81	0.87	0.93	1.00	1.12	1.24	1.36	1.49
700	0.36	0.44	0.51	0.58	0.62	0.65	0.69	0.73	0.76	0.80	0.87	0.94	1.02	1.09	1.16	1.30	1.45	1.59	1.73
800	0.42	0.50	0.58	0.67	0.71	0.75	0.79	0.83	0.87	0.91	1.00	1.08	1.16	1.24	1.32	1.49	1.65	1.81	1.97
900	0.47	0.56	0.65	0.75	0.79	0.84	0.89	0.93	0.98	1.03	1.12	1.21	1.30	1.40	1.49	1.67	1.85	2.03	2.21
1000	0.52	0.62	0.73	0.83	0.88	0.93	0.99	1.04	1.09	1.14	1.24	1.34	1.45	1.55	1.65	1.85	2.05	2.25	2.44
1100	0.57	0.69	0.80	0.91	0.97	1.03	1.08	1.14	1.20	1.25	1.36	1.48	1.59	1.70	1.81	2.03	2.25	2.46	2.67
1200	0.62	0.75	0.87	1.00	1.06	1.12	1.18	1.24	1.30	1.36	1.49	1.61	1.73	1.85	1.97	2.21	2.44	2.67	2.90
1300	0.68	0.81	0.94	1.08	1.14	1.21	1.28	1.34	1.41	1.48	1.61	1.74	1.87	2.00	2.13	2.38	2.63	2.88	3.12
1400	0.73	0.87	1.02	1.16	1.23	1.30	1.37	1.45	1.52	1.59	1.73	1.87	2.01	2.15	2.29	2.56	2.82	3.09	3.34
1500	0.78	0.93	1.09	1.24	1.32	1.40	1.47	1.55	1.62	1.70	1.85	2.00	2.15	2.30	2.44	2.73	3.01	3.29	3.56
1600	0.83	1.00	1.16	1.32	1.41	1.49	1.57	1.65	1.73	1.81	1.97	2.13	2.29	2.44	2.60	2.90	3.20	3.49	3.77
1700	0.88	1.06	1.23	1.41	1.49	1.58	1.66	1.75	1.83	1.92	2.09	2.26	2.42	2.59	2.75	3.07	3.38	3.68	3.97

RPM of Faster Shaft	Rated Horsepower for Small Pulley (Number of Grooves and Pitch Diameter, Inches)																		
	10L 1.194	12L 1.432	14L 1.671	16L 1.910	17L 2.029	18L 2.149	19L 2.268	20L 2.387	21L 2.507	22L 2.626	24L 2.865	26L 3.104	28L 3.342	30L 3.581	32L 3.820	36L 4.297	40L 4.775	44L 5.252	48L 5.730
1800		1.12	1.30	1.49	1.58	1.67	1.76	1.85	1.94	2.03	2.21	2.38	2.56	2.73	2.90	3.23	3.56	3.87	4.17
1900		1.18	1.37	1.57	1.66	1.76	1.85	1.95	2.04	2.14	2.32	2.51	2.69	2.87	3.05	3.40	3.73	4.05	4.36
2000		1.24	1.45	1.65	1.75	1.85	1.95	2.05	2.15	2.25	2.44	2.63	2.82	3.01	3.20	3.56	3.90	4.23	4.54
2200		1.36	1.59	1.81	1.92	2.03	2.14	2.25	2.35	2.46	2.67	2.88	3.09	3.29	3.49	3.87	4.23	4.57	4.89
2400		1.49	1.73	1.97	2.09	2.21	2.32	2.44	2.56	2.67	2.90	3.12	3.34	3.56	3.77	4.17	4.54	4.89	5.21
2600		1.61	1.87	2.13	2.26	2.38	2.51	2.63	2.76	2.88	3.12	3.36	3.59	3.82	4.04	4.45	4.84	5.19	5.50
2800		1.73	2.01	2.29	2.42	2.56	2.69	2.82	2.96	3.09	3.34	3.59	3.83	4.07	4.30	4.72	5.11	5.45	5.75
3000		1.85	2.15	2.44	2.59	2.73	2.87	3.01	3.15	3.29	3.56	3.82	4.07	4.31	4.54	4.98	5.36	5.69	5.97
3200			2.29	2.60	2.75	2.90	3.05	3.20	3.34	3.49	3.77	4.04	4.29	4.54	4.78	5.21	5.59	5.90	6.14
3400			2.42	2.75	2.91	3.07	3.22	3.38	3.53	3.68	3.97	4.25	4.51	4.77	5.00	5.43	5.79	6.07	6.27
3600			2.56	2.90	3.07	3.23	3.40	3.56	3.71	3.87	4.17	4.45	4.72	4.98	5.21	5.63	5.97	6.21	6.35
3800			2.69	3.05	3.22	3.40	3.56	3.73	3.89	4.05	4.36	4.65	4.92	5.17	5.41	5.81	6.11	6.31	6.39
4000			2.82	3.20	3.38	3.56	3.73	3.90	4.07	4.23	4.54	4.84	5.11	5.36	5.59	5.97	6.23	6.37	6.37
4200				3.34	3.53	3.71	3.89	4.07	4.24	4.40	4.72	5.02	5.29	5.53	5.75	6.10	6.32	6.39	6.30
4400				3.49	3.68	3.87	4.05	4.23	4.41	4.57	4.89	5.19	5.45	5.69	5.90	6.21	6.37	6.36	6.17
4600				3.63	3.82	4.02	4.21	4.39	4.57	4.74	5.06	5.35	5.61	5.84	6.03	6.29	6.39	6.29	5.98
4800				3.77	3.97	4.17	4.36	4.54	4.72	4.89	5.21	5.50	5.75	5.97	6.14	6.35	6.37	6.17	5.73
5000				3.90	4.11	4.31	4.51	4.69	4.87	5.04	5.36	5.64	5.88	6.08	6.23	6.38	6.31	6.00	5.41
5200				4.04	4.25	4.45	4.65	4.84	5.02	5.19	5.50	5.77	6.00	6.18	6.30	6.38	6.22	5.78	5.03
5400				4.17	4.38	4.59	4.79	4.98	5.16	5.32	5.63	5.89	6.10	6.25	6.35	6.36	6.08	5.50	4.57
5600				4.30	4.51	4.72	4.92	5.11	5.29	5.45	5.75	6.00	6.19	6.32	6.38	6.30	5.90	5.17	4.05
5800				4.42	4.64	4.85	5.05	5.24	5.41	5.58	5.86	6.09	6.26	6.36	6.39	6.21	5.68	4.77	3.44
6000				4.54	4.76	4.98	5.17	5.36	5.53	5.69	5.97	6.18	6.32	6.38	6.37	6.08	5.41	4.32	2.76

▨ Use this pulley and rpm only if required to obtain speed ratio or to meet diameter limitations. See Page 74.

□ Pulley surface speeds over 6,500 fpm; special pulleys are required.

Source: Courtesy of The Gates Rubber Company.

TABLE 17 Minimum and Maximum Installation Tension for Timing Belts

Pitch		Belt Width											
		1/8″	3/16″	1/4″	3/8″	1/2″	3/4″	1″	1 1/2″	2″	3″	4″	5″
MXL	0.080″ T_s	.99 to 1.2	1.7 to 2.1	2.3 to 2.9									
	Y	.66	.98	1.3									
XL	0.200″ T_s			3.2 to 4.0	5.1 to 6.3								
	Y			.95	1.8								
L	0.375″ T_s					13 to 15	19 to 23	25 to 31					
	Y					10	18	25					
H	0.500″ T_s						54 to 66	72 to 88	110 to 130	140 to 180	220 to 260		
	Y						33	47	73	100	160		
XH	0.875″ T_s									210 to 250	310 to 370	410 to 500	
	Y									200	320	460	
XXH	1.250″ T_s									260 to 320	390 to 470	520 to 640	650 to 790
	Y									320	510	720	930

Source: Courtesy of The Gates Rubber Company.

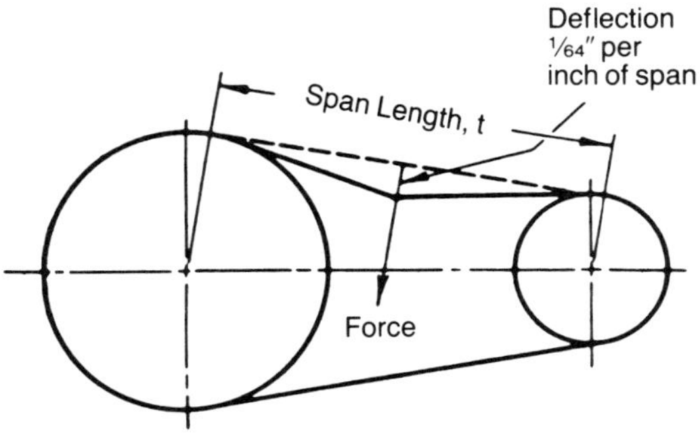

FIGURE 19 Numerical method for tensioning a drive.

7. Determine Installation Tension

The recommended minimum and maximum installation tension are specified in Table 17. The values in the table may be used to determine the numerical method for tensioning the drive as shown in Figure 19.

The recommended deflection forces can be calculated using the following formulas:

$$\text{minimum recommended force} = \frac{T_{st(minimum)} + (t/L)\,Y}{16},\ \text{lbf}$$

$$\text{maximum recommended force} = \frac{T_{st(maximum)} + (t/L)\,Y}{16},\ \text{lbf}$$

where

$T_{st(minimum)}$ = minimum required tension from table, lbf
$T_{st(maximum)}$ = maximum required tension from table, lbf
Y = constant from table
t = span length
L = pitch length of belt

The amount of deflection is determined by dividing the span length by 64.

The force should be applied at the center of the span and perpendicular to the belt. Be sure at least one pulley or sprocket in the drive is free to rotate when measuring installation tension.

11

Belt Tensioning and Bearing Loads

DANIEL L. SCHWARTZ *Industrial Product Application, The Gates Rubber Company, Denver, Colorado*

I. DESCRIPTION AND SIGNIFICANCE OF STATIC TENSION

Belt drive tension is important. Too much tension shortens belt life and can also shorten the life of the other drive components such as bearings. Too little tension allows belts to slip, generating heat, which also reduces the life of drive components. The proper tension is between these two extremes.

Although important, the tension of a belt is not generally thought of as critical. That is, the tension range over which belts will operate satisfactorily is rather wide – particularly with V-belts.

A. Ideal Belt Tension

The ideal tension for a belt drive is the lowest tension at which the belts will not slip at the highest load condition. This ideal tension results in the best belt life and lowest bearing loads for a given job of power transmission. However, the ideal tension is hard to determine and, except for automatically tensioned drives, it is difficult to maintain. Therefore, more practical approaches to tensions and tensioning have been developed, as explained below.

B. Formula for Calculating Static Tension

Running tensions can be calculated quite accurately from load, speed, and the geometry of a drive. Running tensions are very difficult to measure on a drive, however, except by laboratory techniques. Therefore the belt tension when the drive is stopped must be used for establishing tensions for locked center drives. This tension is called static tension (T_{st}), or more commonly, installation tension. Static tension for any belt drive with two sheaves is given by:

$$T_{st} = \frac{T_T + T_S}{2} + mS^2 - \frac{fEI}{L}\left(\frac{1}{d} + \frac{1}{D}\right)\left(\frac{1}{2T_T} + \frac{1}{2T_S} - \frac{1}{T_S}\right) \qquad (1)$$

where

T_{st} = static tension (lb)
T_T = tight side tension (lb)
T_S = slack side tension (lb)
m = belt mass per inch, lb/sec^2
S = belt speed in thousands of ft/min (S = ft/min/1000)
f = belt strain constant (lb)
E = belt modulus of elasticity (lb/in.^{2})
I = belt moment of inertia (in.^{4})
L = belt length (in.)
D = diameter of large sheave (in.)
d = diameter of small sheave (in.)

Note: If the running tensions are per belt, the static tension is per belt. If running tensions are for total drive, the static tension is for total drive.

Of the three terms in the equation:

$$\frac{T_T + T_S}{2} = \text{operating tensions}$$

$$mS^2 = \text{centrifugal tension}$$

$$\frac{fEI}{L} = \text{tension due to the stiffness of the belt (bending tension)}$$

The above equation provides an accurate value for static tension. However, because V-belt tension need not be precisely controlled, the equation is simplified by elimination of the third term, which is primarily concerned with the stiffness of the belt. For a long flexible belt on large-diameter sheaves with high operating tensions (T_T and T_S) the third term will approach zero. Under these conditions static tension can be found from:

$$T_{st} = \frac{T_T + T_S}{2} + mS^2 \tag{2}$$

For stiff, short belts on small-diameter sheaves with low operating tensions, however, the third term does have significance. It will reduce the static tension by as much as 20%. Therefore, if the right side of Equation (2) is reduced by 10% (or multiplied by 0.9), the resulting value of static tension is halfway between that found for the extremes of flexibility. In other words, T_{st} will be accurate to within ±10% for most drives. The equation now reads:

$$T_{st} = 0.9\left(\frac{T_T + T_S}{2} + mS^2\right) \tag{3}$$

Drive design is usually worked in terms of HP ratings found in any V-belt drive design manual. The equation relating $(T_T + T_S)$ to the design HP, the number of belts, N, the belt speed, S, and the arc-of-contact correction factor, G, is:

$$T_T + T_S = \frac{33HP}{SN}\left(\frac{2.5-G}{G}\right) \tag{4}$$

Factor G, in turn, depends on the actual arc of contact on the small sheave (or on the (D-d)/c ratio), where c = center distance between sheaves. Factor G can be obtained from Table 1 (G = 1 for drives with one-to-one speed ratio, 180° arc of contact).

When the value of $(T_T + T_S)$ from Equation (4) is substituted in Equation (3), the static tension equation becomes:

$$T_{st} = 14.85\left(\frac{2.5-G}{G}\right)\left(\frac{HP}{SN}\right) + 0.9mS^2 \tag{5}$$

In the belt industry, the above equation is further simplified by substituting the factor M = 0.9m and by rounding off the multiplier in the first term to two significant figures. Thus,

$$T_{st} = 15\left(\frac{2.5-G}{G}\right)\left(\frac{HP}{SN}\right) + mS^2 \tag{6}$$

Equation (6) is the working equation for determining the static tension necessary for a specific drive. Table 2 gives weight factors, M, for some various sizes (cross sections) of industrial belts. These values may vary from manufacturer to manufacturer.

TABLE 1 Arc of Correction: Factor G

Arc of contact on small sheave, degrees	$\frac{D-d}{C}$	Factor G for V-flat drives	Factor G for V-V drives
180	0.00	0.75	1.00
174	0.10	0.76	0.99
169	0.20	0.78	0.97
163	0.30	0.79	0.96
157	0.40	0.80	0.94
151	0.50	0.81	0.93
145	0.60	0.83	0.91
139	0.70	0.84	0.89
133	0.80	0.85	0.87
127	0.90	0.85	0.85
120	1.00	0.82	0.82
113	1.10	0.80	0.80
106	1.20	0.77	0.77
99	1.30	0.73	0.73
91	1.40	0.70	0.70
83	1.50	0.65	0.65

Source: Courtesy of The Gates Rubber Company.

For more than two sheave drives, running tensions are calculated as part of the drive design procedure. Thus, it is better to put a static tension formula in terms of the running tensions, as:

TABLE 2 Factor M and Factor Y

Cross section	M	Y
Super HC Molded Notch		
3VX	.29	4
5VX	.78	13
Super HC Molded Notch Power Band		
3VX	.39	4
5VX	.98	13
Super HC		
5V	1.0	11
8V	2.6	22
Super HC Power Band		
5V	1.2	11
8V	3.0	22
Hi-Power II		
A	0.51	7
B	0.80	8
C	1.5	18
D	3.0	27
E	4.3	56
Hi-Power II Power Band		
A	0.66	7
B	1.0	9
C	1.8	18
D	3.4	28
E	4.9	60
Tri-Power		
AX	0.47	7
BX	0.76	8
CX	1.31	15
DX	2.56	37

Source: Courtesy of The Gates Rubber Company.

$$T_{st} = 0.9\left(\frac{T_1 t_1 + T_2 t_2 + T_3 t_3 + \text{etc.}}{t_1 + t_2 + t_3 + \text{etc.}} + \frac{mV^2}{10^6}\right)$$

where

T_{st}	= static tension (lb)
T_1, T_2, T_3, etc.	= tension in spans 1, 2, 3, etc. (lb)
t_1, t_2, t_3, etc.	= lengths of spans 1, 2, 3, etc. (in.)
m	= factor for belt weight (lb/in.)
V	= belt speed (ft/min)

Note: If running tensions are per belt, the static tension is per belt. If running tensions are for total drive, the static tension is for total drive.

C. Installation Tension for New Belt

A V-belt that is installed at the static belt tension calculated from the above formulas and is operating at the assumed HP and speed will be operating at the V-belt design tension ratio of approximately 5:1 (referenced to 180° arc). This is called normal installation tension, T_{st}, and is the base factor from which to determine tension for new belts and retensioning for existing drives.

When installing new belts on a drive, installation tension should be set somewhat higher, for example, 1.5 T_{st} to 2.0 T_{st} for industrial multiple belt drives. This is often called 1 1/2 to 2 times normal tension. This higher initial tension is necessary because drive tension drops rapidly during the initial part of the operation of a V-belt drive due to the belt seating in the grooves. This extra initial tension usually does not adversely affect bearings because it decays rather rapidly. For maintenance – retensioning of existing drives – tension can be reset from 1.0 T_{st} to 1.5 T_{st}.

II. METHODS OF APPLYING STATIC TENSION

Knowing the correct static (or installation) tension is only half the job. The other half is having a way to adjust and measure.

A. Movable DriveR or DriveN Unit

Some method must be provided for adjusting the static tension. Most often this is through movement of the driveR or driveN shafts.

1. Methods of Moving Shafts

There are two methods for moving a shaft to adjust the belt tension. With smaller units, a lock bolt is loosened and the shaft is pryed outward while the tension is checked. While holding the shaft in place, the unit is locked in position. The other is through a jackscrew where the shaft can be moved in or out by turning the screw. This may include mounting a large unit on rails and using screws to move it.

B. Shimming Sheaves

Occasionally, the sides of a V-belt sheave can be made movable. For example, shim stock can be placed between the two halves of the sheave. Removing shims forces the belt outward in the groove, effecting some take-up. This also changes the speed ratio unless all sheaves are shimmed precisely.

C. Idlers

Another method is through the use of idlers. Idlers are sheaves or pulleys that are not loaded: they do not transmit power.

D. Spliced Belts

When spliced belts are used, take-up allowances are not absolutely mandatory because the belt length can be changed by resplicing. However, movable shafts or idlers are still desirable for easier installation and for less frequent splicing for take-up.

III. METHOD OF MEASURING STATIC TENSION

After determining the static tension and tensioning procedure, a method of measuring the actual tension is needed for the drive. Accurate measurement is important to determine if it is within the specified range for the drive.

The tension in a belt can be measured by many methods, some of which are:

By measuring the required force with which an adjustable shaft is moved into the drive

By measuring the torque necessary to slip one of the sheaves or pulleys

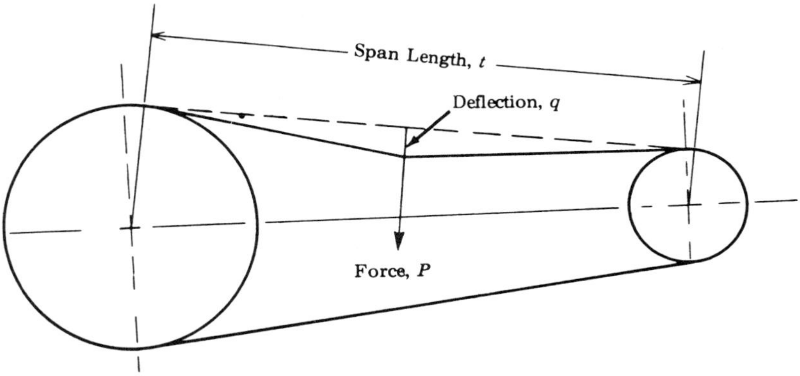

FIGURE 1 Tension measurement by force deflection.

By deflecting a belt, measuring both the amount of deflection and the force necessary to produce it

By using the belt elongation method

Of these, the deflection method has been used more effectively than the others.

A. Force Deflection Method

The easiest way to measure T_{st} is to deflect one belt at the center of the span with a force and then relate the deflection distance and force to a static tension. Figure 1 illustrates this principle.

The basic formulas which relate deflection and the amount of this force to static tension are:

$$T_{st} = \frac{P(t)}{4q} = \frac{2(f)q^2}{t^2} \text{ (for locked sheaves)}$$

$$T_{st} = \frac{P(t)}{4q} = \frac{2(f)q^2}{(L)(t)} \text{ (for sheaves free to rotate)}$$

where

T_{st} = static tension (lb)
P = deflection force applied perpendicular to span at the midpoint (lb)

t = span length (in.)
q = deflection (in.)
f = belt strain constant (lb)
L = belt length (in.)

The difference between thesè two formulas concerns whether the extra tension imposed by the deflection force is distributed only through that span or through the entire belt. This distribution depends upon whether the sheaves or pulleys are locked in place (an example is a multiple-belt drive, since only one belt is being deflected at a time) or are free to rotate (an example might be an automotive accessory drive).

These formulas are quite accurate, but difficult to use in practice. This is because the belt strain constant (which varies between belt types, belt manufacturers, and even throughout the life of the belt) is not known. Also, unless the amount of deflection, q, is kept small in comparison to the span, the assumptions made to derive the formulas are not valid. Therefore, the formulas must generally be simplified by imposing further conditions for use with specific types of belts.

For V-belts, if the deflection amount is fixed at 1/64 in. per inch of span, and if an average value for strain constant is assumed, the formulas can be simplified. In these formulas, the positions of P and T_{st} are rearranged so that the deflection force is the factor calculated.

$$P = \frac{T_{st} + Y}{16} \quad \text{(for locked sheaves)}$$

$$P = \frac{T_{st} + (t/L)}{16} \quad \text{(for sheaves free to rotate)}$$

where

P = deflection force applied perpendicular to span at the midpoint (lb)
T_{st} = static tension (lb)
Y = a constant which is a function of the belt strain constant
t = span length (in.)
L = belt length (in.)

Values of Y are given in Table 2.

Now, the deflection forces can reflect the increased initial values of installation tensions desired, as indicated above (1.5 T_{st} to 2.0 T_{st} for new industrial multiple belts, for example) by substituting the desired multiple of T_{st}.

B. Simplified Tensioning Tables

For V-belt drives that meet certain requirements of speed ratio, speed, geometry, and so forth, an even simpler tensioning method can be used. Most manufacturers have tables which provide deflection force range that should be used to deflect a given belt 1/64 in. per inch of span. An example is Table 3.

C. Precision Tensioning

For laboratory work or demanding applications, more accurate tensioning is sometimes desired. In this case, a third type of formula and method can be derived. It is known as the double deflection method.

When two deflections with two different forces are made, all the variables except the forces, span length, and deflection amount are eliminated. If the second deflection is taken as twice the first, the following accurate formula can be written:

$$T_{st} = \frac{t}{24q_1}(8\,P_1 - P_2)$$

where

T_{st} = static tension (lb)
t = span length (in.)
q_1 = first deflection (second deflection is twice the first) (in.)
P_1 = first force (lb)
P_2 = second force (lb)

D. Elongation Method

The elongation method can also be used to measure static tension. Find the required static tension, T_{st}, as described above. Determine the range of recommended tensions by:

Lower tension = T_{st}

Upper tension = $1.5 \times T_{st}$

Measure the outside circumference of the belt at no tension. This can be done with a belt either on or off the drive. If retensioning a belt already on a drive, slack off on the drive until there is no tension.

Find the appropriate belt length multiplier from Table 4 for the proper belt cross section and the static tensions calculated above.

TABLE 3 Simplified Tensioning Table

V-Belt Cross Section	Small Sheave Diameter Range (Inches)	Small Sheave RPM Range	Speed Ratio Range	Recommended Deflection Force (Pounds)	
				Minimum	Maximum
3VX	2.20	1200-3600	2.00 to 4.00	2.8	4.1
	2.35- 2.50	1200-3600		3.2	4.7
	2.65- 2.80	1200-3600		3.5	5.1
	3.00- 3.15	1200-3600		3.8	5.5
	3.35- 3.65	1200-3600		4.1	6.0
	4.12- 5.00	900-3600		4.8	7.1
	5.30- 6.90	900-3600		5.8	8.6
5VX	4.40- 4.65	1200-3600	2.00 to 4.00	9.0	13
	4.90- 5.50	1200-3600		10	15
	5.90- 6.70	1200-3600		11	17
	7.10- 8.00	600-1800		13	19
	8.50-10.90	600-1800		14	20
	11.80-16.00	400-1200		15	23
5V	7.10- 8.00	600-1800	2.00 to 4.00	11	16
	8.50-10.90	600-1800		13	18
	11.80-16.00	400-1200		14	21
8V	12.50-17.00	600-1200	2.00 to 4.00	28	41
	18.00-24.00	400- 900		32	48

*If replacing 3V or 5V belts with the same belt length and same number of belts in 3VX or 5VX cross section, belt tension does not need to be increased.

NOTE: New drives designed with 3VX or 5VX belts should be tensioned at the respective deflection force value shown in above table.

V-Belt Cross Section	Small Sheave Diameter Range (Inches)	Small Sheave RPM Range	Speed Ratio Range	Recommended Deflection Force (Pounds)			
				Hi-Power II		Tri-Power	
				Minimum	Maximum	Minimum	Maximum
A	3.0	1750 to 3600	2.0 to 4.0	2.7	3.8	3.8	5.4
	3.2			2.9	4.2	3.9	5.6
	3.4- 3.6			3.3	4.8	4.1	5.9
	3.8- 4.2			3.8	5.5	4.3	6.3
	4.6- 7.0			4.9	7.1	4.9	7.1
B	4.6	1160 to 1800	2.0 to 4.0	5.1	7.4	7.1	10
	5.0- 5.2			5.8	8.5	7.3	11
	5.4- 5.6			6.2	9.1	7.4	11
	6.0- 6.8			7.1	10	7.7	11
	7.4- 9.4			8.1	12	7.9	12
C	7.0	870 to 1800	2.0 to 4.0	9.1	13	12	18
	7.5			9.7	14	12	18
	8.0- 8.5			11	16	13	18
	9.0-10.5			12	18	13	19
	11.0-16.0			14	21	13	19
D	12.0-13.0	690 to 1200	2.0 to 4.0	19	27	19	28
	13.5-15.5			21	30	21	31
	16.0-22.0			24	36	25	36
E	21.6-24.0	435 to 900	2.0 to 4.0	32	47	—	—

Source: Courtesy of The Gates Rubber Company.

Multiply the taped outside circumference of the belt at no tension by each of the belt length multipliers. This gives the elongated outside circumferences.

With the belt installed on the drive, tighten it until the taped outside circumference falls between the elongated outside circumferences calculated above.

IV. INSTRUMENTS FOR USE IN MEASURING STATIC TENSION

Gauges are available for measuring the force and deflection for a belt drive. Most belt manufacturers sell simplified, inexpensive tension gauges.

A. Spring Scales

Force is generally measured with a spring scale. The push type is placed against the top of the belt and shows the force required to deflect the belt. Figure 2 shows a push type gauge with sliding O-rings which can be preset to the required deflection force and deflection distance.

Pull-type scales are also used where the belt is pulled down with the scale and the scale then displays the force. Figure 3 shows a pull type scale in use.

A steel scale is used with both to measure the deflection distance. Some spring scales have a steel scale attached, as illustrated in Figure 4.

B. Commercial Gauges

Special gauges are available for field tensioning of belt drives. The most common uses a portion of the span length. The belt is clamped between two fixed points by a spring-loaded plunger. Tension is then measured by the amount of compression in the spring. Figure 5 shows such a device. Variables such as belt strain constant, belt stiffness, and belt thickness reduce the overall accuracy of such gauges, but they can be calibrated for fairly accurate use for specific belt types.

C. Special Measuring Gauges

Special devices such as the one shown in Figure 6 can be used to measure both deflection force and deflection distance for high volume applications such as an assembly line. Such a device would consist of a steel bar that can be laid from one sheave to another with an adjustable guide set so that a gauge and deflecting plunger is at the center

TABLE 4 Belt Length Multipliers

Ts Per Strand (Lbs.)	Cross Section									
	3VX	5VX	5V	8V	A*	B		C		D
					Equal To or Less Than 128″ Length	Equal To or Less Than 128″ Length	Over 128″ Length	Equal To or Less Than 128″ Length	Over 128″ Length	
10	1.00091	1.00033	1.00053	1.00029	1.00048	1.00042	1.00050	1.00025	1.00033	1.00017
12	1.00109	1.00040	1.00063	1.00034	1.00057	1.00050	1.00060	1.00030	1.00040	1.00021
14	1.00127	1.00047	1.00074	1.00040	1.00067	1.00058	1.00070	1.00035	1.00047	1.00024
16	1.00145	1.00053	1.00084	1.00046	1.00076	1.00067	1.00080	1.00040	1.00053	1.00028
18	1.00164	1.00060	1.00095	1.00051	1.00086	1.00075	1.00090	1.00045	1.00060	1.00031
20	1.00182	1.00067	1.00105	1.00057	1.00095	1.00083	1.00100	1.00050	1.00067	1.00034
24	1.00218	1.00080	1.00126	1.00069	1.00114	1.00100	1.00120	1.00060	1.00080	1.00041
28	1.00255	1.00093	1.00147	1.00080	1.00133	1.00117	1.00140	1.00070	1.00093	1.00048
32	1.00291	1.00107	1.00168	1.00091	1.00152	1.00133	1.00160	1.00080	1.00107	1.00055
36	1.00327	1.00120	1.00189	1.00103	1.00171	1.00150	1.00180	1.00090	1.00120	1.00062
40	1.00364	1.00133	1.00211	1.00114	1.00190	1.00167	1.00200	1.00100	1.00133	1.00069
45	1.00409	1.00150	1.00237	1.00129	1.00214	1.00187	1.00225	1.00112	1.00150	1.00078
50	1.00455	1.00167	1.00263	1.00143	1.00238	1.00208	1.00250	1.00125	1.00167	1.00086
55	1.00500	1.00183	1.00289	1.00157	1.00262	1.00229	1.00275	1.00137	1.00183	1.00095
60	1.00545	1.00200	1.00316	1.00171	1.00286	1.00250	1.00300	1.00150	1.00200	1.00103
65	1.00591	1.00217	1.00342	1.00186	1.00309	1.00271	1.00325	1.00162	1.00217	1.00112
70	1.00636	1.00233	1.00368	1.00200	1.00333	1.00292	1.00350	1.00175	1.00233	1.00121
75	1.00682	1.00250	1.00395	1.00214	1.00357	1.00312	1.00375	1.00187	1.00250	1.00129
80	1.00727	1.00267	1.00421	1.00229	1.00381	1.00333	1.00400	1.00200	1.00267	1.00138
85	1.00773	1.00283	1.00447	1.00243	1.00405	1.00354	1.00425	1.00212	1.00283	1.00146
90	1.00818	1.00300	1.00474	1.00257	1.00428	1.00375	1.00450	1.00225	1.00300	1.00155
95	1.00864	1.00317	1.00500	1.00271	1.00452	1.00396	1.00475	1.00237	1.00317	1.00164
100	1.00909	1.00333	1.00526	1.00286	1.00476	1.00417	1.00500	1.00250	1.00333	1.00172
120	1.01091	1.00400	1.00632	1.00343	1.00571	1.00500	1.00600	1.00300	1.00400	1.00207
140	1.01273	1.00467	1.00737	1.00400	1.00667	1.00583	1.00700	1.00350	1.00467	1.00241

Ts Per Strand (Lbs.)	Cross Section									
	3VX	5VX	5V	8V	A*	B		C		D
					Equal To or Less Than 128″ Length	Equal To or Less Than 128″ Length	Over 128″ Length	Equal To or Less Than 128″ Length	Over 128″ Length	
160	1.01455	1.00533	1.00842	1.00457	1.00762	1.00667	1.00800	1.00400	1.00533	1.00276
180	1.01636	1.00600	1.00947	1.00514	1.00857	1.00750	1.00900	1.00450	1.00600	1.00310
200	1.01818	1.00667	1.01053	1.00571	1.00952	1.00833	1.01000	1.00500	1.00667	1.00345
240	1.02182	1.00800	1.01263	1.00686	1.01143	1.01000	1.01200	1.00600	1.00800	1.00414
280	1.02545	1.00933	1.01474	1.00800	1.01333	1.01167	1.01400	1.00700	1.00933	1.00483
320	1.02909	1.01067	1.01684	1.00914	1.01524	1.01333	1.01600	1.00800	1.01067	1.00552
360	1.03273	1.01200	1.01895	1.01029	1.01714	1.01500	1.01800	1.00900	1.01200	1.00621
400	1.03636	1.01333	1.02105	1.01143	1.01905	1.01667	1.02000	1.01000	1.01333	1.00690
450	1.04091	1.01500	1.02368	1.01286	1.02143	1.01875	1.02250	1.01125	1.01500	1.00776
500	1.04545	1.01667	1.02632	1.01429	1.02381	1.02083	1.02500	1.01250	1.01667	1.00862
550	1.05000	1.01833	1.02895	1.01571	1.02619	1.02292	1.02750	1.01375	1.01833	1.00948
600	1.05455	1.02000	1.03158	1.01714	1.02857	1.02500	1.03000	1.01500	1.02000	1.01034
650	1.05909	1.02167	1.03421	1.01857	1.03095	1.02708	1.03250	1.01625	1.02167	1.01121
700	1.06364	1.02333	1.03684	1.02000	1.03333	1.02917	1.03500	1.01750	1.02333	1.01207
750	1.06818	1.02500	1.03947	1.02143	1.03571	1.03125	1.03750	1.01875	1.02500	1.01293
800	1.07273	1.02667	1.04211	1.02286	1.03809	1.03333	1.04000	1.02000	1.02667	1.01379
850	1.07727	1.02833	1.04474	1.02429	1.04048	1.03542	1.04250	1.02125	1.02833	1.01466
900	1.08182	1.03000	1.04737	1.02571	1.04286	1.03750	1.04500	1.02250	1.03000	1.01552
950	1.08636	1.03167	1.05000	1.02714	1.04524	1.03958	1.04750	1.02375	1.03167	1.01638
1000	1.09091	1.03333	1.05263	1.02857	1.04762	1.04167	1.05000	1.02500	1.03333	1.01724

[a] A section powerband belts are not a standard. For availability, check with your local Gates Field Engineer.

Source: Courtesy of The Gates Rubber Company.

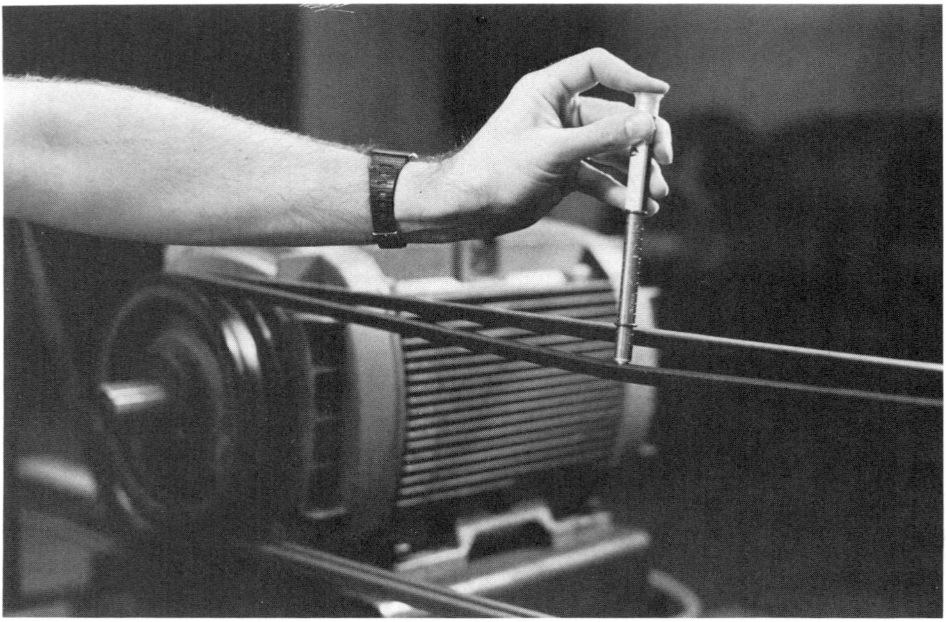

FIGURE 2 Push-type tension tester.

of the span. A collar on the plunger is adjusted to the required deflection (q). The plunger is depressed until this collar hits the top. The deflection force is read on the scale. Once a device like this is set for a particular drive, it can be used to measure tension very quickly on a production line basis.

V. AUTOMATIC TENSION DEVICES

A typical locked-center drive has a sawtoothed decay tension profile. Figure 7 shows this sawtooth effect as belt tension decays and the belt is retensioned. Even though the average tension on a locked-center drive is less than these high initial values, it is greater than can be achieved with automatic tensioning.

A. Advantages of Constant Tension Devices

The major advantage of constant or automatic tensioning is that it makes possible a lower average belt tension, and somewhat longer

FIGURE 3 Pull-type scale for measuring belt tension.

life for the drive components. It also eliminates the need for manual take-up, reducing maintenance costs and ensuring a satisfactory drive tension at all times. This is particularly important for drives that are hard to reach.

In designing constant tensioning methods for belt drives, do not overlook some of the possible disadvantages, depending on the method used. Periodic load fluctuations or extremely high shock loads can create undesirable component vibration, belt instability, or belt breakage unless this is compensated for by additional mechanical components. Reversing-load drives create changing tension requirements that are not easy to handle by most automatic-tensioning methods. The additional automatic-tensioning components add cost to the drive.

B. Spring-Loaded Idlers

Spring-loaded idlers are used extensively on drives that encounter wide load variations along with high peak loads (but not extremely high shock loads). A locked-center drive must be tensioned to

FIGURE 4 Spring scale and steel scale used to measure belt tension.

prevent slip at the peak load; and as the load drops off, the total tension remains nearly constant at the peak-load value. A spring-loaded idler, however, allows the total tension to decrease as the load drops off, and thus increases belt and bearing life.

This effect can be illustrated by an example as shown in Figure 8. A V-belt drive uses a spring-loaded idler on the slack side. Assume there is a ratio of 5:1 between the tight- and slack-side tensions of the drive. The HP load is such that an effective pull (T_T - T_S) of 40 lb is required. The idler spring must be sized, to produce a 10-lb slack-side tension (T_T = 50 lb and T_S = 10 lb). This tension remains constant regardless of whether the load increases or decreases.

If the load were taken completely off the drive shaft, both T_T and T_S would be 10 lb and total tension (T_T + T_S) would be 20 lb. As load increases, T_T goes up while T_S stays at 10 lb. At half the design HP, T_T - T_S is 20 lb and since T_S is a constant 10 lb, T_T would be 30 lb and T_T + T_S would be 40 lb – compared to the 60 lb at full load.

FIGURE 5 Special commercial gauge.

As a comparison, with the same drive using a fixed idler (locked-center drive) at half design HP, effective pull ($T_T - T_S$) is 20 lb, and total tension ($T_T + T_S$) is still 60 lb. T_T will be 40 lb and T_S will be 20 lb. Thus, in the spring-loaded system, T_S is constant and total tension varies with the load, whereas, with a fixed idler, total tension is constant and T_T and T_S vary with the effective pull required.

The recommended location and size of idlers are the same for spring-loaded idlers as they are for conventional idlers. A spring-loaded idler must be placed on the slack side of the drive to effect the reduction in $T_T + T_S$. It should also be as close to the driver as possible. Inside idlers should be no smaller than the smallest loaded sheave and back-side idlers should be at least 1/3 greater than the smallest loaded sheave. Minimum recommended diameters for belt cross sections can be found in the appropriate industry engineering standards.

Spring-loaded idlers should not be used on reversing drives. When the drive reverses, the idler is then on the tight side.

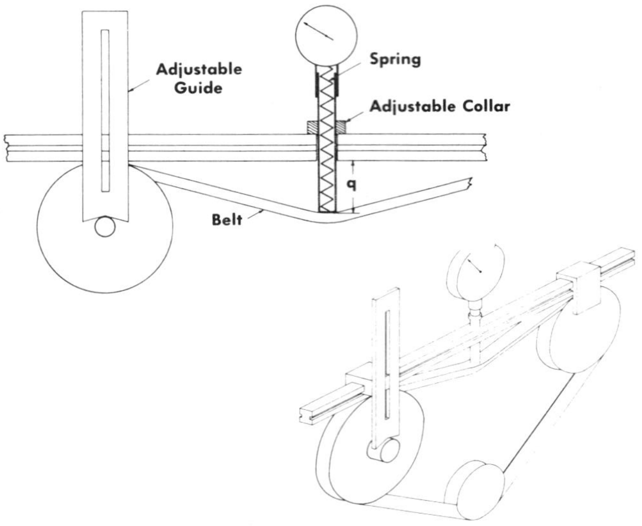

FIGURE 6 Special tension measuring gauge.

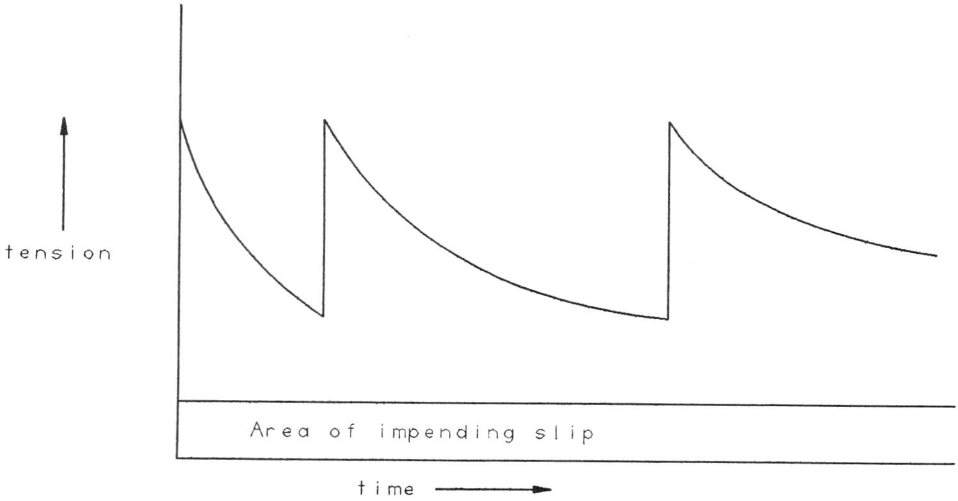

FIGURE 7 Tension vs. time for locked-center drive.

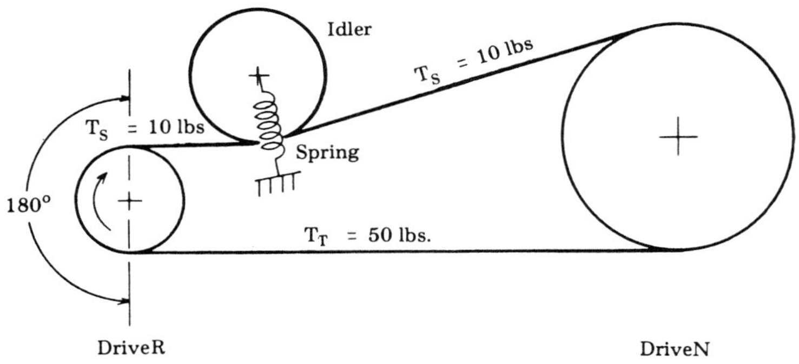

FIGURE 8 Spring-loaded idler.

Idler force is the force that the idler spring and the weight of the idler assembly impose on the belt span through a given type of linkage. This may be quite different from the spring force because of the weight and linkage factors. The span tension, idler force, and linkage assembly system must first be known before determining the actual required spring force.

To determine the required idler force and direction:

1. Lay out the drive to scale and measure the arcs of contact to find the smallest arc.
2. Calculate the span tensions.
3. On the scale drawing, determine the resultant vector force of those tensions. This is the idler force, but in the opposite direction.

For example, in Figure 9, the two T_S vectors are drawn to scale with arrows on them. The resultant vector indicates the idler must push down on the belt with 8.5 lb to impose the 10 lb T_S. The direction of the force is 86° from the drive center-line.

The example in Figure 9 shows the idler in only one position. However, the idler will be in several positions during the life of the drive. The idler force should be calculated with the idler in the new-belt position, and also in the old-belt position (idler further in due to wear and elongation). Since the forces will be slightly different due to different arcs of contact and different angles, use the higher of the two idler forces calculated.

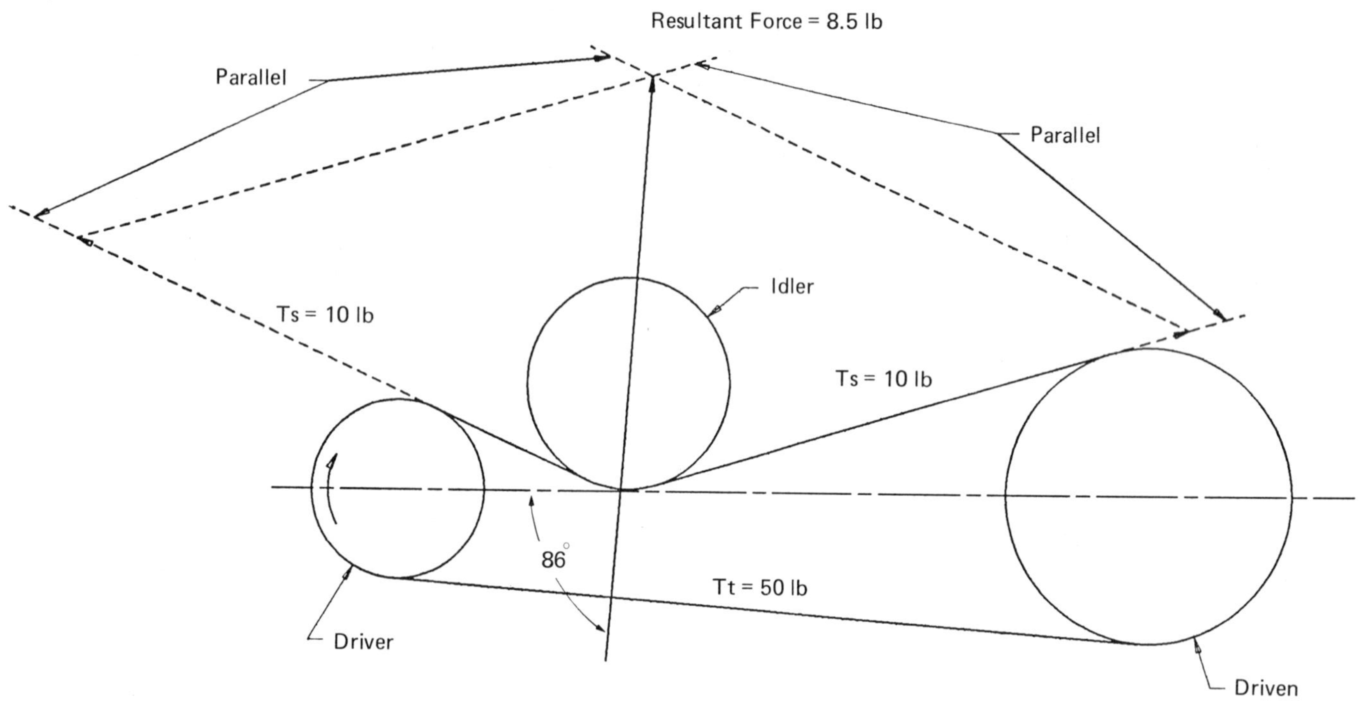

FIGURE 9 Determining idler force.

With automatic tensioning, it is also possible to lower all the tensions somewhat. This is possible because a spring-loaded idler drive does not experience a drop in tension, and therefore the tensions can be somewhat less than for a locked-center drive.

Design of such reduced-tension drives should always be done by the manufacturer, because care must be taken to ensure that the idler force is not reduced to a point where belt slippage occurs. This principle of lowering all the tensions applies also to the following other methods of automatic tensioning.

C. Weighted Idlers

Weighted idlers (sometimes called gravity idlers) are a special case of spring-loaded idler where the weight of the assembly is the idler force and no spring is involved. Belt take-up is accomplished simply by the idler moving in a slot or on a pivot arm by the force of gravity. A drive subject to shock should not use this type of automatic tensioning. Severe idler bounce could result. The advantages are much the same as those for a spring-loaded idler. However, the drive must be a permanent installation to rely on gravity to operate.

D. Spring-Loaded Bases

Spring-loaded prime-mover bases are usually called spring-loaded motor bases because they are used more with electric motors than with other types of prime movers. The operating principle is shown in Figure 10.

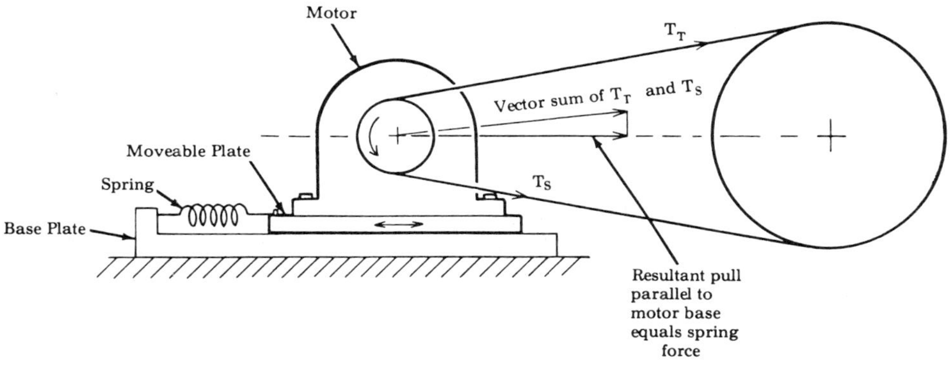

FIGURE 10 Spring-loaded motor base.

Here, the spring force must be such that the motor is pulled back against the drive with a force equal and opposite to the resultant belt pull, parallel to the base. The spring force required can be calculated from the drive data (finding the vector sum of T_T and T_S) as with spring-loaded idlers. The spring must have some extra force, however, to overcome the friction in the base. If the mounting is not horizontal, the effect of motor weight must be taken into account. Spring-loaded motor bases, unlike spring-loaded idlers, provide a nearly constant total drive tension ($T_T + T_S$).

Since spring-loaded motor bases are commercially available, quite often each individual drive is not analyzed for its exact spring force, but relies instead on an adjustment of the spring tension. Some spring force is lost as the belt seats in and elongates, and occasionally spring adjustments may be necessary unless the drive is overtensioned when the belt is new.

Spring-loaded bases are most commonly used on industrial drives with motors from 5 to 10 HP, and occasionally on larger drives. Their main advantage is to maximize the amount of time between tension adjustments.

E. Reactive-Torque Motor Bases

Reactive-torque motor bases use the reactive torque of the motor to help tension the drive. When a motor is delivering power to a sheave or pulley, its frame tries to rotate in a direction opposite to that of the sheave or pulley as illustrated in Figure 11. With the motor in a

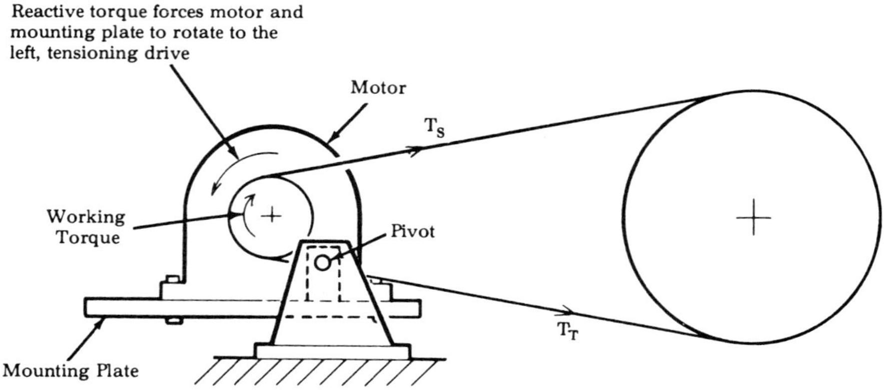

FIGURE 11 Reactive-torque motor base.

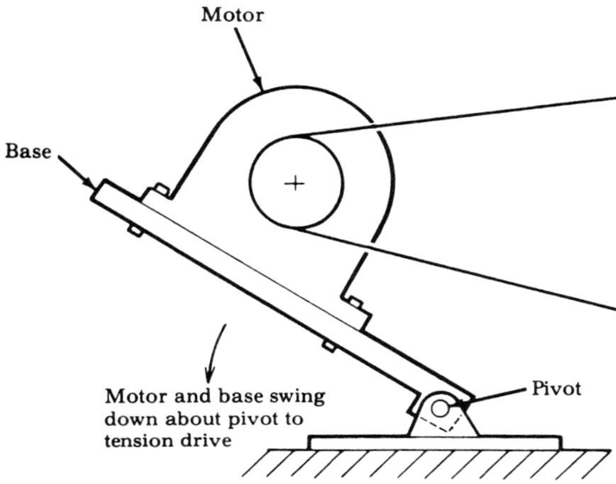

FIGURE 12 Standard gravity-type motor base.

cradle with a pivot point near the motor shaft, the motor can move away from the driven sheave, tensioning the drive. As torque increases, tension increases. This reactive-torque base is the same as the gravity base, except that the pivot point on the gravity base is quite far away from the motor. In the gravity type, reactive torque has little effect on drive tension.

The reactive-torque motor base must be installed so that the direction of rotation is correct. If the motor shaft were rotating in the opposite direction, the reactive torque would actually loosen the drive, rather than tighten it. Operating in the same direction, both the reactive torque and the motor weight act to tension the drive.

F. Standard Gravity Bases

Standard gravity bases use the weight of the motor to tension the drive. The bottom plate of this type base is usually slotted so that the motor can be moved as needed to get the proper tension when the drive is installed. Figure 12 shows how the gravity base operates. Although specification data usually shows a given setting (or a certain specified base) for a certain National Electrical Manufacturers Association (NEMA) motor frame size, it is good practice to check for correct tension after the drive is installed. This is done by imposing the highest load on the motor after the belts are run in, to see if the drive slips. Adjustment of the pivot point can also be made if the tension is not correct. This type of drive should not be used where shock loading could cause motor bounce. It is used primarily on drives such as ventilating fans where drive accessibility is difficult.

VI. BEARING LOADS

The part of a bearing load caused by a belt drive depends upon the shaft load imposed and the location of that shaft load with respect to the bearings. Shaft load depends upon belt pull, sheave or pulley weight, and shaft weight. Belt pull is a function of tensions in the belt. Because the belt tensions under operating conditions are the only ones that affect bearings most of the time, they are used for bearing capacity analysis. These are usually called working tensions.

Some of the above factors depend upon the characteristics of the drive such as the amount of power being transmitted, linear belt speed, and arc of contact. Figure 13 shows the influence of these characteristics on the various bearing load factors.

A. Working Tensions

An assumption must be made to determine working tensions of locked-center belt drives: Centers are locked. Therefore, tensions decrease during operation. The assumption must be made because, as shown in Figure 13, working tensions are a function of the drive characteristics and the installation and maintenance procedures. The assumption is that the ratio of the tensions between the tight and slack sides of a standard two-sheave drive, with sheaves of equal diameter, averages 5:1 during its life. This ratio is the design-tension ratio. The assumption of a value of 5:1 is standard practice in the industry. For drives with unequal sheave diameters (see Figure 14), this ratio must be corrected for arc of contact.

When using standard practice (design-tension ratio equals 5:1) the working tensions of a drive are given by:

$$T_T = 41,250 \left(\frac{HP}{GV}\right)$$

$$T_S = 33,000 \; (1.25 - G) \left(\frac{HP}{GV}\right)$$

where

T_T = tight side tension (lb)
T_S = slack side tension (lb)

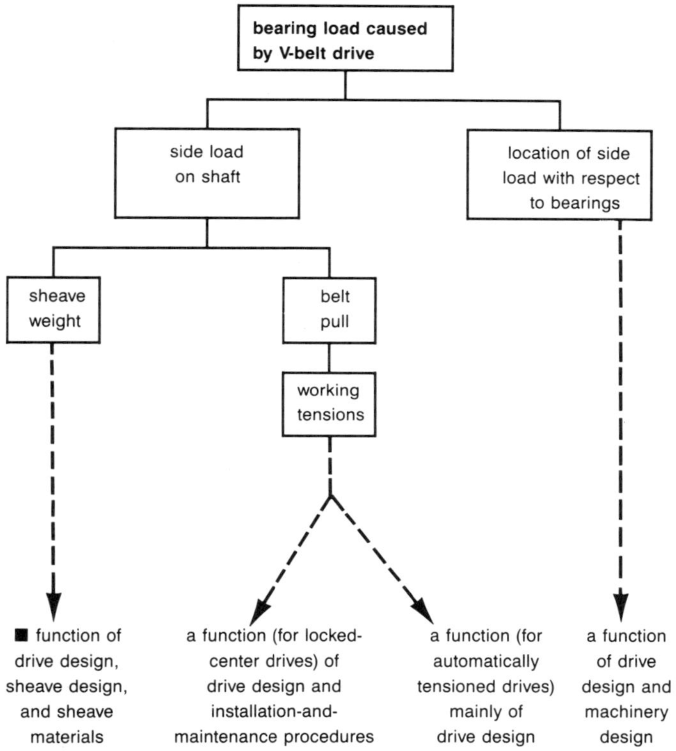

FIGURE 13 Summary of design factors.

HP = design horsepower of drive
G = arc of contact correction factor (see Table 1)
V = belt speed (ft/min)

The T_T and T_S of an automatically tensioned belt drive can be calculated by a different method. For example, to calculate the T_T and T_S of a spring-loaded idler drive, first translate spring force into belt tension for the span tightened by the idler. Then, calculate the tension in the other span from power transmitted. It may be a good idea to have a manufacturer of belts or sheaves check working-tension calculations for automatically tensioned drives.

Working tensions, and therefore, belt pull, are not a function of the number of belts in a drive. However, the number of belts affects the location of the shaft load with respect to the bearings.

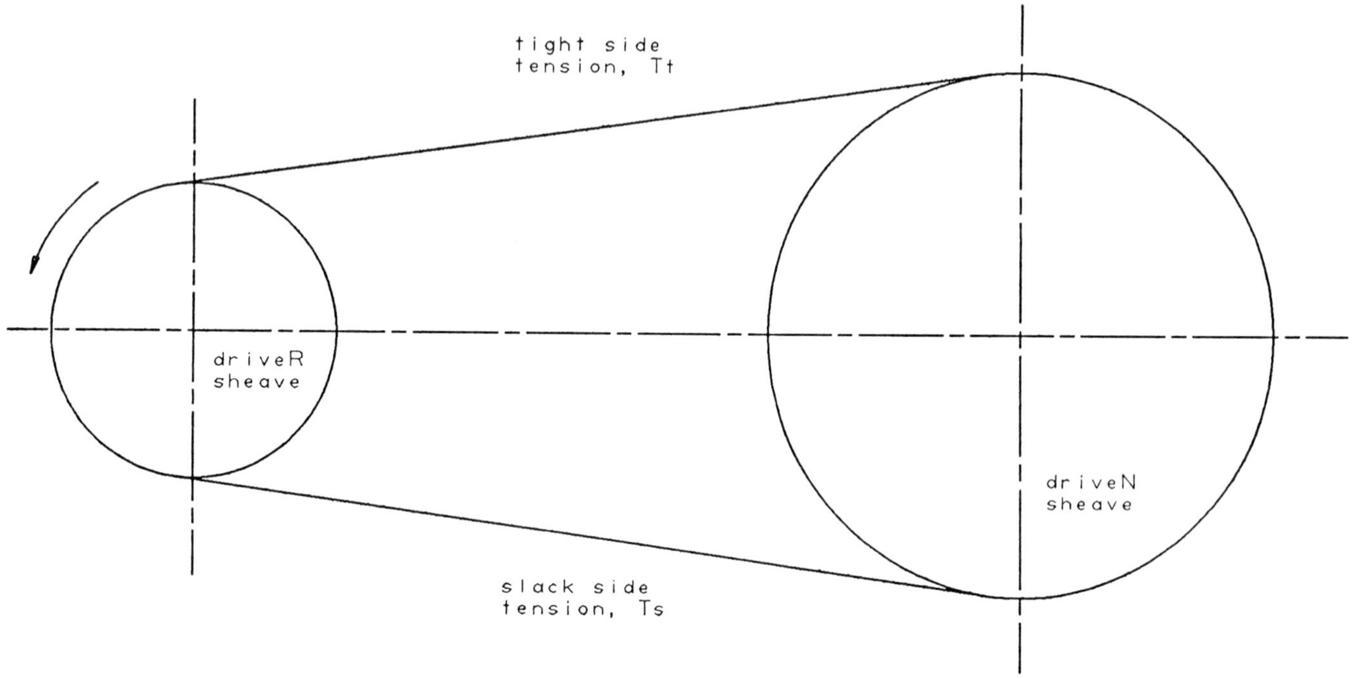

FIGURE 14 Two-sheave drive with unequal diameters.

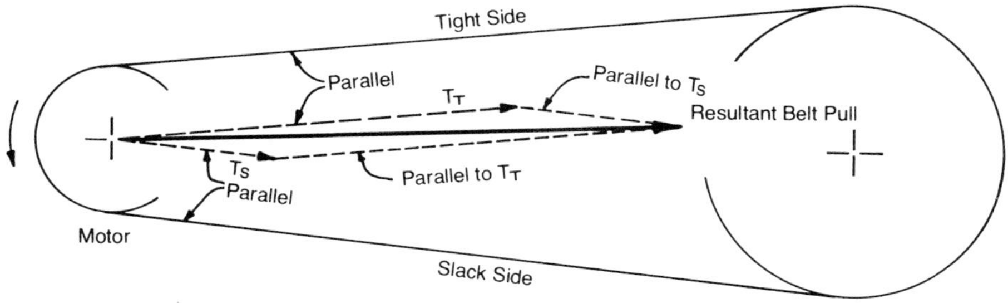

FIGURE 15 Graphic addition of T_T and T_S.

B. Belt Pull

Once T_T and T_S have been determined, it is simple to find their vector sum: the belt pull on the shaft. The vector sum can be found graphically or mathematically.

To use the graphical method, first draw the T_T and T_S vectors to a convenient scale (for example, 1 in. = 100 lb) on a layout, parallel to the tight and slack sides of the drive, respectively. Then add them vectorially, as shown in Figure 15. This figure shows vector addition for belt pull on the driver shaft. If the direction of the pull isn't needed, then use the simpler mathematical method. Add T_T and T_S arithmetically. Then, multiply the sum ($T_T + T_S$) by the vector-sum-correction factor from Figure 16. The result will be belt pull.

For two-sheave drives, belt pulls on the driveR and driveN shafts are equal but opposite. For drives using idlers, both magnitude and direction may be different. *Note*: It is important to recognize that belts can be tensioned between 1.5 to 2.0 times the design tension. This higher tension doesn't exist for the life of the drive, but bearings and shafts must be able to tolerate it, without damage, for a reasonable period of time.

C. Shaft Load

Sometimes it is desirable to neglect sheave weight in a bearing-load analysis for rough checks or if the effect of the drive needs to be isolated. But, if true load on the shaft, including sheave weight (and bushing, if required), is needed, the sheave weight (including bushing, if required) must be added to the belt pull using the same graphical method shown in Figure 15. The sheave weight vector is vertical to the ground. Weights for standard sheaves, sprockets, and bushings are available from belt and sheave manufacturers.

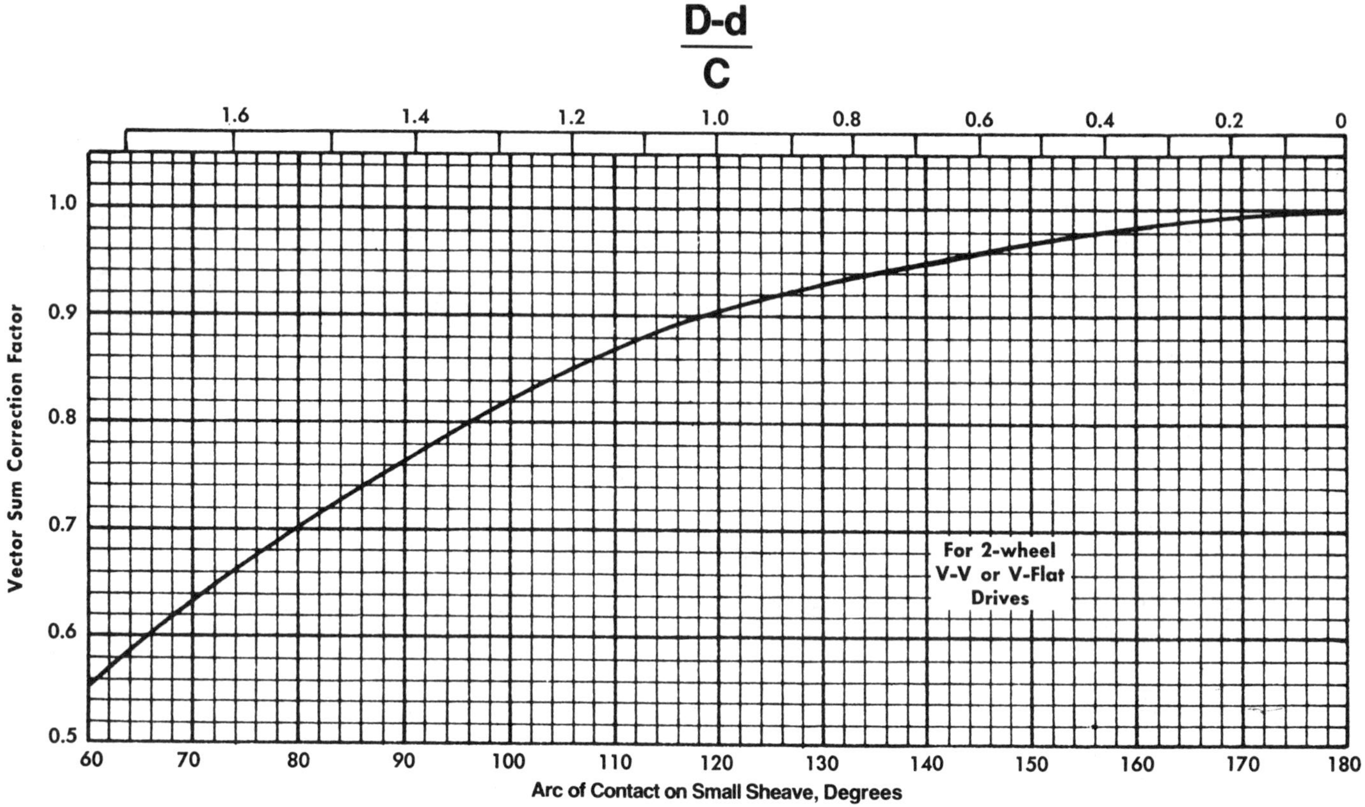

FIGURE 16 Vector sum correction factor.

D. Bearing Load Calculations

In order to find actual bearing loads, it is necessary to know weights of machine components and the value of all other forces contributing to the load. However, as mentioned above, it is sometimes desirable to know the bearing load contributed by the V-belt drive alone. The bearing load due to the drive can be found by knowing the bearing spacing with respect to the sheave center and the shaft load as calculated above. For rough checks, use belt pull alone, ignoring sheave weight. If accuracy is desired, or if the sheave is unusually heavy, actual shaft load including sheave weight should be used. It is usual to assume that the load on the shaft acts at the center of the sheave or pulley.

To determine bearing loads for an overhung sheave, use:

$$\text{load at B (lb)} = \frac{\text{shaft load} \times (a + b)}{a}$$

$$\text{load at A (lb)} = \text{shaft load} \times \frac{b}{a}$$

where

a and b = spacing (in.) per Figure 17.

To determine bearing loads for a sheave between bearings, use:

$$\text{load at D (lb)} = \frac{\text{shaft load} \times c}{(c + d)}$$

$$\text{load at C (lb)} = \frac{\text{shaft load} \times d}{(c + d)}$$

where

c and d = spacing (in.) per Figure 18.

E. Special Procedures

The National Electrical Manufacturers Association (NEMA) publishes recommendations for minimum sheave and sprocket diameters for belt drives applied to electric motors. These recommendations are based on bearing and shaft capabilities and calculation procedures similar to those noted above. Following these recommendations should minimize bearing problems.

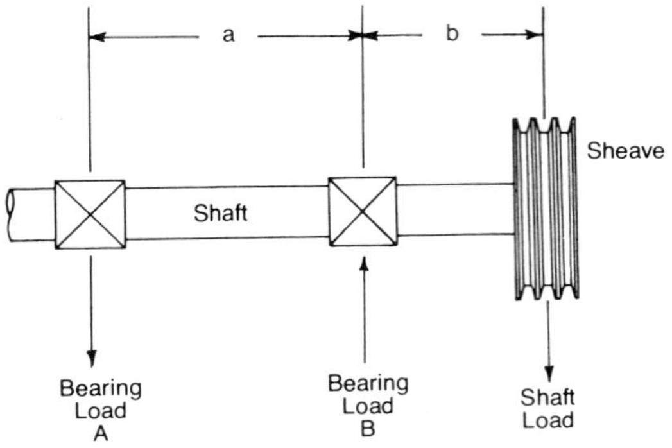

FIGURE 17 Overhung sheave.

Manufacturers of some types of equipment (clutches, speed reducers, and power take-offs, for example) sometimes specify maximum values for belt pull and sheave overhang. Some of their formulas for belt pull differ slightly from those listed above. Usually their formulas differ because they include service factors specific for that equipment. Follow the formulas for the appropriate equipment.

Some handbooks use short cuts, ignoring the effects of weight and other factors or assuming average values for them. Short-cut procedures are usually satisfactory. However, for a given type of machinery, compare them at least once with the fundamental procedures given here, to determine their accuracy.

As noted in the belt pull section (VI.B.), new V-belts are usually installed at tensions of 1.5 to 2.0 times the design tension to provide for initial tension decay. Although this initial overtensioning exists for only a short period, it should be taken into account during final evaluation of the bearing and shaft design.

F. Reduce Bearing Loads

To reduce the part of the bearing load caused by belt drives, reduce at least one of these factors:

1. Working tension, T_T and T_S, by

   - Reducing design HP
   - Increasing linear belt speed
   - Increasing arc of contact
   - Changing from flat belts to V-belts

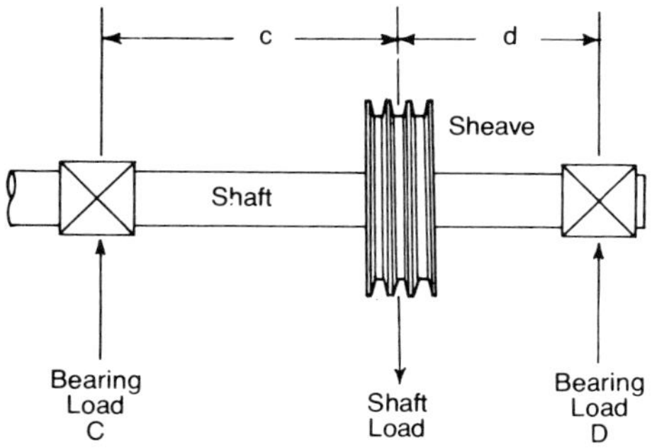

FIGURE 18 Sheave between bearings.

2. Sheave or pulley weight, by
   - Reducing amount of material
   - Using drive of narrower face width
3. Moment arms, by
   - Moving sheave or pulley closer to bearing
   - Using drive of narrower face width

12

Installation, Maintenance, and Replacement of Power Transmission Belts

CHARLES H. ARTUS *V-Belt Product Application, The Gates Rubber Company, Denver, Colorado*

I. INTRODUCTION

V-belt drives require very little maintenance. Usually, periodic belt retensioning is all that is needed. This is one of the major reasons for their wide usage.

There are several simple procedures and precautions that apply to maintenance and replacement of components. These are discussed in this chapter.

II. MAINTENANCE OF OPERATING DRIVES

A. Noise Problems

Properly operating drives make very little noise, but there are exceptions to be noted and corrected.

1. Slapping may indicate the belt is striking a guard or some other obstruction. Usually, it is possible to locate the trouble spot merely by listening. Then, with drive stopped, adequate clearance can be provided.
2. A squeal, or chirp, indicates belts are too loose. Stop the drive and retension the belts.
3. Noise from failed bearings or a loose mechanism in the related machinery may sound as though emanating from the drive. These noises will also have to be investigated and corrected.

B. Visual Evidence

1. Buildup of foreign material inside the guard: This can cause excessive wear on belts and sheaves, or even create a "hot spot." Guards should be inspected regularly to prevent this.
2. Presence of oil on belts or sheaves: Belts will tolerate occasional exposure to oil vapor or slight mist. However, a buildup of oil will eventually deteriorate the belts or cause belt slippage. If excessive oil is present, the source should be eliminated, or additional protection provided for the drive.
3. Excessive vibration: This can result from a number of causes such as loose machine mounts, imbalance in rotating parts, or an intermittent or improperly operating power source. If the cause cannot be readily determined and corrected, it is advisable to contact a belt and sheave supplier for technical assistance. *Note*: See Table 1 for troubleshooting tips.
4. If one belt in a set sags excessively, or seems very loose, it may be necessary to retension the drive, or examine the belt for cord damage.

III. REPLACEMENT OF BELTS AND SHEAVES

A. When to Replace Belts

On many drives, belts are replaced only when they fail to transmit power. This failure can result from belts being broken or a lack of take-up due to stretched belts or worn sheaves. On large machines with multiple-belt drives, it is best to examine drives on a regular basis for evidence of belt wear or sheave wear, and replace them before a breakdown occurs. If one belt fails on a multiple drive, the drive will usually continue to operate, but the entire set of belts should be replaced very soon.

B. Selection of Replacement Belts

When a set of belts fails, or becomes excessively worn and needs replacement, always select the same cross section and length. Replace all of the belts. A mixture of old and new belts will not give good service. Be sure the new set is matched, that is, all the belts must be within certain limits of length tolerance for good drive performance. The supplier will furnish matched sets when specified on the order. All belts should be from the same manufacturer. (See Section III.E. regarding measuring tools which help identify belts.)

It is possible to fill up only some of the grooves by using higher rated belts. However, this will result in wearing those grooves only. If a replacement set of standard belts filling all grooves is installed later, performance may not be satisfactory.

C. Safety Tips

Before starting any maintenance work on drives, all drive sources should be off and all equipment in a safe position. (Refer to Section VI.C.)

D. Removal of Old Belts

Most drives have two sheaves: one on the motor and one on the driveN machine. The motor base usually has slotted holes to allow for adjustment toward and away from the driveN machine. Loosen the motor hold down bolts and move the motor *toward* the driveN machine to the inner end of the adjustment to release tension on the belts. Use a pry bar against the motor base, if needed. The belts should now be quite loose. If the drive has an idler instead of an adjustable motor base, loosen the hold down bolts and swing the idler *away* from the belts, allowing them to hang loose. *Be sure to use all the adjustment in the drive to loosen belts completely.* Now remove

TABLE 1 Troubleshooter's Guide for V-Belt Drives

Trouble area and observation	Cause	Remedy
Belt stretch beyond takeup		
Belts stretch unequally.	Misaligned drive, unequal work done by belts.	Realign and retension drive.
	Belt tensile member broken from improper installation.	Replace all belts with new matched set, properly installed.
All belts stretch about equally.	Insufficient take-up allowance.	Check take-up and follow allowance recommended by belt manufacturer.
	Greatly overloaded or underdesigned drive.	Redesign drive using manuals.
Short belt life		
Relatively rapid failure, no visible reason.	Tensile members damaged through improper installation.	Replace with all new matched set, properly installed.
	Worn sheave grooves (check with groove gauge).	Replace sheaves.
	Underdesigned drive.	Redesign drive using manuals.

Sidewalls soft and sticky. Low adhesive between cover plies. Cross section swollen.	Oil or grease on belts or sheaves.	Remove source of oil or grease. Clean belts and grooves with cloth moistened with nonflammable, nontoxic degreasing agent or commercial detergent and water.
Sidewalls dry and hard. Low adhesion between cover plies. Bottom of belt cracked.	High temperatures.	Remove source of heat. Ventilate drive better.
Deterioration of rubber compounds used in belt.	Belt dressing.	Never use dressing on V-belts. Clean belts and grooves with cloth moistened with nonflammable, nontoxic degreasing agent or commercial detergent and water. Tension drive properly to prevent slip.
Extreme cover wear.	Belts rub against belt guard or other obstruction.	Remove obstruction or align drive to give needed clearance.
Spin burns on belt.	Belts slip under starting or stalling load.	Retension drive until slipping stops.
Bottom of belt cracked.	Sheaves too small.	Redesign for larger sheaves using manuals.

(continued)

TABLE 1 (Continued)

Trouble area and observation	Cause	Remedy
[Short belt life]		
Broken belts.	Object falling into or hitting drive.	Replace with new matched set of belts. Provide shield for drive.
Belt turnover		
	Excess lateral belt slip.	Use joined V-belt.
	Foreign materials in grooves.	Remove material. Shield drive.
	Misaligned sheaves.	Realign the drive.
	Worn sheave grooves (check with groove gauge).	Replace sheave.
	Tensile member broken through improper installation.	Replace with new matched set properly installed.
	Incorrectly placed flat idler pulley.	Carefully align flat idler on slack side of drive as close as possible to driveR sheave.
Belt noise		
	Belt slip.	Retension drive until it stops slipping.

Improper driven speed		
Incorrect driveR-driveN ratio.	Design error.	Use correct sheave.
Spin burns on belt.	Belt slip.	Retension drive until belt stops slipping.
Hot Bearings		
Drive overtensioned.	Worn grooves. Belts bottoming and will not transmit power until overtensioned.	Replace sheaves. Tension drive properly.
	Improper tensioning.	Retension drive.
Sheaves too small.	Motor manufacturer's sheave diameters not followed.	Redesign drive using manuals.
Poor bearing condition.	Underdesigned bearings or poor bearing maintenance.	Observe recommended bearing design and maintenance.
Sheaves out too far on shaft.	Error or obstruction problem.	Place sheaves as close as possible to bearings. Remove any obstruction preventing this.
Drive undertensioned.	Belts slipping and causing heat buildup.	Retension drive.

the belts carefully from each sheave. Do not move motor or machine after belt removal.

Do not pry belts off drive or attempt to roll them off by rotating the sheaves. In doing this, you can severely cut your fingers or hands and may receive permanent injury. This also could damage the belt tensile members.

E. Measuring Tools

The complete belt set includes three belt gauges: one for A, B, and C sections; one for D and E sections; and one for 3V, 5V, and 8V sections. To determine proper belt section, simply place the old belt in the gauge as far as it will go, and read the section marked on the gauge (refer to Figure 1).

The complete set of sheave gauges includes one for each heavy-duty V-belt section (A, B, C, D, E and 3V, 5V, 8V). Each gauge includes a template for all standard angles used with that section and sheave diameter, clearly marked for easy identification. For example: C: 34°, PD 6.00-7.99; 36°, PD 8.00-12.00; 38°, PD over 12.00.

To check sheave grooves for wear, simply select the proper gauge, find the correct template for the sheave diameter used and insert the template into the sheave groove until the rim of the gauge butts against the sheave land. Worn grooves are illustrated in Figure 1. If you can detect 1/16 in. of wear along one side of the groove, poor V-belt life will probably result. Also inspect the sheaves by hand for worn or chipped grooves, hub cracks, or damaged bushings. Wear gloves to avoid cutting fingers on sharp metal edges.

F. Sheave Replacement

If the sheaves need replacing, they should be attached to the motor shaft and driveN machine shaft with bushings as described below:

1. Types

Type QD (Figure 2): Slide QD hub on shaft, flange end first. Assemble key. Position QD hub on shaft. Tighten setscrew over key hand-tight with standard Allen wrench only. Do not use excessive force.

Slide large end of sheave taper bore into position over cone, aligning pullup bolt holes in sheave and tapped holes in flange of hub. Assemble pullup bolts and lock washers.

Tighten bolts alternately and evenly. Do not use extensions on wrench handles. There should be a gap between the face of the sheave hub and the flange of the QD hub to ensure a satisfactory cone grip and press fit. *This gap must not be closed.*

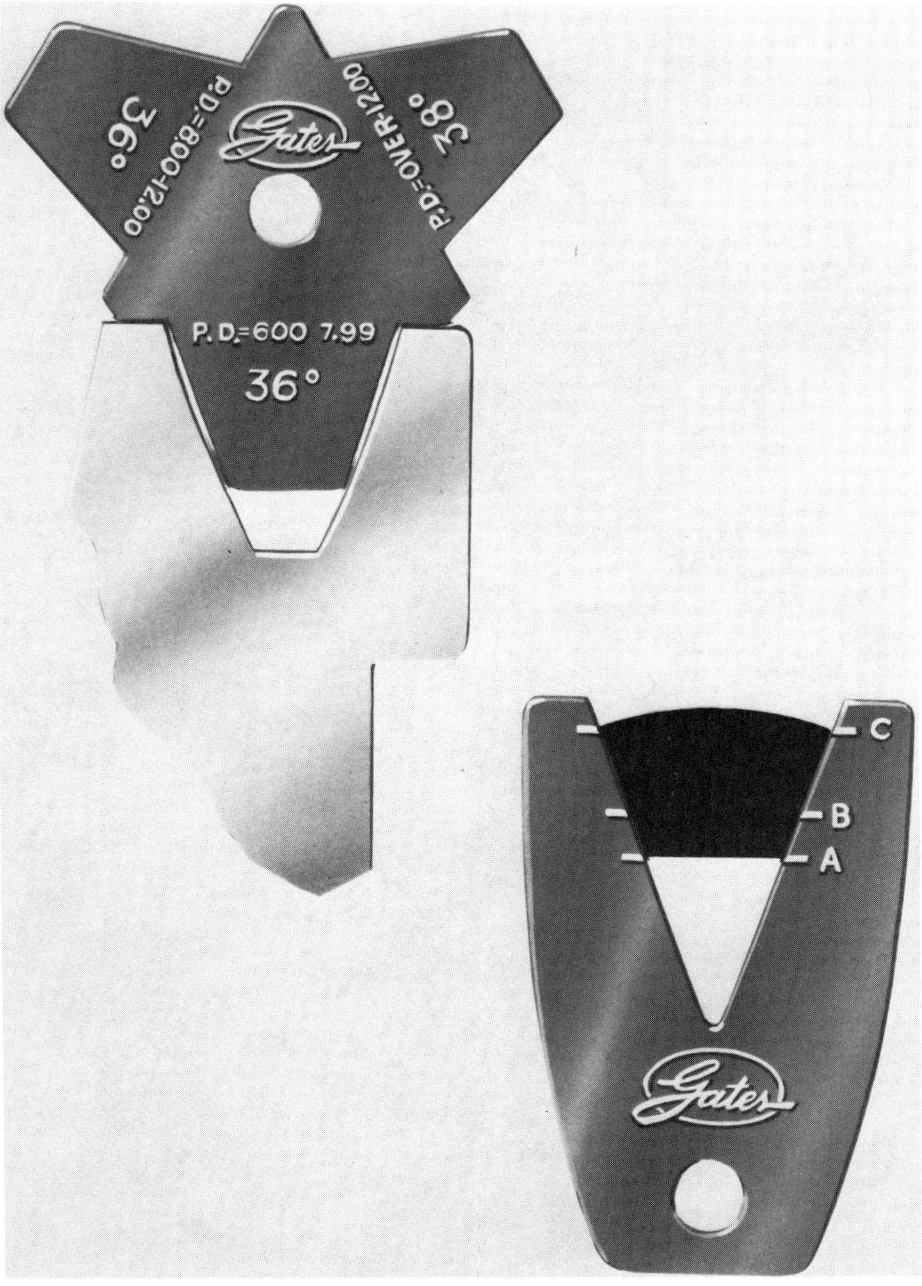

FIGURE 1 Gauges used for measuring sheave groove and belt wear.

FIGURE 2 Type QD sheave.

In addition to this standard mounting arrangement, most sheaves using J size and smaller hubs may have bolts inserted through hub flange into rim to provide alternate hub locations.

Type VT (Figure 3): Insert bushing in sheave. Match holes (not threads) and slip entire unit onto shaft.

Put screws into the holes that are threaded in the sheave only. Align the sheave and tighten screws. As the bushing is wedged inward it contracts and grips the shaft.

Important: With either type of sheave, always make sure that the mating surfaces of the sheave and hub are free of all foreign substances such as dirt, grease, or paint.

After sheaves and bushings are on the shaft, align them before final tightening. Install belts.

2. Alignment

V-belt drive alignment is not as critical as that required for most other types of drives, but unless the belts enter and leave the sheaves in a relatively straight line, wear is accelerated.

The two most common causes of misalignment are shown in Figure 4: (a) the shafts of the driveR and driveN machines are not parallel, and (b) the sheaves are not located properly on the shafts. To check alignment, all you need is a straight-edge or, for drives with longer centers, a steel tape. If these aren't available, you can, as a last resort, use heavy string. Just line the straight-edge or tape along the outside face of both sheaves as shown in the illustration. Misalignment will show up as a gap between the sheave face and straight-edge, or perhaps as a "break" in the tape or string. Make sure that the width of the outside land is equal on both sheaves when using this method.

A third cause of misalignment is tilted or improperly mounted sheaves. The best way to prevent this is to follow the mounting steps shown in Section III.F. Sheaves that are already installed can be checked for tilt by using a spirit level.

G. Tensioning

When the new set of belts has been installed and the sheaves aligned, the drive should be properly tensioned. The methods for V-belts are discussed in Chapter 11 and those for timing belts in Chapter 10.

IV. STORAGE

If extra belts are kept on hand, they should be stored:

In cartons as received
Coiled up on shelves
Hung on racks

FIGURE 3 Type VT sheave.

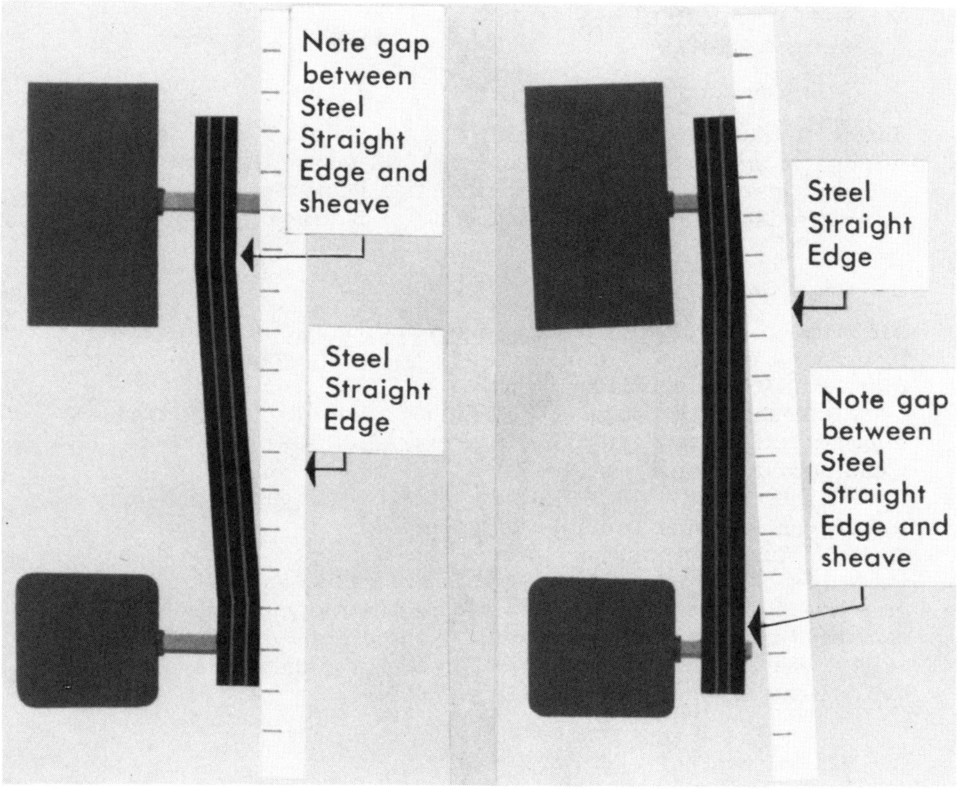

FIGURE 4 Determining alignment.

When hanging belts, wide supports or multiple hooks should be used to keep the bend diameter as large as possible. The storage area should be as cool as possible and not in direct sunlight. Stored sheaves and pulleys should be kept in the original carton whenever possible. If received unwrapped or on pallets, they should be protected with a light coating of oil or antirust agent. They should be stored away from excessive amounts of moisture. Sheaves and pulleys should not be allowed to bump together or roll on concrete floors. This could form burrs or nicks on the lands of the groove that are sharp enough to damage belts.

V. SAFETY PROCEDURES AND REQUIREMENTS

A. Safety Around Drives

Safety around V-belt drives calls for safe action. There are a number of positive steps that can be taken that will insure that belts can be operated in a safe manner and that routine inspection procedures can be accomplished safely and with a minimum of time and trouble.

B. Drive Guards

Virtually all regulatory agencies, including OSHA, API, and MSHA, insurance companies, and other safety authorities recommend or require that V-belt drives have guards.

A makeshift, partial guard can often be more dangerous than no guard at all. Such a guard could give a false sense of security and could encourage unsafe action.

OSHA Standard 1910.219 covers general requirements for guards for mechanical power transmission apparatus.

Drive guards should have provisions for proper ventilation and inspection of the drive through the use of grills, inspection doors, or removable panels. The guard should have no gap where workers can reach inside and become caught in the drive. Besides being a safety asset, a proper drive guard protects the drive from weather, debris, and tampering (Figure 5).

C. Maintenance Tips

When working near drives that are running, personnel should be alert to loose clothing — neckties, shirt tails, and sleeves. Even if drives are properly guarded, loose clothing can cause serious injury to personnel if it becomes entangled in a drive that is operating. Also, personnel should carefully watch the placement of their hands around drives that are running.

When drives are to be serviced, make sure equipment is turned off — even if the maintenance procedure is simple and brief. It is recommended that a warning sign be used to indicate that the machine is "down for maintenance." This is so that someone else won't turn the machine on while the drive is being serviced. Padlock the controls if possible, or remove the key from the starter if there is one.

Make sure all machine components are in a safe position or condition before servicing the drive. For example, weights should be put on an oilfield pumping unit in the "down" position so that shafts won't turn while the drive is being maintained. The same type of safe action should be taken with flywheels, counterweights, gears, and clutch mechanisms on other types of machinery. Always follow

FIGURE 5 Example of a vented belt guard.

equipment manufacturers' recommendations for safe maintenance procedures.

D. Avoid Overloading Sheaves

Sheaves are designed to carry a certain HP and torque. If drives are to carry more load than the normal belt HP design level, the sheave supplier should verify load carrying level. This also applies to serious shock loads that might occur.

E. Observe Speed Limitations

The standard upper rim speed limitation for cast iron sheaves is 6500 ft/min for both narrow cross sections and classical sections. Above these speeds, the stresses set up by centrifugal forces may become critical. Special sheaves are available for the higher rim speeds. *Note*: These limits do not apply to V-belts – just to sheaves. Design manuals show upper limits based on HP ratings.

F. Hazardous Atmospheres

Various groups, including the Rubber Manufacturers Association (RMA), International Standards Organization, National Fire Protection Association, American Petroleum Institute, Mine Safety and Health Administration, and OSHA, have published standards concerning requirements for V-belt drives operating in a potentially explosive atmosphere. The following is a general description of the standard as published by the RMA (IP-3-3).

Under certain conditions of temperature and humidity, a V-belt drive may generate a significant potential of static electricity. Belts intended for operation in a potentially dangerous atmosphere can be constructed with a relatively low resistance characteristic. It has become common practice to specify and refer to such belts as "static conducting."

The standard method of measuring electrical resistivity consists of two 5/8 in. diameter flat contacts, 8 1/2 in. apart on centers, moistened with water, and pressed against the belt with a force of 12 1/2 lb per contact.

The resistivity (conductivity) between the contacts is measured with an ohmmeter operating at a potential of 500 V and having an accuracy range from zero to 10 MΩ.

Experience has demonstrated that the dissipation provided by belts having a resistivity of 6 MΩ or less (when new and measured as above) is satisfactory for operation in hazardous atmosphere. This provides only for the static charge generated by the belts and pulleys.

Belts on such drives should be kept reasonably free from an encrusted accumulation of nonconducting extraneous material. In addition, all elements of the drive must be interconnected and grounded to earth.

No explosion has been known or suspected to have been caused by the static generated by a V-belt drive. Neither the Rubber Manufacturers Association or its participating members can accept responsibility beyond that of furnishing belts within the limits heretofore found satisfactory as described herein.

13

Belt Testing

H. L. COX *Belt Test Laboratory, The Gates Rubber Company, Denver, Colorado*

I. INTRODUCTION

In any industry, the proof of the product is in the testing. What might appear to be a very simple process of running belts to failure is actually a rather complicated process which consists of: establishing standard tests that correlate with appropriate industry standards, carefully controlling test conditions, recording pertinent data (including failure patterns as the belt progresses to failure), establishing and relating to large volumes of historical data, and performing statistical analyses of the test results.

II. PURPOSE OF TESTING AND TYPES OF TEST PROGRAMS

In a typical test facility, various test programs are established to measure product performance. For example, test programs to measure belt quality and performance include the following.

A. Production Control Testing

One of the major objectives of a test laboratory is to monitor product quality trends of various manufactured products. Many in-process tests are made and final inspections are performed before belts are submitted to the test laboratory for dynamic testing, however, quality trend testing in the belt laboratory is actually the final test of belt products to assure established quality levels. The philosophy of quality control tests will be discussed later in this section.

B. Product Development Testing

These tests evaluate new materials, new processes, new belt constructions, and designs. This type of testing is often performed with comparisons to standard control tests, as well as to historical tests under equivalent test conditions applied to standard products.

C. Special Engineering and Application Testing

An extremely important function of the test laboratory is to evaluate basic engineering characteristics of the product involved. In the case of belts, this includes:

Testing to determine the load-life capacity of the belt so that power ratings may be established and published in various catalogs.
Evaluating belt performance under unusual operating conditions, including such items as shock loading, misalignment, and abnormal ambient conditions (excessive moisture, dust, and temperatures).
Special testing to evaluate belt performance characteristics under extraordinary customer application conditions. (For example, measuring clutching performance of belts applied to lawn-and-garden equipment.)

D. Competitive Testing

Competitive belts are continually tested and analyzed in order to develop comparisons against equivalent constructions. These belts are subjected to selected standard tests and the results are compared to production and developmental belts by using various types of statistical analysis.

E. Field Testing

After an experimental product has achieved the desired performance level in the laboratory, it is often field tested under actual application conditions so that a final determination of the product's acceptability can be made. Each product engineer is responsible for the field testing of his belts. During the field testing program, special attention is paid to the performance and failure modes of the belts so that future laboratory testing can be designed to duplicate field failure patterns.

III. COMPARISON OF DESTRUCTIVE AND NONDESTRUCTIVE TESTS

Tests for any of the above programs can be selected from many test conditions that are divided broadly into two groups: (1) destructive and (2) nondestructive. In the first group, tests are usually designed to subject the belts to repetitive fatiguing conditions which eventually exceed the ultimate capabilities of the belt and destroy it. The second group generally utilizes test conditions which explore characteristics of belts, or components, within expected performance parameters and do not necessarily cause failure.

A. What Can Be Learned from Each Type of Test

1. *Destructive Tests*

Destructive tests are used most commonly to establish the performance capacity of a belt. A running, or fatigue-type, test is the most typical of destructive tests; specifically, a test belt is run to failure to measure durability -- or hours of life -- under chosen test conditions.

By mapping out tests with different loads or tensions, pulley diameters, speeds, and belt lengths, belt capacity data can be recorded as the basis for belt development, production control, drive design, and other programs.

The tensile, or pull, test is another destructive belt test, but of very limited value. The only data it provides would be typical of manufacturing process control, and that data is available by other means.

2. Nondestructive Tests

Nondestructive tests can be further divided into measuring (geometric) inspection and engineering characteristics determination.

Measuring (geometric inspection): Belt measuring techniques most generally involve a measuring fixture (Figure 1) that uses two pulleys of equal diameter, a method of applying force, and a means of measuring the center distance between the two pulleys as the belt rotates. One of the two pulleys is fixed in position, while the other is movable along a graduated scale. Pulley dimensions and measuring forces can be found in appropriate industry standards.

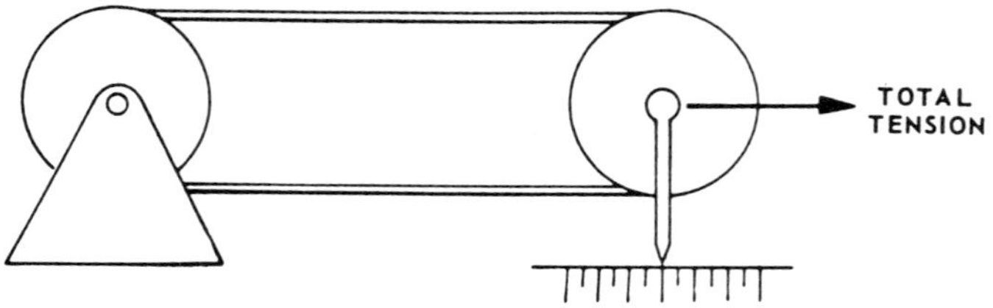

FIGURE 1 Diagram of a fixture for measuring belts.

This same technique can be useful in measuring belt cross section, position in pulleys, and center distance change (or smooth running) as the belt rotates. Suitable fixturing and instrumentation would be required.

Engineering characteristics: Other nondestructive tests are used to characterize belt behavior or performance on application set-ups. Included would be such tests as:

Tension decay: Tension change in a belt on a locked-centered drive, as it occurs over time, with regard to wear, compression-set of the undercord, or tensile cord change.
Efficiency: Relationship of belt losses.
Tension ratio/slip: Looking at slip, or driveN pulley speed loss, as a function of varied total tensions while holding constant torque.
Noise: Running noise levels as affected by materials, section design and test conditions.
Stability: The effects of axial and/or angular misalignment of pulleys, span lengths, and other vibration factors, loads, speeds, and belt construction characteristics.
Clutching: The ability of a belt to act as a clutch – such as on a lawn mower or a riding garden tractor – without "grabbing"; variations in belt sidewall characteristics, dimensions, and specific test conditions are involved.

B. Limitations of Nondestructive Tests

From the comments above, it should be obvious that nondestructive tests, properly applied, provide useful and necessary information on belt performance. The most common application of that type of testing is dimensional inspection in final inspection areas, generally where production is sampled, inspected, and returned for stocking without loss.

While such a use is valuable in manufacturing process control functions, nondestructive testing tells essentially nothing about the power transmission capacity of the belts – the primary characteristic of belt quality. Therefore, to measure that characteristic, fatigue (destructive) tests are used universally in the analysis of belt quality.

IV. FATIGUE-TYPE TESTS

A. The Need for Basic Standard Tests

Basic tests used in a laboratory should be simple, repeatable, reliable, and should follow industry standards, if applicable. By using standardized tests, it is possible to maximize use of the test data obtained.

Whenever possible, standard tests should duplicate, or closely follow, tests which are recommended by other organizations, such as The Society of Automotive Engineers and The International Standards Organization.

1. Standard Test Considerations

Many years of testing experience have shown that a good standard test should cover the following factors.

Repeatability: The test should lend itself to repeatability at any given time or location. Such an approach enables a laboratory to reproduce procedures and test conditions so that test results can be compared directly. This is particularly helpful when suppliers and customers need to verify results or when comparisons are necessary between laboratories of multiplant companies. Furthermore, it allows large test databases to be built for analytical work on specific belt constructions and test conditions.

Optimum duration: The test should maintain test lives in an optimum average life range. For belts, the rule of thumb might be a range of 100 to 300 hr for tests with power loading and 300 to 500 hr on deadweight testers. Too severe a test will give an erroneous picture of the quality of the product, since abnormal failure patterns may be encountered on tests where belt lives average less than 100 hr. A test that consistently results in lives greater than 300 hr on loaded tests does not contribute significantly to data obtained and is inefficient and uneconomical for the laboratory.

Controlled ambient conditions: A standard test should control ambient conditions. Variability of ambient temperatures, air velocity, humidities, and so forth, can introduce undesirable test variables. Most standards recommend ambient temperatures be controlled to 85°F ± 5°F, combined with low air velocities and moderate humidities.

Run to failure: The test should run belts to failure. Removal of a belt from a test before failure does not give a complete picture of the life of the belt. Failure is usually defined as the point at which a belt will no longer run or transmit power in an acceptable, uniform manner.

2. Industry and International Standards

As mentioned before, major standards-writing bodies have developed specific standards, or recommended practices, for belt testing. For example, some existing recommendations for automotive belts are:

The Society of Automotive Engineers (SAE J637)
The International Standards Organization (ISO 5287)

3. Advantages of Standards

The advantages of meeting these or other established test recommendations cannot be stressed too strongly; belt testing benefits from consolidation of sizes, standardization of test conditions and procedures, and correlation of results.

Consolidation of sizes: Consolidation of test sizes is important since each industry has a wide range and large number of belt sizes to serve its applications. A test laboratory would be hard pressed to cover all and provide useful data. Consolidation establishes preferred sizes and types, and it allows laboratories to concentrate their equipment needs accordingly.

Standardization of test conditions and procedures: Even with sizes consolidated, there are many conceivable test conditions that could be run. Belts could be run with load or no load, high or low tension, fast or slow, large or small diameters, two pulleys or three, or more, low or high ambient temperature, and even rightside up or upside down! Through standard test recommendations, industries can focus all laboratory efforts on specific, chosen test conditions and procedures for efficient use of equipment and reliable results.

Correlation of results: Not only is reliability important, but correlation of results within a lab or between labs is essential. The same test conditions, perhaps the same equipment, running belts from the same lots may give significantly different results. However, through standardization at the industry level, the variables of test sizes, conditions, and procedures are minimized; problems with correlation of results can be reduced to everpresent factors of operator training or technique and equipment maintenance.

4. Precautions for Standard Testing

In the section on Consolidation of Sizes, it was indicated that the objective of standard tests is to provide duplication at any location. While the objective is extremely important, experience has indicated that there is commonly more average variation between test laboratories than between products tested in a given laboratory. This is attributed to the additive effect of slight variables in the equipment, ambient conditions, and operation of the test equipment by trained personnel. Variances between laboratories should be monitored continually; whenever a laboratory appears to be out of control, measures should be taken to bring the laboratory back to established tolerances. One typical difference between laboratories involves control of ambient temperature. Several studies have been made concerning the affect of ambient temperatures on belt life (see the section on Controlled Ambient Temperatures, above). A recent study indicates that each 2°F change in ambient temperature produces a 1°F change in the internal belt temperature, and in turn, an 18°F reduction in internal belt temperature doubles the life of a belt; conversely, an 18°F increase cuts belt life in half (within reasonable temperature ranges). Thus it is evident why the need exists for closely controlled ambient conditions in a test facility.

It was noted that ultimate belt life was used to measure belt performance (see the section, Run to Failure, above). While ultimate hours of life is one measure of belt performance, the failure patterns and belt growth that occurs during the testing of a belt are equally important and should be reported at daily intervals. In this manner, the failure modes that occur during progression to ultimate failure are recorded for use by the product engineer. This allows the engineer to evaluate the performance of the belt during its test life in terms of appearance, growth, and specific failure patterns. If, during the test of the belt, unusual failure modes are observed, the product engineer can establish corrective action to eliminate, or minimize, the problems which might produce user warranty failures.

B. Types of Belt Testers

The majority of the test equipment used in a belt test laboratory can be divided into the three following conventional types.

1. Deadweight Testers (see Figure 2)

The deadweight tester is a very simple and inexpensive test station that uses constant tension weights, relatively high speeds (3,600 to 10,500 rpm) and minimum diameters to test belts. No power is transmitted, thus the name "dead weight." This type of tester does not give the full picture of belt performance, but is a useful and inexpensive tool used frequently in production control and development work for automotive, light-duty, and some industrial belt cross sections.

2. SAE-Type Testers (see Figure 3)

The SAE-type testers are designed to duplicate exactly the industry standard test specified by the Society of Automotive Engineers Test Recommendation SAE J637. The belt configuration is standardized and is similar to a typical engine front-end drive with the driveR shaft equivalent to the crankshaft, the idler equivalent to an alternator or a generator, and the driveN equivalent to the fan/water pump shaft. The installation and operation of the test belt is completely specified by the SAE standard. These testers typically use an AC motor with a belt jackshaft drive to obtain the desired driveR

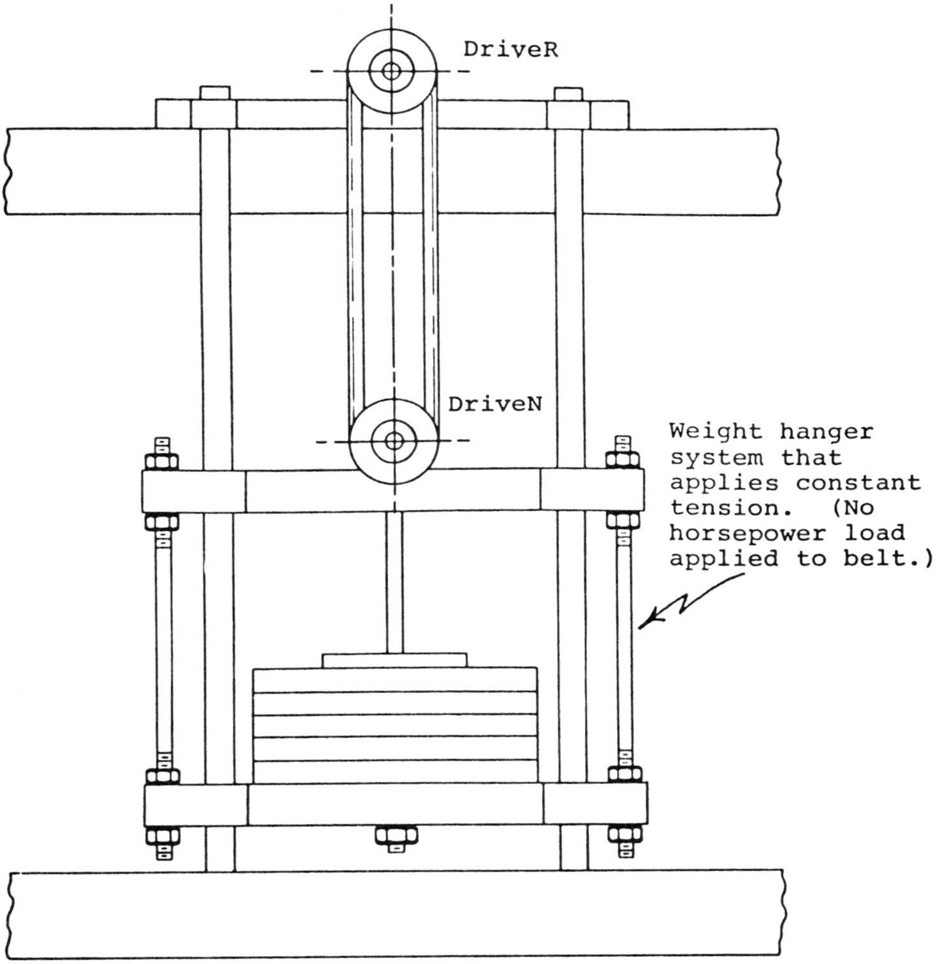

FIGURE 2 Deadweight belt tester.

rpm and a water-brake, or electrical absorption unit, on the driveN shaft. Special electronic monitoring equipment is used to shut the testers down at 8% slip, per SAE J637. The belts are run under a locked tension condition which duplicates the application. This type of tester is used principally for automotive types of belts.

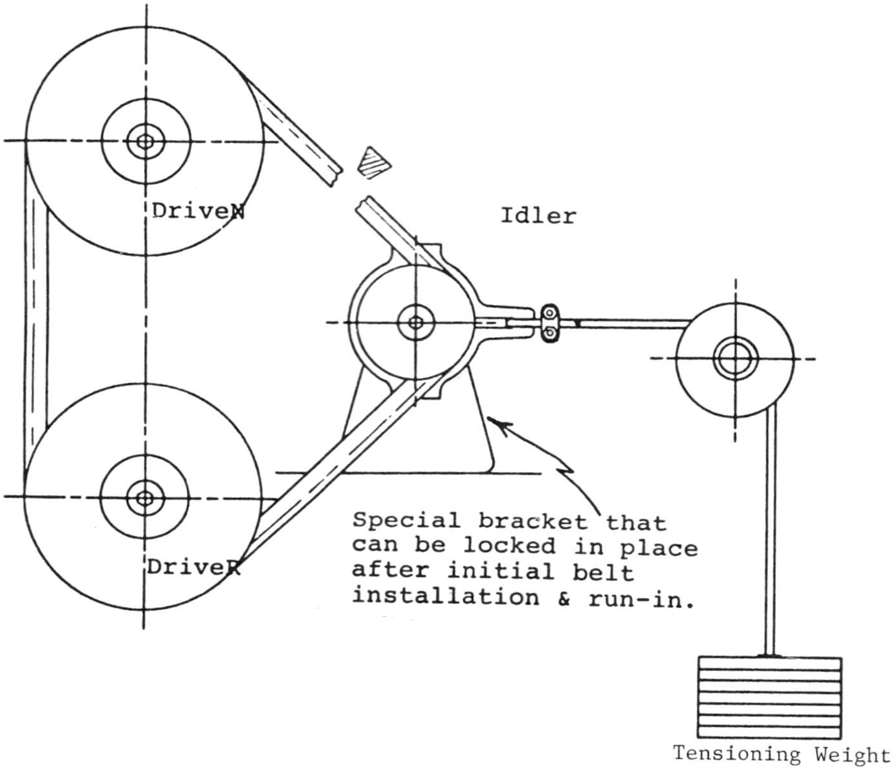

FIGURE 3 SAE automotive belt tester.

3. Electrical Dynamometer Testers (see Figure 4)

The electrical dynamometer testers are used for all of the test programs listed above and are used for virtually all cross sections of belts produced. Typical electrical dynamometers use a DC motor to drive the belt, and a DC generator is used to absorb the power transmitted by the belt. Figure 4 shows the specially designed cradle system for the absorption unit which allows accurate load measurement and carefully controlled applied tension. The power absorbed is measured through torque reaction of the generator which is mounted in large, low friction bearings. A scale is used to measure the reactive force of the generator. Most V-belts applied to this type of tester are subjected to constant, controlled tension through the special weight and lever arm system. The system is designed using a parallelogram principle so that as the belt elongates, the tension applied

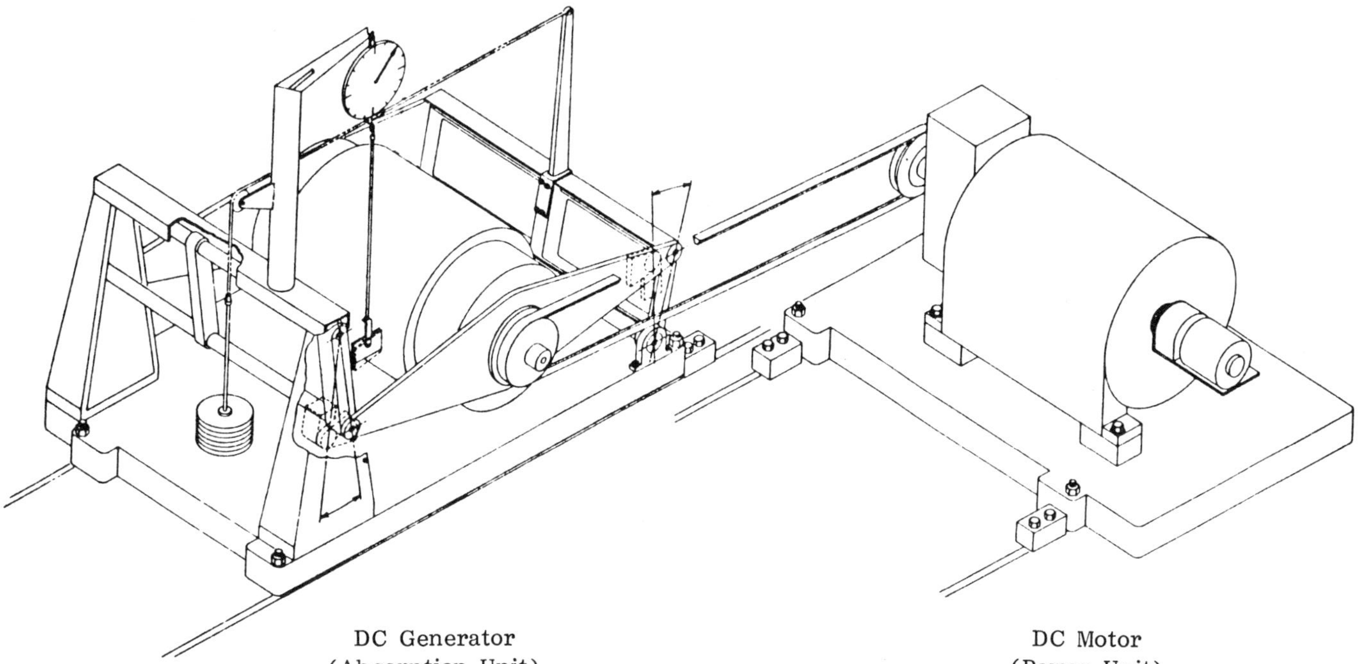

FIGURE 4 Electrical dynamometer belt tester. Generator is mounted in low friction bearings which allows limited rotation about its shaft, thus reactive torque (test torque) can be measured with a calibrated scale. Generator is mounted in specially designed and balanced cradle system to maintain constant belt tension over several inches of center distance movement. Motor and generator are track mounted to accommodate various length belts.

to the belt remains constant over a relatively large center distance movement. Synchronous belts and some V-belts are tested under fixed, or locked, center conditions by locking the cradle in a fixed position.

Because of the DC system, the loads and speeds are easily adjusted and lend themselves to feedback control systems which very accurately control the speed and loads to within 1% of preset values.

As a matter of interest, this type of system is relatively efficient in terms of electrical power required, since the power generated in the DC generator absorption unit is returned to the system, thus requiring only a small amount of power to make up system losses.

C. Methods of Accelerating Belt Test Lives

As explained previously, it is desirable to achieve standard tests that result in average test lives in the range of 100 to 300 hr as compared to the actual application usage which can run anywhere from 1,000 to 25,000 or more hr. In order to better understand how test conditions are accelerated, it is necessary to understand the process by which belts fail in normal service: stress fatigue. This process is described in more detail in Chapter 3.

As shown in Figure 5, belts are subjected to repetitive cyclic stresses which are a combination of the following three forms of tensions:

Tight side tension (T_T): This is often referred to as working tension and is a function of the torque being transmitted.
Bending tension (T_B): A tension created in the belt as it is bent around the pulleys of the drive.
Centrifugal tension (T_C): A tension created in the belt due to its mass and rotational speed.

The combination of these three tensions is commonly referred to as the peak tension in the drive and can be related to cycles to failure as shown in Figure 6.

By carefully balancing the use of increased power loads, small diameter pulleys, and increased rotating speeds (thus increasing the tight side, bending, and centrifugal tensions), the life of the belt can be reduced to acceptable test levels and still measure the necessary performance characteristics of the belt. In addition, short belts of standardized lengths are frequently used so that the number of cycles per unit of time are increased. (If nonstandard lengths are tested, the test lives are corrected by the logarithmic relationship specified in SAE J637.) It is extremely important to balance the accelerating variables of loads, pulley diameter, and speed in order to match the laboratory belt failure modes with those which occur in the actual field application of the belt. A test which would unnecssarily stress

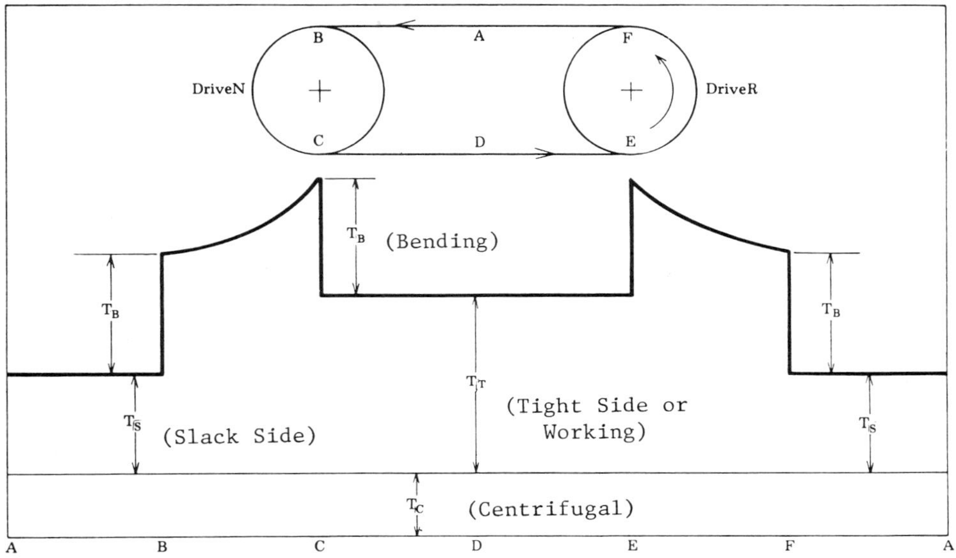

FIGURE 5 Diagram of working, bending, and centrifugal tensions in a belt drive.

one of the above contributors to peak tension could produce unrealistic test results, which in turn could lead the product engineer to improper conclusions about the performance characteristics of the test belts.

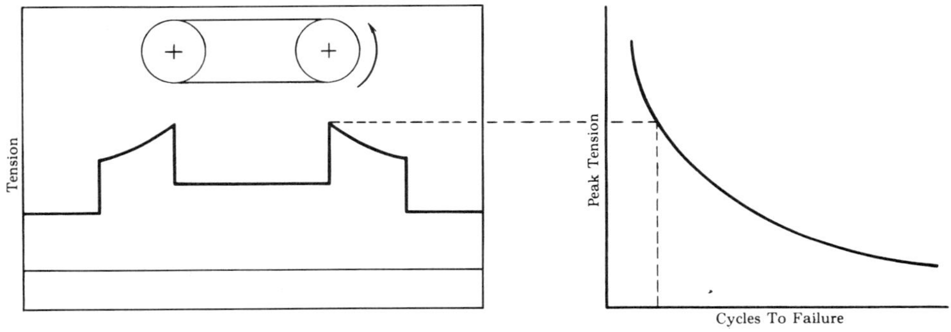

FIGURE 6 Peak tension versus cycles to failure.

D. Reliability and Interpretation of Test Results

1. Product Reliability

Consideration of product reliability is an important part of a product engineer's task today. The reliability of a product is related directly to product quality, product liability, warranty costs, and often to government mandated performance standards.

Manufacturers today must be confident of the reliability of their product before it reaches the marketplace. One way of attaining this confidence is testing the finished product. It is impossible and impractical to test every product; the cost of testing is high and is increasing daily. Preferred destructive tests can only be performed on a minimum number of components. Therefore, through the use of statistics and the projection of quality levels, the manufacturer can assess the reliability of a product with minimum sample sizes.

The reliability of a product can be defined as the measured or predicted performance capacity of that product on a given application or test together with the expected variability of the total population of the product under these conditions. In other words, reliability involves both performance capacity and variability.

2. Interpretation of Test Results: Weibull Distribution Function

The Weibull distribution function has come to the fore in recent years as a means of analyzing fatigue-type tests. Using special probability paper, it enables the user to determine the capacity and distribution pattern for the product, as well as the correctness of fit of the data to the plotted line. All these criteria are important.

One of the main reasons for the acceptance of Weibull graphs as applied to fatigue testing is the ability to determine reliability (capacity/variability) at a glance.

This function enables the plotting of a straight line graph to represent cumulative percentages of failures by means of appropriate coordinate scales – a clear graphic picture of the fatigue life distribution pattern. Product reliability is depicted by the placement of the curve and its slope which represent performance capacity and product variability, respectively.

Significance of curve placement: capacity indicator: Capacity is defined by the position of the curve on the abscissa in miles/cycles/time/etc. This curve tells us the life (with 50% confidence) that would be expected if an infinite number of samples were tested on a given condition at any percent failure point. The two values most commonly used are B50 (the point at which 50% of the samples have failed) and B10 (at which 10% of the samples have failed). The capacity of a product on a laboratory test must be such that it will give satisfactory service on an application. Laboratory tests should not be accelerated to the extent that they will be influenced by factors which will never be encountered on an application such as excessive loads, speeds, temperatures, or diameters.

Significance of slope: variability indicator: The variability (distribution pattern) is defined by the slope of the curve. It must be analyzed along with the capacity of the product. When the Weibull distribution function is used, the slope of the line defines this value. A normal distribution pattern, in which the median and mean are equal, would give a slope of 3.5. Fatigue-type tests generally produce slopes lower than this. Each manufacturer of different products should perform sufficient tests to determine an expected distribution pattern for his product. Fatigue tests run on rubber products have established the criteria that 10% of the samples will fail in the range of 35 to 50% of average life.

To illustrate the use of Weibull graphs for determining product reliability, refer to Figure 7. In this illustration, both curves represent products which have the same capacity, i.e., the same median fatigue life, yet their variability is quite different. The product with the slope = 1 could produce warranty failures, and perhaps result in product liability problems.

Significance of sample size: No discussion of fatigue testing/statistical analyses is complete without mentioning the effect of sample size on predicted results. Until a criterion is established to define the variability of a product, along with its capacity, all tests should be run to failure on a controlled test condition. This permits the assimilation of a larger history of test data to be used as a guide in screening samples at a later date.

The confidence that can be placed in a Weibull slope denoting the distribution pattern is dependent on sample size. Figure 8 shows a plotting of data from two separate days' production (lines C and D) with a sample size of three. A probable error in slope of ±27% can exist with this sample size. Line B represents 18 samples from one month's production (including the samples from lines C and D). A probable error in slope of ±18% can exist in this line. Line A represents a sample size of 566. The probable error in slope of this line is less than ±2%, yet it includes data from the other three groups.

Significance of fit of data to line: In analyzing fatigue-type data for rubber products, a deviation from the least square fit line has been noticed in the upper and lower regions (see Figure 9). The cause for these deviations is under investigation. The best theory to date is that they represent a different fatigue failure mode from

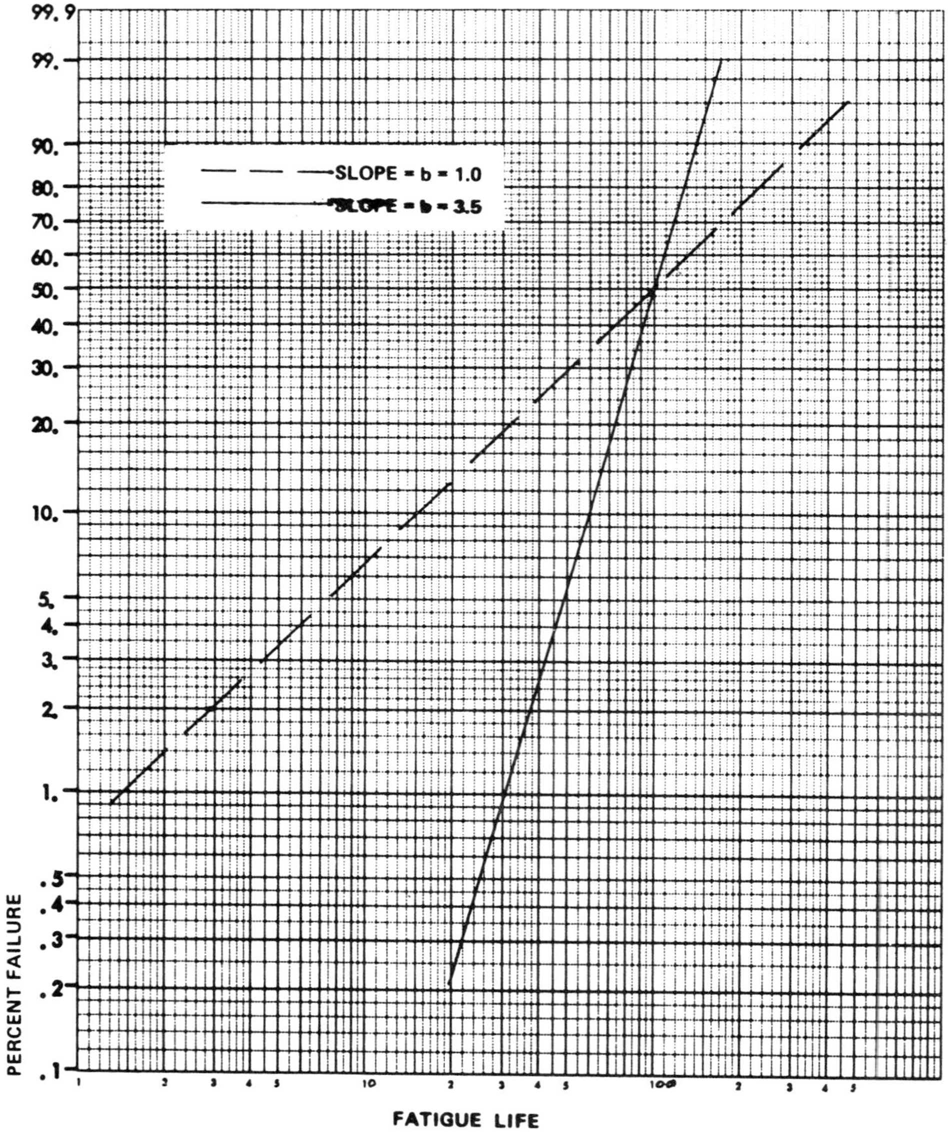

FIGURE 7 Illustration of the effect of variability for products having the same median life capacity.

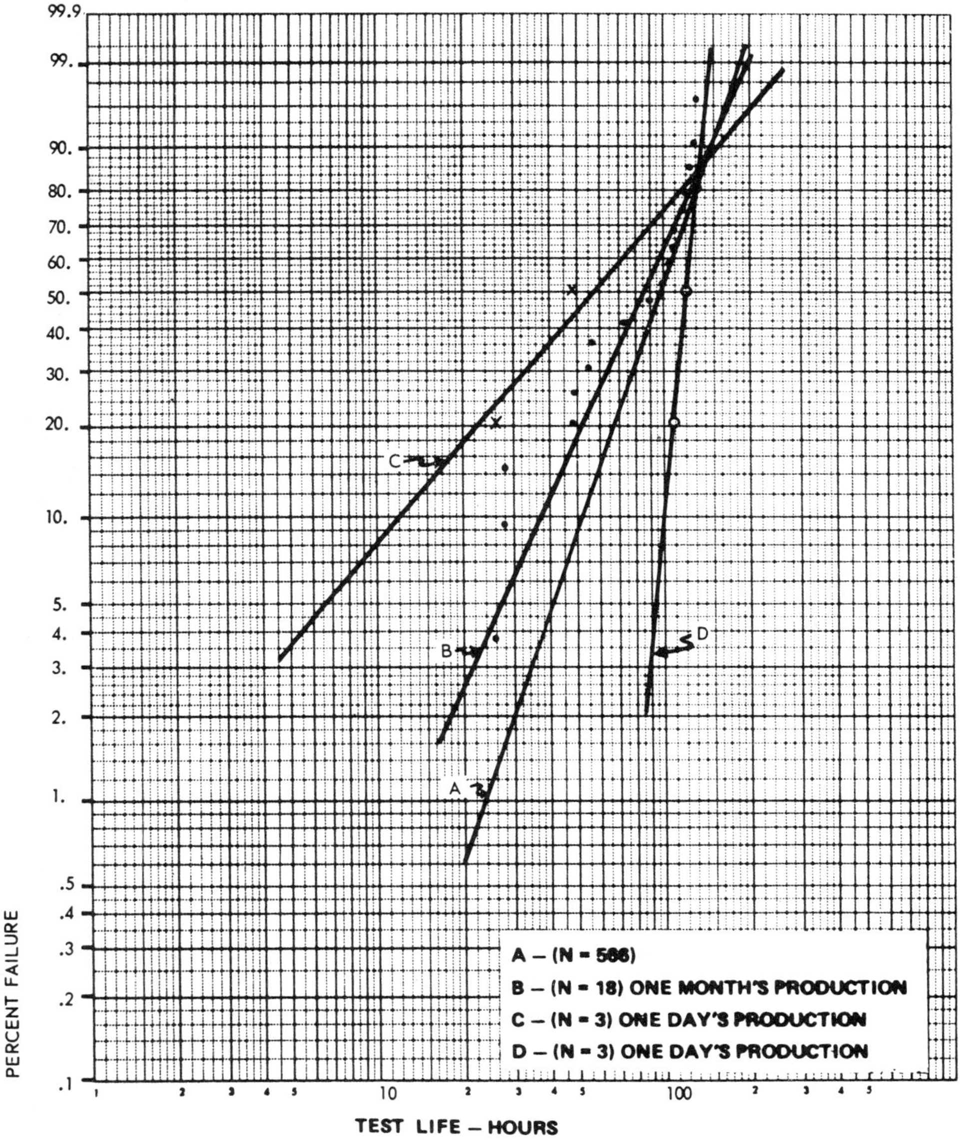

FIGURE 8 Effect of sample size on test life distribution.

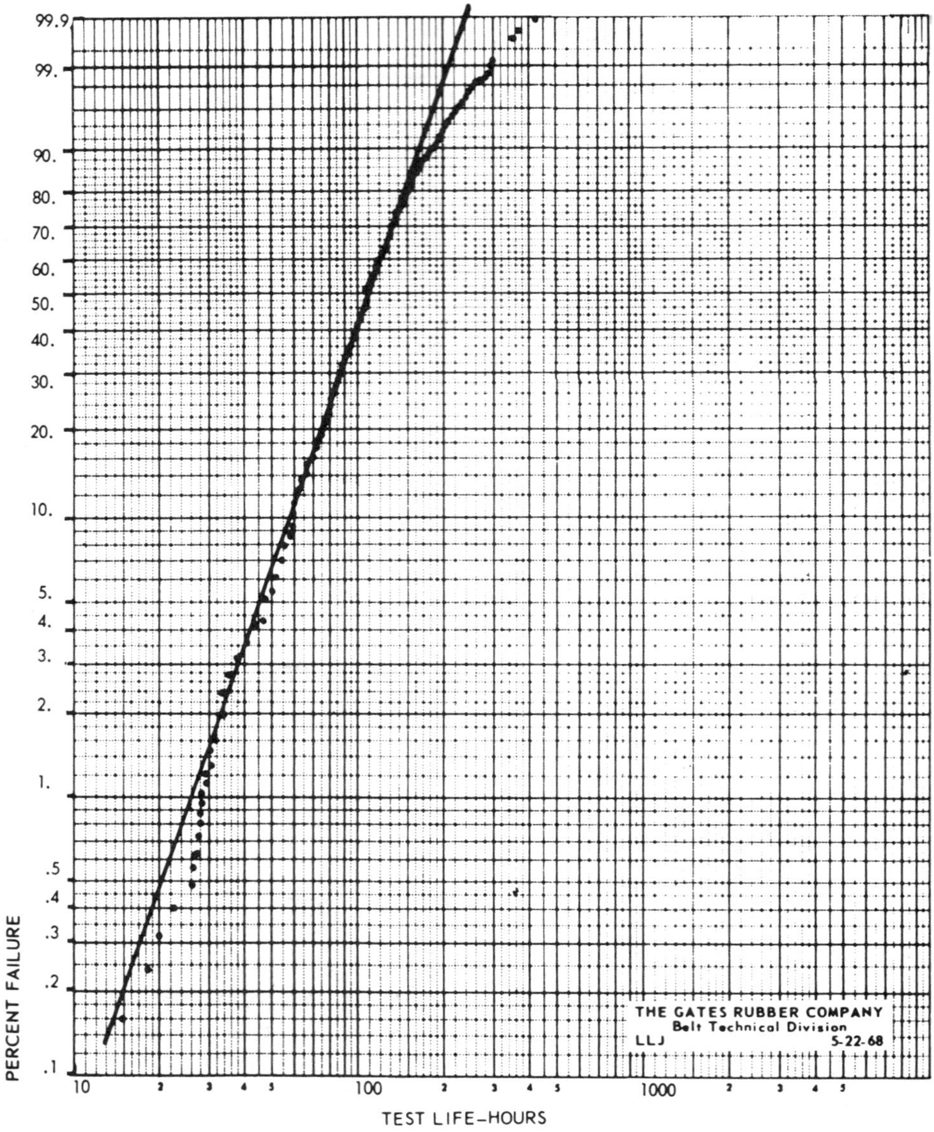

FIGURE 9 Weibull graph from original light-duty industrial belt fatigue test with representative test results plotted illustrating non-linear fit at the upper and lower portion of the curve.

the majority of the population. However, the deviation appears to occur in less than 10% of the population and does not appear to be sufficiently significant to cause concern.

V. QUALITY CONTROL TESTS

All quality control tests performed in the laboratory are standardized tests that relate to large amounts of case history. Because of the extremely large quantity of belts produced each day in combination with the numerous constructions and cross sections involved, it is not practical to perform actual "lot" testing. Rather, a "family system" is used. In this system, belts are grouped by construction, cross section, and manufacturing methods. From these families, belts are selected on a scheduled basis for production control testing. Because lot testing is not performed, the term "quality trend testing" or "quality auditing" is often substituted for the title "quality control testing."

It is the objective of this testing to measure long term trends in belt quality. At the same time, careful observation is made of the immediate test results to make sure a particular family of belts is not out of control.

The test results are also measured against published load-life capacity ratings. The lower control limit of these tests is set so that the test results must meet or exceed these ratings. This assures the user that the belt he receives will perform as rated when properly applied.

Testing, in itself, cannot assure high quality belts: the quality must be built into the product during the manufacturing phases. A comprehensive raw material and in-process control program must be used so that in-process controls and tests are performed during each manufacturing operation to ensure that the final product meets the standards of the company. The dynamic quality trend testing performed in the laboratory is merely one final check of these controls which assures the user of quality products.

Frequently, the laboratory is requested to perform "quality control tests" on actual user equipment. This type of application testing should be avoided whenever possible. It can be generally stated that this type of testing produces minimal data. Frequently, the equipment on which the belt is tested is not sufficiently strong to withstand the rigors of accelerated testing. In other words, the test often ends up measuring the performance of the mechanical parts rather than the belt. Through various drive design manuals, it is possible to relate standardized testing to the user's application and thus it is not necessary, or desirable, to test on the actual equipment.

VI. FIELD TESTING

A. Description and Explanation of Need

The term "field testing" most generally refers to belt tests run on actual applications and in the typical environment of the application. For example, automotive belts might be installed on operating cab fleets, industrial heavy-duty belts on oil drilling rigs, and agricultural belts on custom-cutting combines during the wheat or soybean harvests. All of these tests are of the real world, and they reflect the service for which a given belt is designed and manufactured.

Such tests are essential to product engineers for the development of belts which meet the actual needs of a user. Field tests must be used to supplement laboratory testing in order to evaluate the many and varied environmental factors that can not be duplicated reasonably in a laboratory.

B. Conducting Field Tests

For the most part, OEM and/or operators of equipment encourage field testing and work cooperatively in the tests. The objectives of the test must be understood mutually to maximize the amount and avilability of data generated, and to control the variables involved in any series of tests.

1. Recognizing Field Test Variables

In most cases, field test belts are installed by product engineers or under their supervision, preliminary records are made, and instructions given to operators. However, unless the tests are followed "real time" for a specific result, the belts may be subjected to abnormal operating conditions, unrecorded environmental variations, accidental abuse by operators or other machine components, and general lack of attention to desired test conditions.

In some cases, belts may fail or be removed with vague or no records of time. Furthermore, such belts may not be saved. To minimize loss of important data, the largest practical number of test samples should be run.

Most field test programs, however, are run with only a few of these problems and the results provide essential data in the evaluation of the belts.

2. Procedure of Field Testing

Once the program is initiated and the belts installed, periodic inspections must be made to record use of the belts and to observe progressive or ultimate failure patterns as they occur. At the conclusion of the tests, all field data should be analyzed and, perhaps, correlated with applicable laboratory test data for a complete evaluation of test belt performance.

Test operators benefit from knowing the purpose and scope of the testing, and sometimes make valuable suggestions as to the conduct of the program. Special operating conditions, observations, or procedures should be highlighted for maximum cooperation.

Periodic follow-up by the development engineer is mandatory in most field test programs. That approach assures necessary test progression data that may not be furnished by the test operator. Such data is invaluable in the final analysis of belt performance and should be taken, as a minimum, at 1/3 and at 2/3 expected life. A last observation must be made to get final results and test belts if available.

C. Analysis of Field Test Results

Correlation of laboratory and field testing is always a difficult task. While the laboratory test is performed under ideal, controlled conditions, the field test variables are quite significant. The objective of the laboratory test is to define fatigue failures. Other criteria, such as progression of failure pattern, elongation, temperature, and the like are also observed and recorded during the fatigue test. Field tests are often set up to evaluate one specific belt characteristic such as belt wear, noise, or stability with less emphasis on ultimate fatigue life.

In an attempt to tie laboratory testing in with field testing, test data on automotive belts as tested in the laboratory and in the field are included in Figures 10 and 11.

Laboratory tests were conducted on SAE 3-point tests using accelerated conditions. Field tests were run on an alternator and a water pump drive on several large commercial trucks. Two different belt sections were involved.

Field data represent belt failures. However, it is often difficult to determine if the failures are actually fatigue oriented, or the result of damage from equipment problems, improper maintenance, or another factor. Even considering the variables involved, the variability of the results compares favorably with the laboratory results.

Laboratory tests should be closely correlated with field testing. Some products proving superior in the laboratory may not out-perform others in the field. This could be due to the uncontrolled conditions of the field application. It could also be attributed to the use of highly accelerated conditions in laboratories necessitated by the high cost of testing.

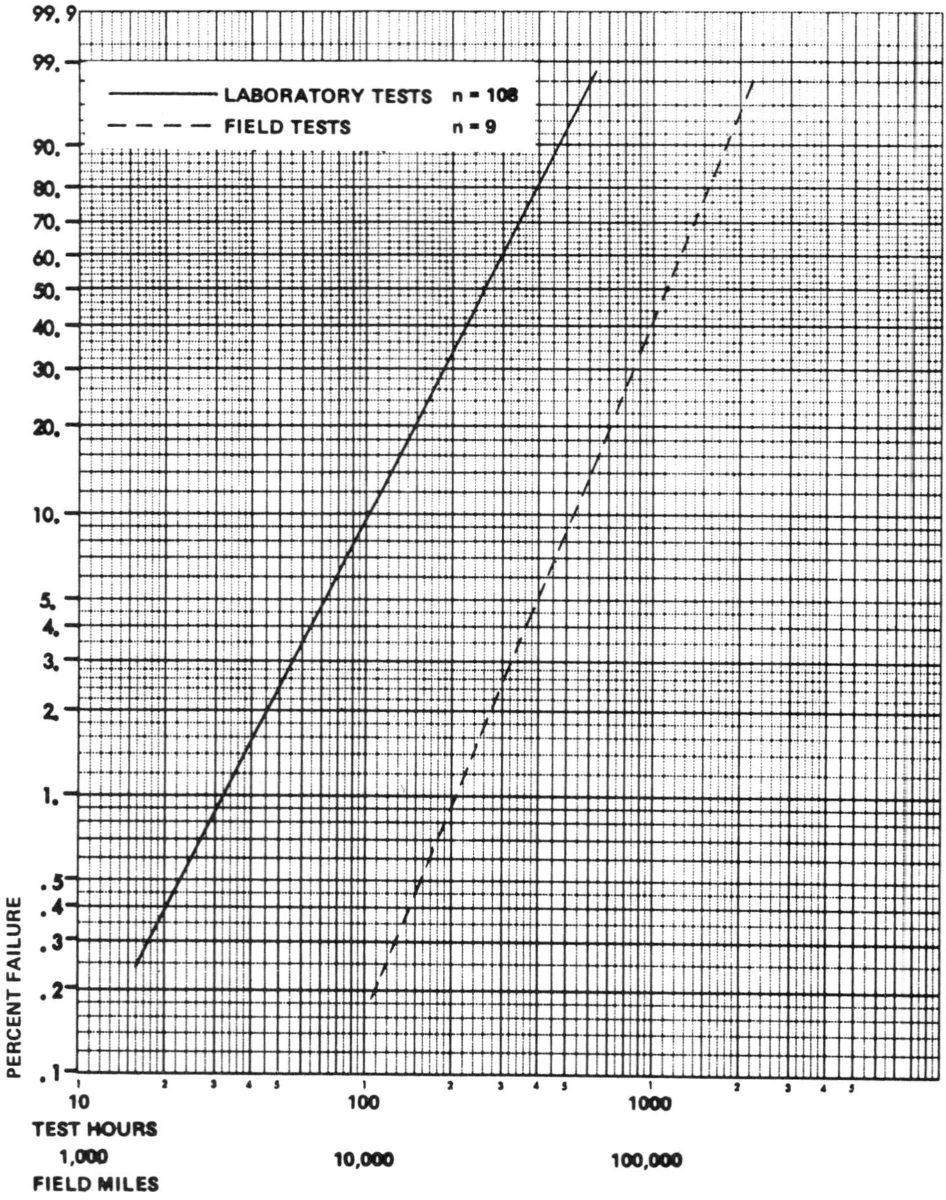

FIGURE 10 Comparison of data distribution for automotive belts subjected to laboratory and field tests (alternator drive).

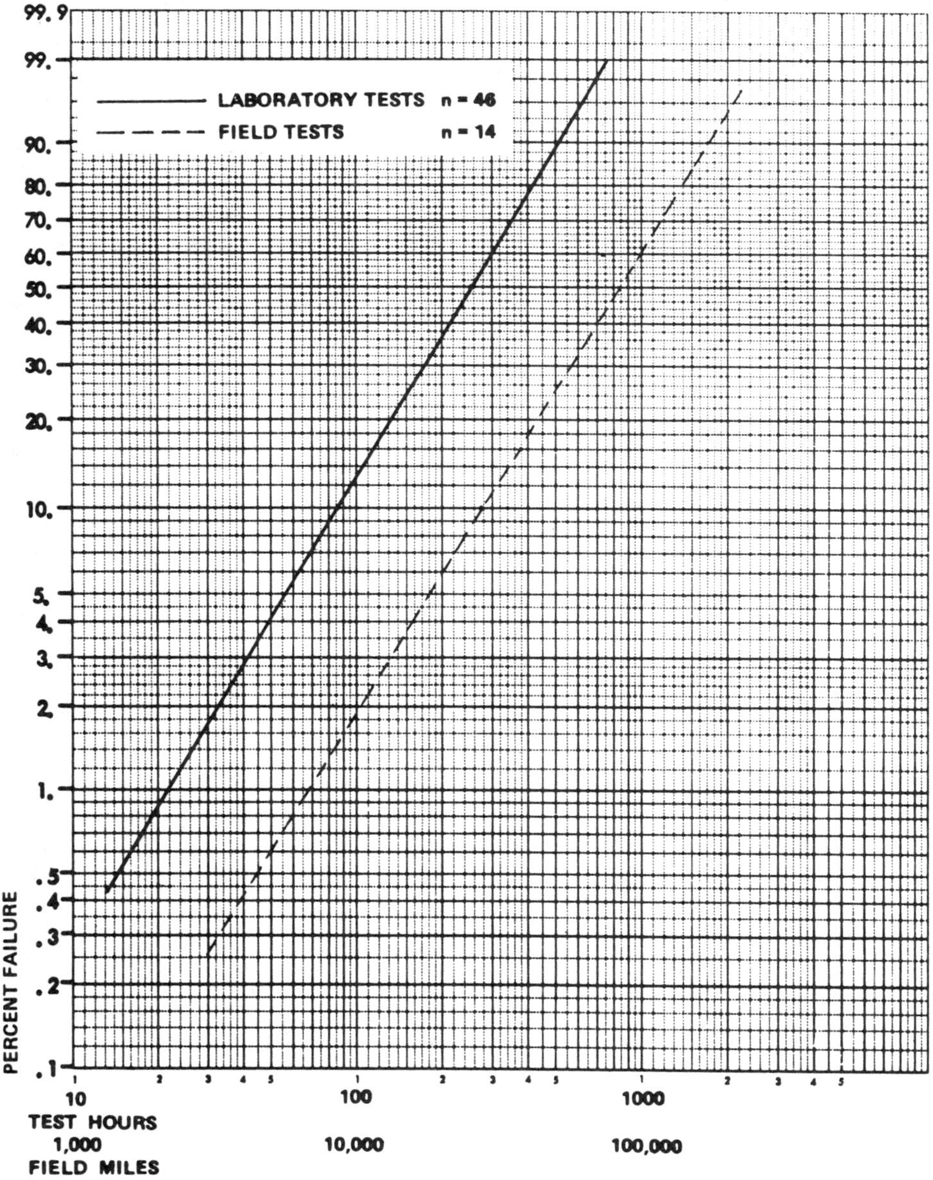

FIGURE 11 Comparison of data distribution for automotive belts subjected to laboratory and field tests (water pump drive).

VII. CONCLUSION

This section has discussed the high degree of control and standardization that is required to establish meaningful belt test programs. By continually relating the laboratory test results to actual belt performance on applications in the field, it is assured that important belt parameters will be monitored.

In the end, this approach assures both the belt manufacturer and the ultimate user that belts are developed, produced, and measured against rigid standards of acceptance.

14

Sheaves and Pulleys

RAY ELDER II and ROY DAVID BAYNE *Automotive Products Technical and Development Division, The Bates Rubber Company, Denver, Colorado*

I. GENERAL DESCRIPTION

A. Introduction

In the normal sequence of events, belt drives are often the last item addressed by the designer and sheaves are usually the last item considered in a drive design. It is assumed the sheaves or pulleys will be readily available at a reasonable cost. This assumption is typically true for industrial V-belt drives but sheave availability at a reasonable cost becomes less certain when the design requires state-of-the-art belt performance. If the drive will often experience unusual load conditions such as heavy shock loads, or if power must be taken from or supplied to prime movers or mechanical equipment that is coupled by specialized mechanisms other than belts, i.e., high HP traction motors used in drill rig applications, sheaves may be difficult to acquire. This chapter will not attempt to specifically address any of these potential problem areas, but it will cover the basic engineering considerations for sheaves and pulleys and will point out methods for avoiding some of the more common pitfalls that may be encountered.

What is the difference between a sheave and a pulley: These names are often used interchangeably but there is a technical distinction. A pulley is any type of simple device over which a belt is run, regardless of the configuration. The two requirements for being a pulley are rotation and change in direction of applied force. A sheave is a special type of pulley. It is a disk or wheel with a circumferentially grooved rim. Regardless of the technical distinction, this chapter will conform to common usage and use "pulley" and "sheave" interchangeably.

B. Manufacturing Methods

1. V-Belt Sheaves

The typical V-belt groove is most often formed in the sheave or pulley either by a machining operation or mecahnical deformation of the sheave material. Where close tolerances are required or certain materials are needed for strength or long life, machining is the preferred production method.

Sheaves with roll-formed grooves or of split construction, i.e., joining mirror-image parts to form the pulley, are usually used in instances for which (1) long equipment service life in the form of actual operating time is not required, (2) long service life is desired but drive loads are light, and (3) wider tolerances on dimensional accuracy are acceptable. Pulleys formed by one of these methods are usually one-groove with certain automotive pulleys being notable exceptions. Parts produced by roll forming are generally below 12 in. in diameter and normally have larger tolerances applied to the various groove profile dimensions, i.e., groove angle, groove topwidth, and so forth.

Grooves may also be cast or molded in the pulley. Such pulleys require very minimal finishing before they are ready for use. As-cast pulleys are also used in applications which do not require high levels of performance. The pulleys that are die cast or injection molded are commonly found on lightly-loaded drives such as ventilating fan drives and larger household electrical appliances. High production volumes are usually needed to offset the initial tooling costs for either of these methods. Nonmetallic pulleys can also be molded or cast and used in the as-cast condition. These pulleys are usually not available in diameters over 12 in.

2. Synchronous Sprockets

Before discussing the methods of synchronous tooth formation, it should be understood that the tooth of a synchronous pulley requires that the tooth profile be very accurately formed regardless of the particular tooth shape. The profile of the tooth and the pitch-to-pitch spacing of the teeth around the circumference of the pulley must be accurate so that the meshing of the synchronous belt with the pulley teeth is free-running and smooth. Pulleys that do not meet this criterion can cause excessive tooth wear, premature belt failure, and high levels of noise during operation.

The following three methods are most often used to produce the teeth on synchronous pulleys.

1. Machined teeth – produced by either shaping or hobbing. Both of these processes generate the tooth profile on the pulley and both methods require their own unique machine to accomplish the tooth generating process. A brief description of each method follows.

Shaping – a pinion-type cutting tool is used to generate the tooth form on the pulley. This tool reciprocates parallel to the longitudinal axis of the pulley blank and is slowly fed into the blank until the required pulley diameter is reached. Once this occurs, the cutter and the blank rotate slightly after each stroke of the cutter across the face of the pulley blank. Each tooth of the cutter does the machining as it rotates into position and is drawn across the pulley face. Once the blank has made one full revolution on the machine, pulley tooth formation is complete.

Hobbing – a hob cutter is simply a cutting tool with rows of cutting teeth. The rows are in the shape of a worm (as in worm and worm gear). The cutting teeth on the hob have straight sides, as in a rack, but the axis of the hob is turned relative to the axis of the pulley blank through the lead angle of the hob. The hob is fed slowly across the face of the blank as the blank and hob are rotated about their own axes. It is important that the hob and blank rotate at the proper angular velocity ratio as the hob travels across the face of the blank.

2. Molded teeth – with this method a very accurate mold or die must be made for each diameter part. This method lends itself to parts produced from plastics or sintered metal processes. Parts produced with this method are generally 6 in. or less in diameter.
3. Extruded teeth – this process requires that a very accurate die be machined for each part diameter. Part materials are limited to plastics and malleable metals such as brass and aluminum. Part diameters produced by this method generally fall in the range of 0.25 to 5.0 in.

C. Materials: General Requirements

1. Introduction

The type of material chosen when manufacturing a sheave or pulley is determined by the end use of the product and its total cost. The two primary factors that influence material selection are: sheave or pulley diameter and the power transmission requirement of the drive, i.e., product strength.

The secondary design considerations of high speed application, corrosion resistance, heat conduction, static conductivity, wear resistance, and noise reduction will also enter into the decision and will be discussed. Depending upon the application, some materials are better suited than others. This section will cover the more general material consideration. Additional discussion of materials will be included in each application section.

2. Primary Material Design Factors

Attaining rough form: Being able to get the material in a circular blank or round molded form at the lowest cost is an overriding consideration that goes hand-in-hand with forming the sheave groove or pulley teeth at the chosen product diameter. Many manufacturing methods can be used to form grooves or teeth when part diameters are below 12 in. as discussed above. For diameters over 12 in., the foremost method of groove or tooth formation is machining.

Casting is the most economical method for producing blanks for parts over 12 in. in diameter. Casting allows part blank dimensions such as rim thickness, spoke or web area, and the hub features to be formed to dimensions that are only slightly larger than their finished (machined) measurements. Since the general dimensions of the various parts of the sheave or pulley are cast to shape, the cost of excess material and material removal (machining) are saved.

When producing parts below 12 in. in diameter, the end use of the part, i.e., power transmission application and service environment, will generally dictate the type of material to be used and the accuracy needed for the groove or tooth forming process.

Strength: The majority of heavy duty industrial sheaves are made of cast iron. The larger sheaves are usually made of at least class 30 gray cast iron. Ductile or nodular iron is used in some of the smaller sheaves where its additional strength is used to compensate for reduced material thickness. The added strength resists the radial forces exerted by the bushing or setscrew when the sheave is tightened on the shaft. In general, sheave designs using standard materials have sufficient strength to resist all of the forces exerted by the belt drive and the drive designer need not be concerned with material strength. In some instances, stress does become a consideration, and there will be further discussion of this later in this chapter.

High-speed applications: Material strength becomes a consideration when unusually high rim speeds are encountered. Industry practice is to use grey cast iron sheaves on industrial drives where the rim speed is 6,500 ft/min or less. When the rim speed is between 6,500 ft/min and 10,000 ft/min, ductile iron is the commonly recommended sheave material. For rim speeds above 10,000 ft/min, steel is generally the recommended material, but in extreme high-speed applications, exotic materials may be used. It is the responsibility of the drive designer to specify the sheave material in these cases.

Material stresses that a sheave is subjected to in operation are not well defined. Considerable safety factors have been built into sheave designs in order to compensate for uncertain levels of stress. Indeed, some of the limitations on sheave rim speed may be traced back to the limits established by insurance companies at a time when calculations of material stress levels were in their infancy.

Residual thermal stresses in cast sheaves are very difficult to determine, and their magnitudes and locations cannot be generalized across a product line. Actual burst testing of cast iron flywheels has demonstrated that residual tensions may play an important role in rim failures. Eight separate constructions (1 solid casting and 7 fabricated constructions) were run to destruction. All constructions failed at a calculated rim tension below the assumed material tensile strength. From these experiments, it can be seen that unknown stresses significantly affect sheave strength [1].

3. Corrosion Resistance

Introduction: The typical industrial sheave materials – iron or steel -- are not corrosion resistant. To prevent corrosion during storage, the sheaves are usually chemically treated and/or painted during manufacture.

After the sheave is installed, the user may apply additional corrosion prevention treatments. Often, however, surface rust is considered acceptable and no action is taken to prevent rust in service.

Rusting of bores can cause difficulty in installation and removal of the sheaves and may contribute to catastrophic drive failure. While difficulty with installation and removal of the sheave may be obvious, the more serious consequences of equipment destruction may not be. If the sheave bore rusts, the shaft will probably rust also. The resulting blemish on the shaft becomes a starting point for fatigue failure of the shaft. Many instances of "broken shafts" have been reported, but, under close examination, it would become apparent that the majority of these breaks were fatigue-related breaks rather than stress-related breaks. The massive overloads that are needed to break even a small shaft would almost certainly cause serious damage to other parts of the machinery. Many times the only damaged part is the shaft.

Typical coatings: Sheave manufacturers use phosphating or other chemical conversion coatings on the smaller sheaves. Alone, these coatings are not adequate barriers to corrosion so additional corrosion resistance is achieved by the application of oil or paint to the treated surface where they are absorbed into the surface coatings. Larger sheaves present a handling problem and may not be phosphated or otherwise treated. Larger sheaves may only be painted to provide corrosion resistance.

Corrosion protection needed:

Normal operation: Resistance to corrosion during actual service is a minor concern for the majority of applications. Because the sheave is used as part of a piece of equipment, an additional protective coating of paint is applied when the finished product is painted. In most cases a coat of paint is adequate because equipment is often sheltered in service and is rarely exposed to corrosive conditions. Certain types of equipment are normally left outdoors where the sheaves can be exposed to the elements. While rain and snow are not beneficial to either iron or steel, surface painting will prevent most corrosion, and the rust that does develop is usually not detrimental to equipment operation.

Severe operation: A small proportion of sheaves receive intermittent exposure to highly corrosive conditions or are required to operate continuously in a corrosive environment. There are two areas on a sheave that can be seriously damaged by exposure to weather: the groove area and the shaft bore. If the equipment is in continuous or regular operation at short intervals, serious rusting will not occur in the sheave grooves. Any small amount of surface rust that does occur will be polished out shortly after the drive starts up. Serious rusting can occur when equipment sits idle for long periods. Extensive rusting can cause the groove sidewalls to become rough and pitted. Belt life is reduced because an abrasive rust can destroy the belt in a relatively short time. A practical solution to this problem is to enclose the drive and to operate the drive at more frequent intervals in order to maintain groove condition.

When the drive must operate in a corrosive environment such as in a chemical plant or a flocculator drive in a sewage plant, additional corrosion resistance may be needed. This additional protection may involve surface treatment such as special organic paints and coatings or one of many types of plating. The effectiveness of surface treatments will depend upon the base sheave material.

Fretting corrosion: Corrosion on ferrous parts -- usually generalized as rust -- may be the result of either chemical or mechanical causes. The above discussion implied chemically induced reactions. Fretting corrosion, however, is induced by microscopic relative motion between the sheave and the shaft: the sheave is loose on the shaft. Fretting corrosion is usually indicated by the presence of a very fine red powder around the shaft and the bore – rust. Removal

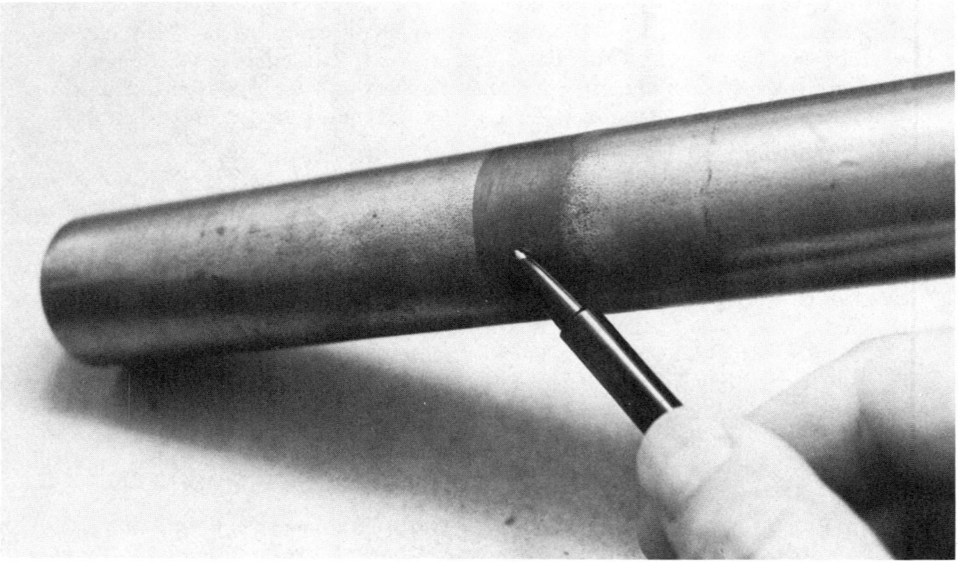

FIGURE 1 Illustration of fretting corrosion.

of the sheave and inspection of the shaft and bore will reveal matching areas of corrosion on the shaft and in the bore (see Figure 1). The sheave can actually become welded to the shaft or, in other cases, fatigue failures originate in the area of fretting corrosion and eventually result in a broken shaft.

Since fretting corrosion is not chemically induced, there is little that can be done to chemically limit its progression. Oiling the shaft, greasing, or keeping the drive dry rarely prove effective. The solution to this problem is by and large mechanical in nature: the sheave must be tightened on the shaft.

Tightness is a situational measure. Since fretting corrosion is due to microscopic motion between bore and shaft, a "tight" fit may not be tight enough to prevent fretting if the motion-inducing force is of the right magnitude and frequency. On drives where fretting is a factor, force-producing mechanical properties such as radial and axial runout, shaft alignment, balance, and natural drive harmonics must be controlled in order to reduce relative motion. The sheave-to-shaft attachment system is also important since the more secure fastening methods will reduce the possibility of fretting.

Special materials: In rare instances, special sheave materials are needed. Sheaves made of materials such as stainless steel, zinc, aluminum, or plastics are either commercially available or fairly easily fabricated in smaller diameters and groove cross sections. However, it may be difficult to find a supplier of large-diameter, multiple-groove sheaves in these same materials. Fabrication by a local machine shop may be more economical and convenient than trying to find a supplier who specializes in producing sheaves made from these materials.

Corrosion can be very complex and involved for each application, and there are many different methods available to prevent corrosion depending upon the type of corrosion encountered. The choice of special materials and prevention methods will be left to the equipment designer and corrosion engineer. However, the equipment designer should remember that sheaves of exotic materials may not be available at an acceptable price.

4. Heat Conduction

Introduction: Heat builds up in all belts as they run. The ability of the belt to dissipate this excess heat depends upon various factors such as air flow, ambient temperature, foreign material in the environment, and sheave material. The majority of belt drives do not have a problem with excessive heat.

On very few drives, extreme conditions combine to cause the belt to become overheated. The belt becomes overheated when its heat cannot be dissipated by the pulley or when excess heat is transmitted to the belt from the pulley. In these situations, excessive heat buildup may be prevented by changing the sheave material to transfer heat away from the belt faster – changing the sheave material from cast iron to aluminum, for example. The sheave can also be redesigned so that the sheave is composed of thinner, more conductive cross sections or fins (Figure 2) can be added or air flow redirected to increase convection losses. Careful consideration should be given to the effect of each option on cost, performance, and serviceability.

Factors affecting heat conduction: Experiments performed by The Gates Rubber Company have revealed that an interaction of factors influence belt temperature and life. A knowledge of these factors can suggest some solutions to belt overheating problems.

Ambient temperature is a logical factor to influence belt temperature. Belt life is inversely proportional to belt internal temperature. For each change in internal belt temperature of about 18°F the belt fatigue life will change by a factor of 2:1. A 2°F change in ambient temperature will produce 1°F change in internal belt temperature.

Air velocity also influences internal belt temperature. Internal temperature declines approximately exponentially with velocity. By

FIGURE 2 Example of a sheave with fins.

varying only air flow over the belt during one set of experiments, the belt temperature was varied by as much as 100°F.

The difference between belt internal temperature and surface temperature varies depending on the belt construction. A typical difference in temperature is 10–15°F higher for internal temperature [2].

5. Wear Resistance

Cast iron: The standard industrial sheave material is cast iron: either grey iron or ductile iron. It is ideally suited for the application because of its low cost and excellent resistance to wear. Most common materials do not wear as well as cast iron and it becomes very expensive to enhance their wear characteristics.

Steel: Steel is used for sheaves when long-time operation is not a consideration or where high strength is required. Although steel is generally less resistant to wear than cast iron, it is often cost effective in applications where the drive is expected to run for relatively short periods of time, say, 3,000 hr. Lawn-and-garden equipment drives and a majority of automotive drives fall into this category. (If the average speed of operation for an automobile is 40 mph, then 3,000 hr is 120,000 miles.) By taking advantage of material savings and mass production methods, steel can have a distinct cost advantage over cast iron.

When long life and/or high strength is required, specialty steels become the material of choice. The variety of chemical and physical properties that are readily available with steel is unmatched and at the fingertips of even the most average designer. High-strength steels are available from most suppliers, and many of these steels can be either heat treated or surface coated to increase hardness and resistance to wear. This flexibility comes at a cost and specially treated steels are not often used for high volume production.

Other materials: Other sheave materials can be processed to resist wear. Aluminum can be anodized and longer wearing zinc-aluminum alloys are commercially available. Although various treatments and alloying have increased the wear resistance of zinc and aluminum, these still fall short of the wear characteristics of grey iron. At this time, there is no general rule-of-thumb for choosing alternate materials. The acceptability of alternate materials must be determined experimentally for each application. Typically, sheave manufacturers have no store of information about alternate sheave materials. There has been little economic justification for the commercial manufacturer to investigate materials other than those usable in his production processes. The more exotic materials have not gotten close scrutiny unless the equipment manufacturer has chosen to do his own investigating.

6. Noise

A belt power transmission drive that creates excessive noise is usually a result of the misapplication of the belt on the drive. Sheaves or pulleys that are misaligned or drives on which the belt is improperly tensioned can develop excessive noise problems. Drive tension and misalignment should always be checked before changing the sheave materials or the general design of the part. The design of a sheave may have an amplifying effect and cause excessive noise to emanate from the drive. Parts that are made from sheet metal such as spun steel sheaves are more susceptible to noise generation because steel is a better sound transmitter than cast iron and the relatively thin material is more likely to vibrate and generate noise as the belt and the sheave interact. The general design of spun steel sheaves, concave from rim to hub, can also act like a megaphone to amplify sound.

If a drive is noisy due to the sheave or pulley design and not traceable to the belt, a change of materials or design may reduce the noise level [3]. Before changing the sheave, verify that the correct belt is being used for the sheave groove, that it is properly tensioned, and inspect the sheave for damage such as cracks or broken welds. Also, check the sheave for proper installation and tightness on the shaft.

Since grey cast iron is a nonhomogeneous material, it generally has good sound-deadening characteristics when compared to most steels. Making a part from cast iron in block form (without concavities or spokes) and as massive as is practical is one of the better approaches to eliminating sheave noise. If noise persists, a thorough analysis of the belt-machine system will be needed to isolate and eliminate the source of the noise.

D. Manufacturing Tolerances and Inspection Methods

1. Industry Standards

Introduction: Commercial tolerances and inspection methods for most industrial sheaves are specified in standards published by the Rubber Manufacturers' Association in conjunction with the Mechanical Power Transmission Association and the Rubber Association of Canada. Some of these have also been adopted as an American National Standard. Groove information can be found in the following Engineering Standards:

Classical multiple V-belts A, B, C, D, and E cross sections	IP-20
Narrow multiple V-belts 3V, 5V, and 8V cross sections	IP-22
Specifications for drives using single V-belts 2L, 3L, 4L, and 5L cross sections	IP-23
Synchronous belts MXL, XL, L, H, XH, and XXH belt sections	IP-24
V-ribbed belts H, J, K, L, and M cross sections	IP-26

The International Standards Organization also publishes belt standards similar to those above. Refer to Chapter 18 for more details.

Groove dimensions: Belt drives are usually specified in either a pitch system or an effective length system. Both systems establish a datum point from which drive design can begin. These systems establish groove dimensions and sheave diameters relative to the groove. It is important to let the sheave manufacturer know which system applies when speaking of sheave diameters – there is a 0.4-in. difference in the pitch diameter and the outside (effective diameter) of a standard C cross section industrial sheave. The reference system is especially important when a nonstandard sheave is being specified. Once the reference system has been established, the sheave manufacturer can supply the appropriate diameter and standard groove.

There are two fundamental methods for establishing groove dimensions. The groove dimensions remain the same for both the pitch system and the effective length system. These methods are:

1. By fixing the outside diameter, the groove angle, and the groove topwidth
2. By fixing a diameter over balls or rods, and the outside diameter, and either the groove angle or groove top width [4].

Figures 3 and 4 illustrate how sheave dimensions are specified by each method. Figure 4 gives dimensions for production sheaves in the narrow cross sections (3V, 5V, and 8V). Figure 3 illustrates dimensioning using the diameter-over-balls method for measuring sheaves. Measuring sheaves are used in quality control inspections to measure belt parameters such as rideout and effective length. Measuring sheaves are normally made to much closer tolerances than are the production sheaves shown in Figure 4.

2. Other Methods

From a practical standpoint, the above inspection methods cannot be used by the average customer. Probably, the most convenient method for groove inspection is a set of groove gauges. These gauges are manufactured in the profile of the cross section groove and are used by inserting the gauge into the sheave groove. Acceptability of the groove is determined by the amount of clearance between the gauge and groove sidewall and the ride of the gauge in the groove. A set of these gauges can be very helpful in determining groove cross section. Geometric and optical inspection methods can be used in a machine shop environment. If the sheave is small enough, the groove can be inspected using an optical comparator. When the sheaves are large and accurate groove dimensions are desired, two geometric methods can be used: Sine plate inspection or comparison of the rides of two known diameter balls in groove.

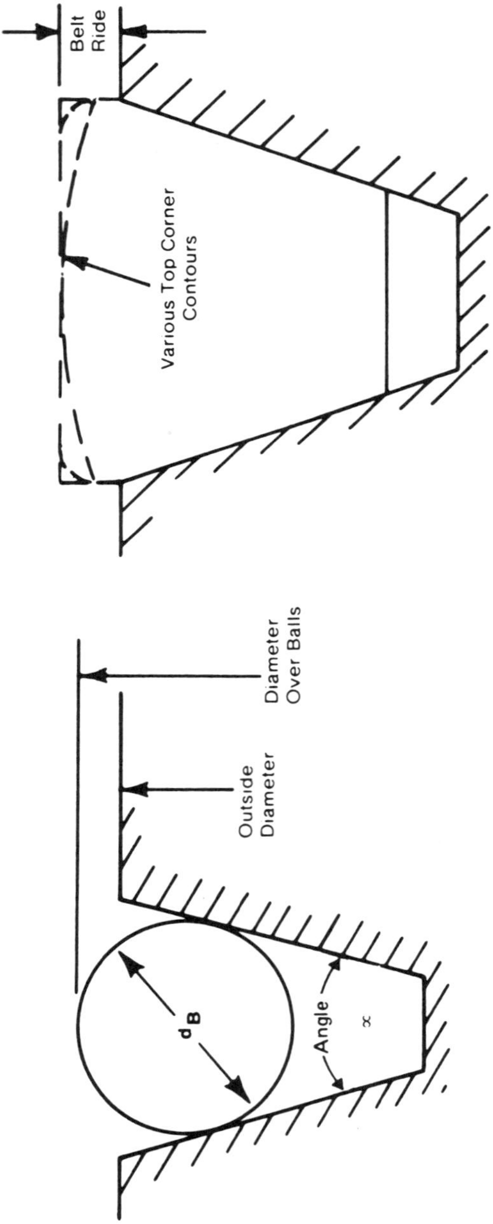
Belt Ride
Various Top Corner Contours
Diameter Over Balls
Outside Diameter
d_B
Angle
α

Cross Section	Outside Diameter +0.005 (Inches)	Effective Outside Circum-ference (Inches)	α Groove Angle ±0.25 (Degrees)	b_e Groove Effective Width (Reference) (Inches)	d_B Ball or Rod Diameter ±0.0005 (Inches)	Diameter-Over-Balls Or Rods ±0.005 (Inches)	h_g Groove Depth (Minimum) (Inches)	Total Measuring Force Per Belt (Pounds)	Maximum Ride, Position Of Belt With Respect To Top Of Groove (Inches)	
									Not Joined	Joined
3V	3.820	12.000	38	0.350	0.3438	4.203	0.340	100	+0.10	+0.20
5V	7.958	25.000	38	0.600	0.5938	8.633	0.590	225	+0.12	+0.25
8V	15.916	50.000	38	1.000	1.0000	17.083	0.990	500	+0.16	+0.30

FIGURE 3 Sheave dimensions using the diameter-over-balls method.

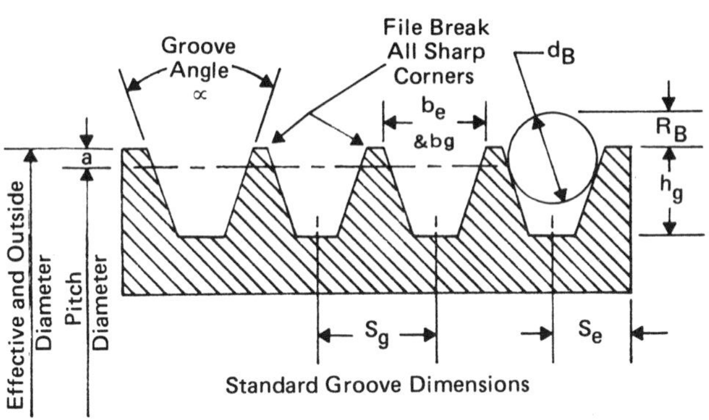

Standard Groove Dimensions

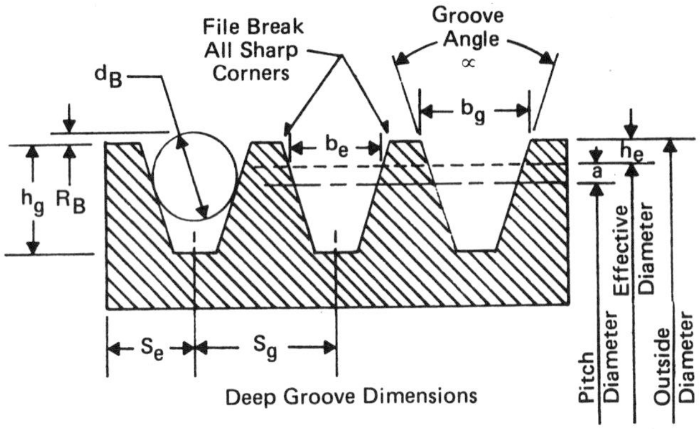

Deep Groove Dimensions

Face Width of Standard and Deep Groove Sheaves

$$\text{Face Width} = S_g (N_g - 1) + 2S_e$$

Where: N_g = Number of Grooves

STANDARD GROOVE

Cross Section	Standard Groove Outside Diameter	Standard Groove Dimensions: Groove Angle ± 0.25 Degrees	b_g ±0.005	b_e Ref.	h_g Min.	R_B Min.	d_B ±0.0005	S_g ±0.015	S_e	Design Factors: Minimum Recommended Outside Diameter	2a
3V	Up Through 3.49	36	0.350	0.350	0.340	0.181	0.3438	0.406	0.344 +0.094 -0.031	2.65	0.050
	Over 3.49 to And including 6.00	38				0.183					
	Over 6.00 To And including 12.00	40				0.186					
	Over 12.00	42				0.188					
5V	Up Through 9.99	38	0.600	0.600	0.590	0.329	0.5938	0.688	0.500 +0.125 -0.047	7.10	0.100
	Over 9.99 To And including 16.00	40				0.332					
	Over 16.00	42				0.336					
8V	Up through 15.99	38	1.000	1.000	0.990	0.575	1.0000	1.125	0.750 +0.250 -0.062	12.50	0.200
	Over 15.99 To And Including 22.40	40				0.580					
	Over 22.40	42				0.585					

DEEP GROOVE

Cross Section	Deep Groove Outside Diameter	Deep Groove Dimensions: Groove Angle ± 0.25 Degrees	b_g ±0.005	b_e Ref.	h_g Min.	R_B Min.	d_B ±0.0005	S_g ±0.015	S_e	Design Factors: Minimum Recommended Outside Diameter	2a	$2h_e$
3V	Up through 3.71	36	0.421	0.350	0.449	0.070	0.3438	0.500	0.375 +0.094 -0.031	2.87	0.050	0.218
	Over 3.71 To And Including 6.22	38	0.425			0.073						
	Over 6.22 To And Including 12.22	40	0.429			0.076						
	Over 12.22	42	0.434			0.078						
5V	Up Through 10.31	38	0.710	0.600	0.750	0.168	0.5938	0.812	0.562 +0.125 -0.047	7.42	0.100	0.320
	Over 10.31 To And Including 16.32	40	0.716			0.172						
	Over 16.32	42	0.723			0.175						
8V	Up Through 16.51	38	1.180	1.000	1.252	0.312	1.0000	1.312	0.844 +0.250 -0.062	13.02	0.200	0.524
	Over 16.51 To And Including 22.92	40	1.191			0.316						
	Over 22.92	42	1.201			0.321						

Summation of the deviations from "S_g" for all grooves in any one sheave shall not exceed ±0.031 inch. The variations in pitch diameter between the grooves in any one sheave must be within the following limits:

Up through 19.9 inches outside diameter and up through 6 grooves—0.010 inches (Add 0.0005 inches for each additional groove)

20.0 inches and over on outside diameter and up through 10 grooves—0.015 inches (Add 0.0005 inches for each additional groove)

This variation can easily be obtained by measuring the distance across two measuring balls or rods placed in the grooves diametrically opposite each other. Comparing this "diameter over balls or rods" measurement between grooves will give the variation in pitch diameter.

Deep groove sheaves are intended for drives with belt offset such as quarter-turn or vertical shaft drives. (See Power Transmission Belt Technical information Bulletin No. 10). They may also be necessary where oscillations in the center distance may occur. **Joined belts will not operate in deep groove sheaves.**

OTHER SHEAVE TOLERANCES

OUTSIDE DIAMETER

Up Through 8.0 inches outside diameter . . . . ±0.020 inches

For each additional inch of outside diameter add . . . ±0.0025 inches

RADIAL RUNOUT*

Up Through 10.0 inches outside diameter . . . . 0.010 inches

For each additional inch of outside diameter add . . . . 0.0005 inches

AXIAL RUNOUT*

Up Through 5.0 inches outside diameter . . . . 0.005 inches

For each additional inch of outside diameter add . . . . . 0.001 inches

*Total Indicator Reading

FIGURE 4 Sheave dimensions measured in inches.

E. Methods of Mounting Sheaves

1. Types of Mounting

Although there are many different methods of attaching sheaves to shafts, the two methods most commonly used in the United States to mount industrial sheaves are tapered bushing and setscrew with key (bored-to-size).

Other sheave mounting methods are occasionally used. Press fits may be used when it is desirable not to mar the shaft, collett mechanisms other than taper bushings are commercially available, and mounting flanges are also used with certain types of equipment.

2. Setscrew with Key

A setscrew with key is the least expensive method for positively attaching a sheave to a shaft. It is easily understood by installers and usually does not require special tools or instructions.

There are several disadvantages to setscrew with key mounting. When servicing a low-volume general market, because of the costs involved with stocking a selection of standard bore sheaves, this method has a very high cost. The sheaves may loosen on the shafts because of shock loads, reversing loads, and vibrations because of improper installation. Worn shafts that need an undersized bore may also be encountered. Because of performance problems and bore availability, the use of fixed bore sheaves is usually limited to high volume OEM applications and lightly loaded drives.

3. Tapered Bushings

Description: The majority of industrial sheaves are attached to the shaft with a tapered bushing. This system consists of a bushing, a sheave, and often a setscrew and key. The bushing is stocked in a range of standard bore diameters. The outside of the bushing is tapered and fits a mating tapered bore in the sheave. The same bushing can be used with a number of different sheave diameters, sheaves with different numbers of grooves, and sheaves grooved for different belt cross sections. Thus, the size and value of inventories can be reduced for both the supplier and the user.

Types of taper bushings: A variety of different taper bushing designs is available. Most of these designs are not interchangeable because of differences in taper angles and bolt patterns. Figure 5 illustrates a taper bushing that utilizes a mounting flange and Figure 6 illustrates a "headless" or flush mounting arrangement.

FIGURE 5 Flanged bushing.

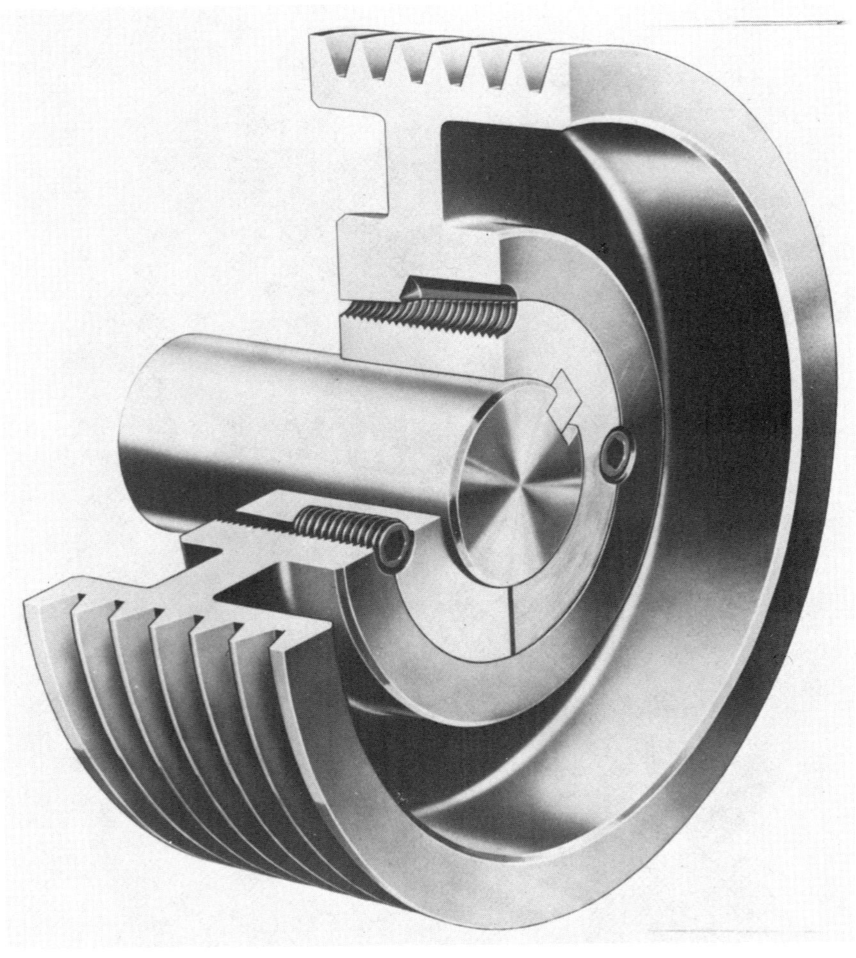

FIGURE 6 Headless bushing.

All of the taper bushing designs rely primarily on the pressure generated by forcing the mating tapers together to transmit power from the shaft to the sheave. The mating tapers are drawn together by tightening the bolts or Allen screws. Very high radial forces are generated between the bushing and sheave and the bushing and the shaft. The radial force clamps the bushing to the shaft with a force equivalent to a pressed fit. The high normal forces thus generated can resist considerable torque before slippage. The bushing is usually keyed to the shaft in standard practice, but for the most part, the key acts only as a safety device should the sheave become loose from improper installation or severe torque overload.

Precautions when installing: Taper bushings require more skill to assemble than bored-to-size sheaves. Instruction sheets are usually included with either the bushing or sheave and, unless these instructions are followed, the sheave, the bushing, and even the equipment can be seriously damaged. (A story is related in the industry about a customer returning a sheave and bushing to the manufacturer with a section of 4 in. shaft still in the broken bushing. The user had not followed disassembly instructions and had damaged the bushing to the point that it could not be removed. In the end, he had cut the shaft in order to remove the sheave from the equipment.)

Regardless of which type taper bushing is used, there are some precautions that should be taken during installation. Always clean the shaft bore in the bushing, the tapered bore in the sheave, and the taper on the bushing to assure proper mating. Never lubricate a taper bushing or mating bore. The reduced friction may allow excessive radial pressures to be generated between the tapers and the hub in the sheave may split. If a setscrew is provided in the bushings, carefully tighten the setscrew until the bushing is snug on the shaft. Overtightening may result in a misaligned or loose sheave. Progressively torque installation bolts to the recommended torque value.

4. Maintenance

Sheaves should be periodically inspected to ensure that they are properly attached to the shaft. With setscrew and key mounting, sheave looseness can usually be remedied by tightening the setscrew(s). However, if the loose condition has existed for some time, the bore in the sheave may be wallowed out or the shaft may have been damaged by the setscrew. If so, the sheave or the shaft may have to be replaced.

Sheaves mounted with taper bushings rarely loosen in service. Improper installation or extreme overload are the usual causes of looseness. If a tapered bushing and sheave are found to be loose, (1) verify that the bushing is bored to the correct diameter for the shaft diameter, and (2) carefully inspect the sheave for cracks in the hub, the mating tapers for proper mating, and any signs of damage to the sheave, shaft, bushing, or assembly bolts. When taper bushings with mounting flanges and sheave are assembled on the shaft, there should always be a space between the flange of the bushing and the hub of the sheave.

II. HEAVY-DUTY INDUSTRIAL SHEAVES

A. Availability

Heavy-duty industrial sheaves are manufactured by a number of different companies, but relatively few companies are full line manufacturers. Most companies do not make and stock sheaves for all standard belt cross sections or in a full range of diameters. The greater the number of grooves, the more difficult it will be to find the sheave. Even the full line companies have minor differences between their standard lines. The larger diameter standard sheaves with the greater number of grooves are the more difficult to obtain. The designer should not assume that a sheave diameter found in a belt manufacturer's drive design table is available from his sheave supplier even if they are the same company.

B. Standard Materials

Stock heavy-duty industrial sheaves are made of an intermediate strength class of grey cast iron – class 30 or class 30,000 psi tensile strength cast iron. The actual strength of the iron in the cast product can vary substantially because of residual stresses or self-annealing of larger sheave cross sections. The usual nondestructive method for determining the class of grey iron is by determining the Brinell hardness. If the hardness falls within a specific range on the Brinell scale, the iron is considered to be class 30.

Some sheaves may also be made of ductile iron or steel at the option of the manufacturer. Steel or ductile iron is used because of additional strength requirements or because production economies can be realized with the alternate material.

C. Construction

1. Types of Construction

Depending on the sheave diameter and number of grooves, industrial sheaves are made in three configurations: block, web, and arm (see Figure 7). When there is no general thinning in material between the hub and the rim, the configuration is called block. If there is substantial thinning but no sizeable longitudinal holes between the hub and the rim, the sheave is of web configuration. Arm or spoke construction should need no explanation. The number of arms or spokes for a given diameter can vary between manufacturers.

2. Design Considerations

The majority of heavy-duty industrial sheaves has a hub that is machined for a taper bushing. Although it is desirable from a stress standpoint to have the bushing mating hub faces located at equal distances from the outside faces of the sheave, standardization in the length of electric motor shafts makes this approach impractical with wide-faced sheaves. The hubs on many sheaves have been offset so that standard length motor shafts will extend through the hub when the sheave is normally mounted. (Some types of bushings may be mounted with the bushing flange near the motor case – normal mounting – or with the hub between the bushing flange and the motor case – reverse mounting.) With short shafts, the rule-of-thumb is that the shaft should reach at least 2/3 of the way through the length of the bore in the hub of the sheave. While this general rule is applicable to most situations, highly fluctuating loads or reversing loads may loosen the grip of the bushing on the shaft. The motor or equipment shaft should extend through the length of the bore in the hub of the sheave whenever possible.

When designing a drive, the offset of the hub may be a critical consideration. If there is not enough offset, the shafts may be too short to properly engage the bushing. If there is too much offset, bearings may fail because of overloading. Therefore, careful consideration should be given to drive parameters whenever there is a deviation from standard sheaves.

III. LIGHT-DUTY OR FRACTIONAL HORSEPOWER (FHP) SHEAVES

A. FHP Sheave

1. Construction and Materials

FHP sheaves are made in a wide variety of materials and configurations. The most common of these materials and constructions are: split steel, zinc die cast, cast iron, and plastic. The choice of sheave material should be determined based upon economics and drive performance requirements.

Because of the wide variety of materials from which FHP sheaves are made and production methods, construction types are not as easily classified as heavy duty sheaves (Figure 8). Consult the sheave supplier for his construction.

2. Groove Cross Sections

Fractional HP (FHP) sheaves are usually one- or two-groove sheaves, although some three-groove sheaves can be found. The sheaves are often made with a combination groove that will accommodate at least two of the light-duty belt cross sections. Light duty belt cross sections are 2L, 3L, 4L, or 5L.

B. 2L Sheaves

The 2L cross section is so infrequently used that most of the sheave manufacturers do not advertise that 2L belts will run in their sheaves.

FIGURE 7 Sheave types (above) QD, (right, top) VT, (right, bottom), spoke.

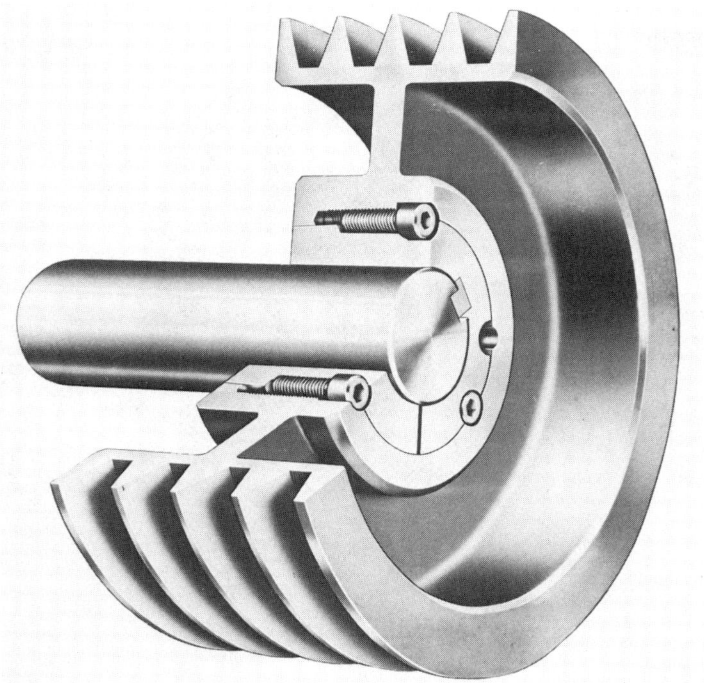

FIGURE 8 Examples of fractional horsepower sheaves.

The 2L cross section belt will fit in a standard 3L groove although it will ride deep into the groove. If the groove is not deeper than the minimum recommended depth for a 3L groove, there will be less than the minimum recommended clearance between the bottom of the 2L belt and the bottom of the groove. Consult the sheave supplier before using a 2L belt in a 3L groove.

C. 3L, 4L, and 5L Combination Groove Sheaves

1. Common Practice

Although sheaves for individual light-duty cross section belts are available, most manufacturers find it economically advantageous to produce a groove that can be used with more than one belt cross section. This is accomplished by deepening the groove and allowing the smaller belt cross section to ride below the top of the groove. The typical combination grooves are 3L-4L and 4L-5L combination grooves.

Quite often, light-duty sheaves are also proclaimed to have A-B combination grooves in recognition of conventional practice. The A groove is dimensionally similar to the 4L groove and the B groove is dimensionally similar to the 5L groove. On many fractional HP drives, A or B heavy-duty belts are used out of convenience instead of the light-duty belts. In low HP applications the use of an A or B belt is an overbelting of the drive and an increase in cost, but is usually not noticeably detrimental to drive service life or pose a safety problem.

2. Danger of Mixing FHP and Heavy-Duty Sheaves in Application

Using FHP sheaves in heavy-duty (A or B design HP) applications is not recommended. The use of FHP sheaves in heavy-duty applications may endanger the safety of personnel and equipment. Because light-duty cross section belts and heavy-duty cross section belts may be used in the same sheaves, most sheave manufacturers produce FHP sheaves with spokes strong enough to meet MPTA minimum recommended stress levels. However, since this practice is not universal, there is the possibility that MPTA minimum recommended spoke stress may be exceeded on some drives and spoke breakage could result. Since the FHP sheaves were intended to transmit low HP loads, the methods of securing the sheave to the shaft may not be adequate for higher HP loads. The sheave may work loose, damage the shaft, or come off the shaft.

D. Methods of Attaching FHP Sheaves to Shafts

FHP sheaves are attached to shafts by three methods: setscrew, setscrew with key, and tapered bushing for some cast iron or steel FHP sheaves. Fixed bores are available up to 1 7/16 in. and tapered bushings are available with bore diameters up to 1 1/2 in. Taper bushings are also available with metric bores. Stocked bore sizes vary between manufacturers.

IV. AUTOMOTIVE AND AGRICULTURAL EQUIPMENT SHEAVES

A. General Information

Sheaves used on automotive and agricultural equipment are not included in RMA-MPTA standards. Dimensions and tolerances for these sheaves can be found in the Society of Automotive Engineers Standard SAE J636c, SAE J1459 (V-ribbed), SAE J1313 (synchronous drives) or in the American Society of Agricultural Engineers Standard ASAE S211.3. Although many of the belt cross sections used on automotive and agricultural drives appear very similar to standard industrial belts, the belts are often very different. They vary in materials, construction, and load carrying capability. The sheaves for these belts are also made differently from their industrial look-alikes so that optimum performance can be obtained within the special operating environments. The belt manufacturer should be consulted before using industrial belts in automotive or agricultural applications.

B. Availability

Many automotive and agricultural sheaves are made for a specific application and are not easily adapted to other applications. Indeed, most of these sheaves are only available through the original equipment manufacturer, and in many cases, the equipment manufacturer is also the manufacturer of the sheave. The equipment manufacturer may also elect to use industrial sheaves and in those cases, replacements will be available from an industrial supply house. Consult the manufacturer if in doubt about the belt or sheave to be used on automotive or agricultural drives.

C. Standard Materials

Automotive and agricultural sheaves can be made of any suitable sheave material but the majority of these sheaves is made from either steel or cast iron. Steel is used more often for automotive sheaves because sizeable economies can be realized through high-volume production. Agricultural sheaves are usually made of either cast iron or steel. Iron is used because of its durability in an abrasive agricultural environment and castings are usually more economical in small quantities. Steel is used for agricultural sheaves when additional fabrication is required.

V. V-RIBBED PULLEYS

A. Materials and Production Considerations

1. Materials

V-ribbed sheaves can be made from a variety of materials and in various configurations; however, this variety is generally confined to K-section sheaves where high production volumes can justify the initial tooling costs associated with the different manufacturing techniques. J, L, and M cross section sheaves are most commonly available in cast iron.

Automotive sheaves are made of either steel, cast iron, plastic, or aluminum, but it can be expected that new materials will be added to the list if they have a cost-to-performance advantage over the present day materials.

2. Groove Formation

The grooves may be cast into the sheave, roll formed, or machined. Regardless of the method of production, V-ribbed grooves are hard to produce correctly. The small size of the groove, the less distinctly defined profile, and closer tolerances combine to increase manufacturing difficulties. It is very difficult to correctly produce multiple grooves with a single point tool. When multiple grooves are to be machined in the pulley, a tool capable of producing several grooves at a time can be obtained from a cutting tool supplier.

The H cross sections has very few power transfer applications. It is used predominantly in very light-duty applications such as office equipment. In these applications, the power transmitting capability of the belt far exceeds any possible demands that the equipment can make on the belt and the main concern is the transfer of motion within the mechanism. Consequently, each application is unique to the equipment and the pulleys are designed for that piece of equipment.

B. Cross Sections Available

1. Industrial Cross Sections

V-ribbed belt drives have been enjoying increased popularity in recent years, especially in automotive and some appliance applications. The J cross section V-ribbed belt is the most popular industrial cross section. L and M cross section belts and sheaves are seldom used and may only be available from a limited number of suppliers. The K cross section is normally thought of as an automotive cross section, but it too has found some limited industrial usage.

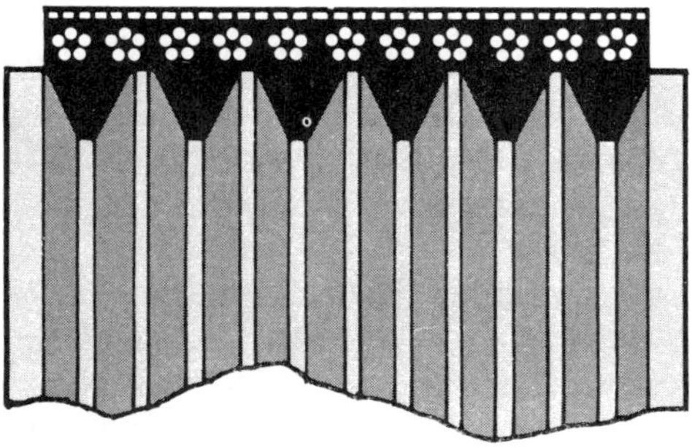

FIGURE 9 V-ribbed belt with pulley cross section.

2. Automotive Cross Sections: K and L

K-section V-ribbed belts have gone from nonexistence only a few years ago to the preferred belt for new passenger car applications. The K-section is used predominantly in automotive engine accessory applications and virtually all available sheaves are designed for use on a particular automotive type application.

The L-cross section V-ribbed belts have also had some limited automotive usage. Groove dimensions are available in SAE J1459 for both L and K cross section sheaves (Figure 9). When working in the metric system, the cross sections are designated as PL and PK respectively by SAE.

VI. SYNCHRONOUS PULLEYS

A. Sprocket Materials and Tolerances

Synchronous sprockets are made from a variety of materials. Cast iron and steel are the most common materials for industrial pulleys. Other metals and plastics are also used for certain applications. Powdered metal has also been used to make sprockets in smaller diameters but not in larger diameters because of manufacturing (press) limitations.

Unlike V-belts, synchronous belts depend upon the shear strength of the belt tooth material to transmit load. Belts and sprockets must be held to close manufacturing tolerances on pitch to be sure that the maximum amount of material is in shear when transmitting load. Close

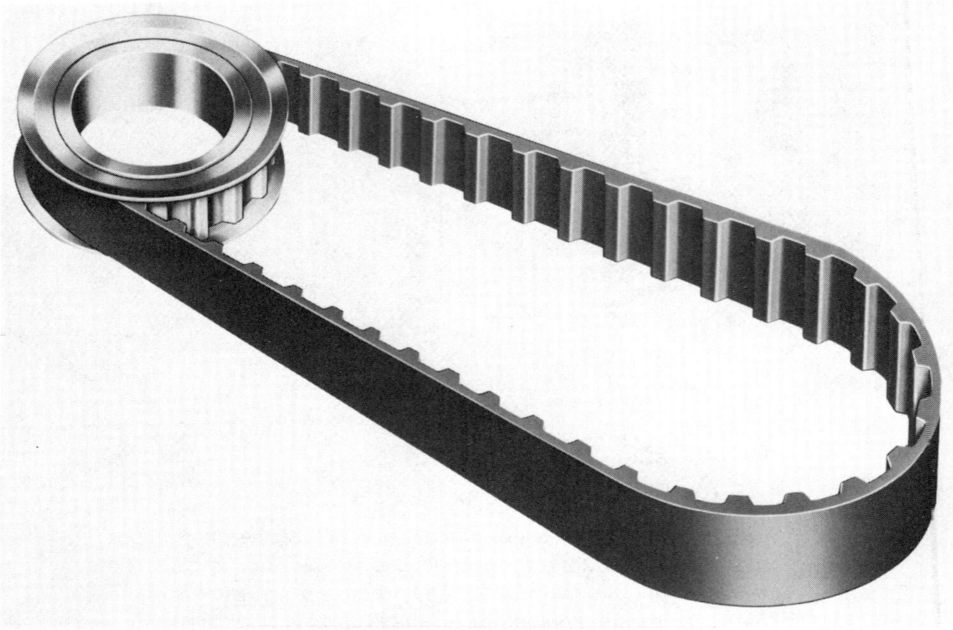

FIGURE 10 Synchronous belt with sprocket.

adherence to recommended machining tolerances is a must in order to ensure satisfactory drive performance (Figure 10).

In addition, the sprocket and belt must not interfere with each other. The belts are made with high modulus cords so that length change under load will be very small. This ensures that the belt teeth will not interfere with the sprocket teeth. The sprockets are machined to very close tolerances for smooth, predictable engagement and disengagement. The profile and diameter of the sprocket must allow the belt to engage smoothly and positively transmit the load. The teeth on the sprocket must be maintained within close tolerances on the outside diameter (OD) and remain perpendicular to the plane of the drive.

B. Synchronous Tooth Profiles

1. Standard Tooth Profiles

Industrial standards do not exist for the all synchronous belt profiles. Industrial standard (RMA-MPTA IP-24) is for standard trapezoidal-shaped teeth in six standard pitches. Belts and pulleys for these profiles can be obtained from several sources as stock items.

Light-duty profiles: MXL and XL profiles are usually used on drives for synchronization with lesser emphasis on transmission of power. As with the H cross section V-ribbed belts, many of these applications can be found in office- and computer-type equipment, although there are also industrial applications in areas such as textile machinery. Because most of the applications are at the original equipment manufacturer level, a limited number of companies make and stock sprockets for these profiles.

Heavy-duty profiles: Synchronous belt tooth profiles L, H, XH, and XXH are commonly used for industrial applications. These belts have a trapezoidal tooth profile and tooth pitches of 0.375, 0.500, 0.875, and 1.250 in., respectively. There are standardized HP-rating tables for XL, L, H, XH, and XXH in the RMA/MPTA Engineering Standard IP-24. These are the only tooth profiles for which a published industry standard exists. This standard also includes drive design information.

Sprocket machining information is also included in this standard for the above cross sections as well as MXL. Sprocket machining is particularly important for synchronous belts in order to ensure even loading across the width of the belt. Drastic reductions in belt life can occur as the result of uneven belt loading due to poorly machined sprockets. Sprocket dimensions such as diameter, runout, bore tolerances, and pitch-to-pitch must be closely adhered to for maximum belt service life.

Many sheave manufacturers and machine shops are capable of producing these synchronous belt profiles. In smaller diameters and tooth profiles, these sprockets may be extruded or stamped.

C. Other Synchronous Tooth Profiles

The world of synchronous belts is in a state of transition. There are a large number of competing tooth profiles and very little standardization. The designer should be aware of some of the difficulties and expense associated with obtaining sprockets for many of these nonstandard tooth profiles. Almost without exception, these sprockets/pulleys will be special order from a limited number of potential suppliers. Often the belt manufacturer will neither stock nor supply pulleys for his special profile but will recommend a job-shop that can make the hardware.

VII. SHEAVE BALANCE

A. Reason for Balancing

Sheaves are balanced in order to limit the transfer of damaging energy fluctuations to the driving or driven machine. For most industrial drives, a precision degree of balance is not required. Consequently, most industrial sheaves receive only a static balance. When it is known that a more precise degree of balance will be required, the more precise balance requirement must be communicated to the sheave supplier. The sheave supplier can then give the sheave a close static balance or dynamic balance as needed. When dynamic balancing is required, the tapered bushing that is to be used on the drive should be balanced with the sheave. Differences in concentricity between different bushings can influence the balance of the sheave-bushing system.

Quite often there is a misunderstanding of why dynamic balancing is recommended. Dynamic balancing is recommended to reduce the amount of vibrational energy transmitted to the drive because of unbalanced couples. Excessive vibration may damage the equipment or result in an unsatisfactory product. Dynamic balancing does not increase the recommended rim speed for grey iron or ductile iron.

B. Balancing Standards

1. MPTA Standard Practice

There are many different balancing standards used in industry. Usually varying grades of balance are specified. Guidelines may be given as to which grade of balance to use for particular types of components. The designer must then use his judgement as to which grade of balance to apply to his particular component. The MPTA has published "Standard Practice for Balancing of Power Transmission Pulleys." At present, this standard sets guidelines for static balancing of pulleys. The equipment designer must still determine the need for and degree of dynamic balance. The MPTA standard is being revised to include recommended guidelines for dynamic balance.

Heavy-duty sheave balancing:

Closeness of balance: Balance guidelines are divided into two groups: guidelines for heavy-duty pulleys and guidelines for light-duty pulleys. Heavy-duty sheaves should be balanced so that the "Maximum static unbalance at the rim should be limited to 0.1% or 0.001 × the pulley mass (1 gram per kilogram of pulley weight) or 5 grams whichever is greater."*

The bottom limit of 5 g was established so that extremely difficult or unattainable values of balance would not be specified routinely.

---

*Standard Practice for Balancing of Power Transmission Pulleys, No. 80, 1980. Mechanical Power Transmission Association.

This limit also makes allowance for inaccuracies inherent in measuring methods and balancing equipment.

Equivalent grade of balance: Certain standards such as the International Organization for Standardization (ISO) Standard 1940 define balance quality grades. When the above MPTA guideline is evaluated at the maximum recommended rim speed for cast iron pulleys, this recommendation is equivalent to a balance quality grade of G33. This grade of balance is better than would generally be required for engine crankshafts or car wheels but not as good as for propeller shafts. If the pulley does not run at 6500 ft/min, the grade of balance grade improves because balance grade is inversely proportional to speed [5].

Light-duty pulleys: Light-duty pulleys should be balanced so that the "maximum static unbalance at the rim should be limited to 0.1% or 0.001 × the pulley mass (1 gram per kilogram of pulley weight) or 15 grams, whichever is greater."*

Light-duty sheaves (FHP) are generally much lighter in weight and construction than heavy-duty sheaves. The space available for drilling balance holes is limited by this lighter construction, thus, the higher unbalance floor limit.

C. Types of Balance

1. Static vs. Dynamic

There are two basic types of balancing: static and dynamic. A perfect static balance would require that the pulley mass be distributed around the rotational axis such that it can cause no rotational moment about the "axis of rotation." Perfect dynamic balance would have the additional requirement that the mass distribution can cause no rotational moment about an axis that passes through the center of mass and perpendicular to the axis of rotation, i.e., the sheave would then be balanced in two perpendicular planes.

2. Dynamic Balance Guidelines

It is not within the scope of this chapter to provide a detailed discussion of dynamic balancing so that every application will be addressed. However, satisfactory balance can usually be obtained by applying the rules of conventional practice. Many different sources

---

*Standard Practice for Balancing of Power Transmission Pulleys, No. 80, 1980. Mechanical Power Transmission Association.

RECOMMENDED BALANCING REQUIREMENTS
STATIC VS DYNAMIC

For **HEAVY MACHINERY APPLICATIONS,** where noise and vibration is not critical, calculate on formula

$$\text{RPM} = \frac{20{,}000}{\sqrt{\text{DF}}}$$

For **LIGHT MACHINERY APPLICATIONS,** where noise and vibration are critical, calculate on formula

$$\text{RPM} = \frac{10{,}000}{\sqrt{\text{DF}}}$$

The nature of your driven machinery is the real determining factor on the need for dynamic balance. Where vibration is a critical problem, dynamic balance is recommended regardless of operating speed.

As a guide, this graph shows the recommended maximum speed limits in RPM for statically balanced wheels for **GENERAL MACHINERY APPLICATIONS,** when exceeding these speeds dynamic balance is recommended. Graph is based on formula

$$\text{RPM} = \frac{15{,}500}{\sqrt{\text{DF}}}$$

where D = Diameter
F = Face Width

EXAMPLE: Using a 30.0" PD, 14 groove, "C" Sheave (14⅜" face), heavy machinery application.

$$\text{RPM} = \frac{20{,}000}{\sqrt{30 \times 14\tfrac{3}{8}}} = \frac{20{,}000}{20.77} = 963$$

963 RPM max. for static balance
(Ductile or steel required at this speed)

EXAMPLE: Using a 30.0" PD, 14 groove, "C" Sheave (14⅜" face), light machinery application.

$$\text{RPM} = \frac{10{,}000}{\sqrt{30 \times 14\tfrac{3}{8}}} = \frac{10{,}000}{20.77} = 481 \text{ RPM max. for static balance}$$

CAUTION: USE DUCTILE IRON OR STEEL WHEN RIM SPEEDS EXCEED 6500 FEET PER MINUTE.

FIGURE 11 Sample guideline for determining dynamic balancing.

of tables and charts illustrating these guidelines can be found in industrial literature. One such chart is illustrated in Figure 11. This chart is only a guideline and the designer should carefully consider the performance requirements before settling on a guideline.

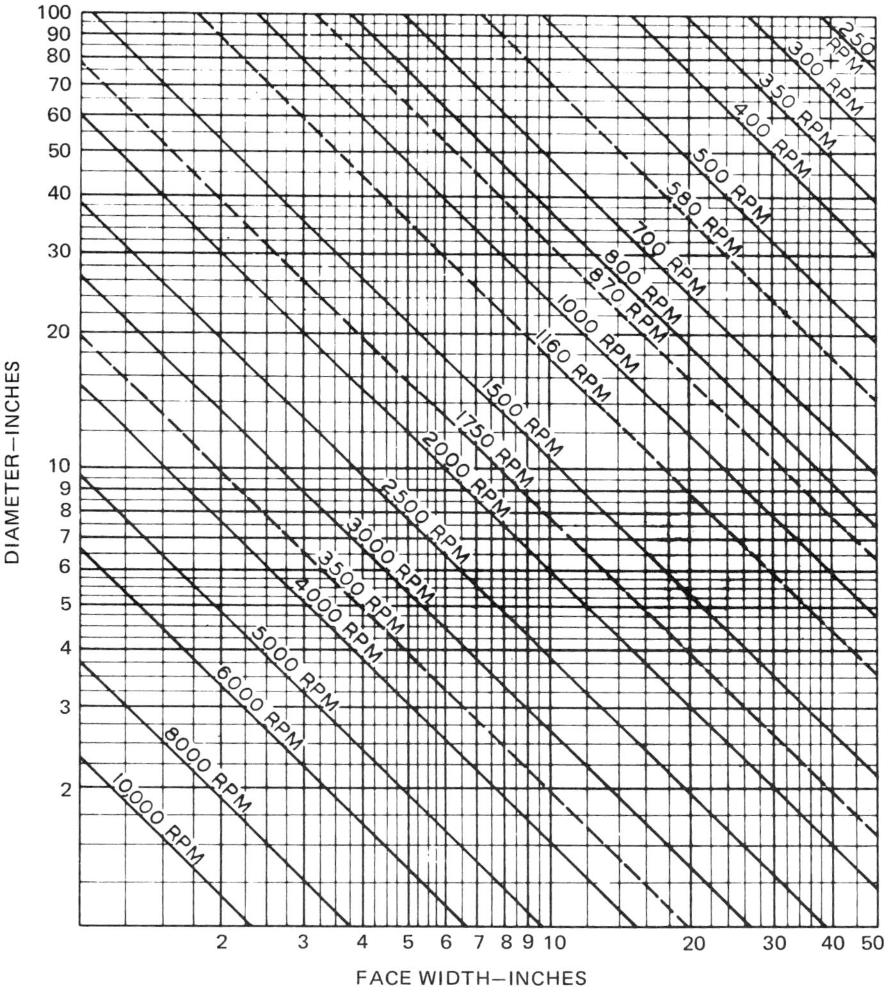

FIGURE 11 (Continued) How to use this graph: Find wheel face at bottom and diameter at left. Read up to the point where the face width (vertical) line intersects the diameter (horizontal) line. If the intersecting point is above the required rpm line, dynamic balance is recommended.

Example: 14-C 30.0" sheave (14 3/8 face) at 800 rpm—dynamic balance required.
8-C 30.0" sheave (8 3/8 face) at 800 rpm—static balance acceptable.

Rim speed = .262 × D × rpm
= .262 × 30.0 × 800 = 6288 fpm—cast iron acceptable.

REFERENCES

1. *Machinery's Handbook*, 15th Edition (1977). Industrial Press, New York, p. 336.
2. Shock, E. E., and Lundstrum, J. F. A Knowledge of Temperature Effects Can Help Improve V-Belt Drive Designs. Gates Rubber Company Belt Technical Division, Denver.
3. Jones, P. E., et al. (1983). The Design and Development of the B Series Diesel Engine. Society of Automotive Engineers, Volume 831344.
4. Narrow Multiple V-Belts (1977). In *Engineering Standard*, Volume IP-22, Rubber Manufacturers Association, p. 6.
5. International Standard 1940 (1973). International Organization for Standardization, Switzerland.

15

Computerized Drive Design

MICHAEL S. COLEMAN *V-Belt Product Application Department, The Gates Rubber Company, Denver, Colorado*

I. BASIC PRINCIPLES AND ADVANTAGES

The computer revolution has taken tedious, repetitive tasks out of the hands of the designer. As with any iterative process involving numerical analysis, V-belt drive design lends itself nicely to computerization. From a main frame computer to a programmable hand-held calculator, these machines can be made to handle all or part of the drive design process.

Computers are programmed to follow the same design procedure as a human designer. The two will ultimately, therefore, arrive at the same end. The difference comes in the speed at which that final design is arrived. The time saved by using the rapid number-crunching ability of the computer can be used to play "what if" games with input data. The result is a more thorough engineering study in less time.

II. LIFE-IN-HOURS DRIVE DESIGN

A. Use

As was noted in previous chapters, there are two basic types of drive design. The life-in-hours method is used to design drives for automotive, agricultural, and light-duty belt applications. This method was introduced by The Gates Rubber Company in the early 1950s.

In life-in-hours design, a fatigue rate is calculated based on drive geometry, belt speed, and predicted loads. This fatigue rate is then used to estimate average belt life. The procedure and relevant equations are outlined in Chapters 4 and 5. This procedure can be considered an algorithm and used to code a life-in-hours design program.

B. Multiple Condition Drives

Computerized life-in-hours design is most useful for those drives with multiple operating conditions. When analyzing a multiple condition drive, the designer must repeat a majority of the steps for each condition. To do the job properly, the tension from the worst condition should also be imposed on the other conditions. This adds even more calculations to an already time-consuming and laborious task.

As an example, consider the drive in Figure 1, a typical layout for automotive accessory drives. Assume that the arcs of contact, span lengths, and belt length have been previously determined to be as indicated. For demonstration purposes, only two operating conditions will be considered. Many agricultural and automotive applications have five or more operating conditions. Horsepower loads for two crankshaft speeds are estimated as shown in Table 1.

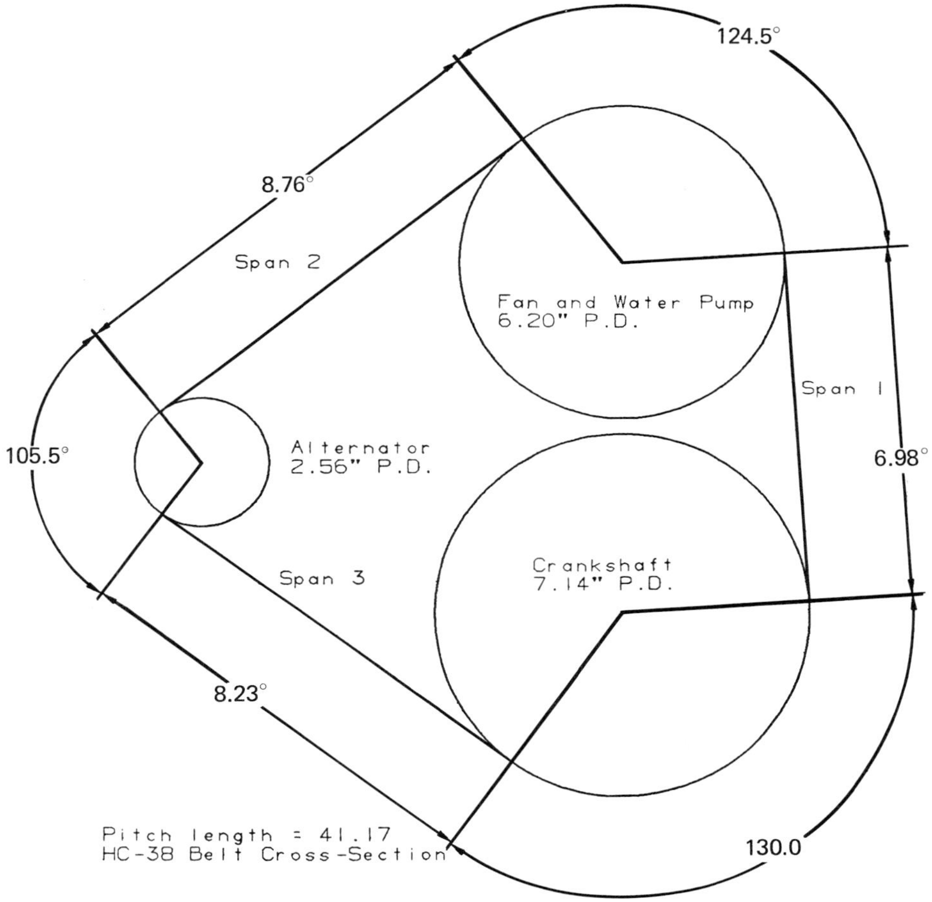

FIGURE 1 Typical layout for an automotive accessory drive.

TABLE 1

Crankshaft rpm	1000	2000
HP loads		
fan and water pump	0.8	1.9
alternator	1.7	2.3
Percent of total operating time	70	30

The first step is to calculate belt speed for each condition using Equation (1).

$$\text{belt speed} = V = \frac{(\text{rpm})(\text{PD})}{3.82} \quad (1)$$

Condition A	Condition B
$V_a = \frac{(100)(7.14)}{3.82}$	$V_b = \frac{(2000)(7.14)}{3.82}$
$= 1869$ ft/min	$= 3738$ ft/min

Next, calculate belt effective pull for each sheave and each drive condition using Equation (2).

$$\text{effective pull} = EP = \frac{(\text{Design HP})(33{,}000)}{\text{belt speed}} \quad (2)$$

	Condition A	Condition B
Crankshaft	$EP_c = \frac{(2.5)(33{,}000)}{1{,}869}$	$EP_c = \frac{(4.2)(33{,}000)}{3{,}738}$
	$= 44.14$ lb	$= 37.08$ lb
Fan and water pump	$EP_f = \frac{(0.8)(33{,}000)}{1{,}869}$	$EP_f = \frac{(1.9)(33{,}000)}{3{,}738}$
	$= 14.12$ lb	$= 16.77$ lb
Alternator	$EP_a = \frac{(1.7)(33{,}000)}{1{,}869}$	$EP_a = \frac{(2.3)(33{,}000)}{3{,}738}$
	$= 30.02$ lb	$= 20.31$ lb

Now that the effective pull is known for each sheave, the span tensions can be calculated using Equations (3) and (4)

$$T_1 = \frac{(1.25)(EP)}{G} \quad (3)$$

where

T_1 = tight span tension (lb)
G = arc of contact correction factor from Figure 2.

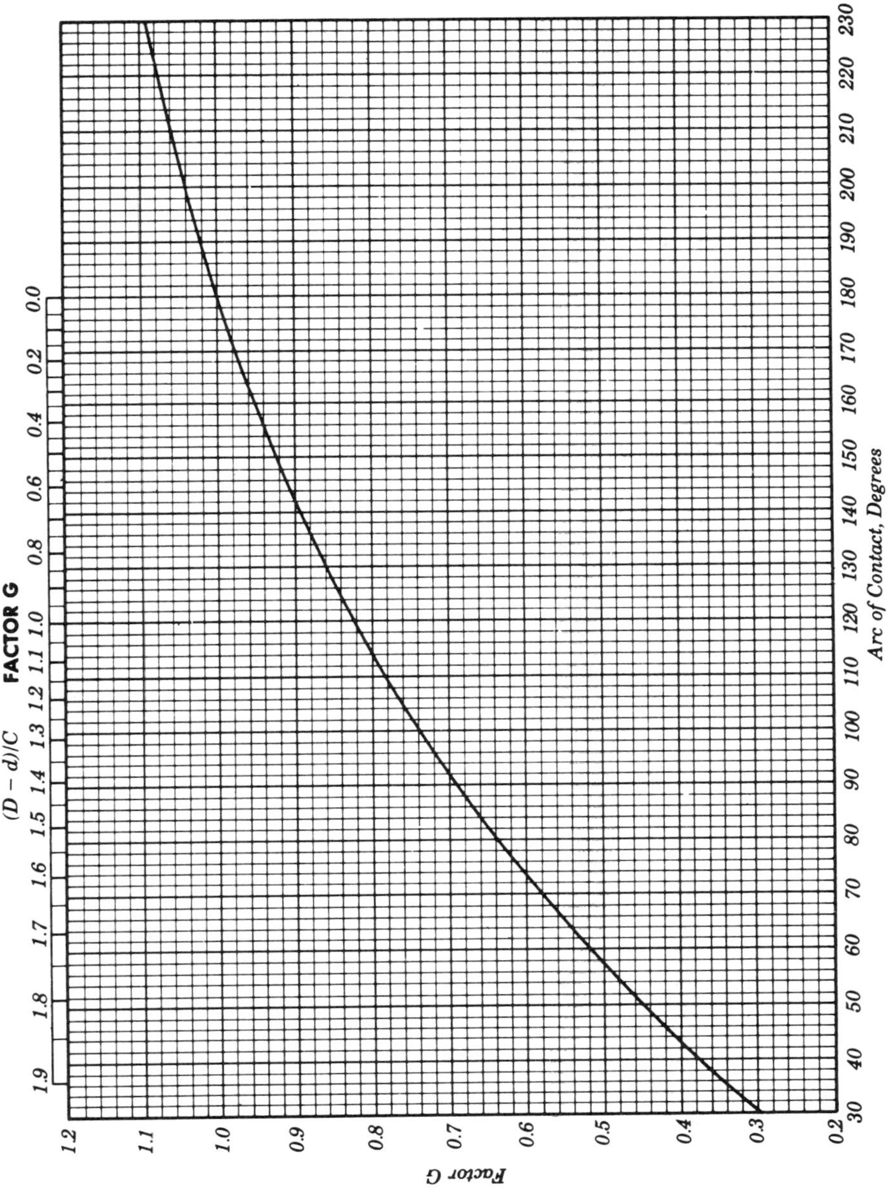

FIGURE 2 Graph of Factor G.

$$T_2, T_3, \text{etc.} = T_1, T_2, \text{etc.} - \text{EP of } dN_1, dN_2, \text{etc.} \tag{4}$$

where

T_2, T_3, etc. = tension of span 2, 3, etc. (lb)
T_1, T_2, etc. = tension of span 1, 2, etc. (lb)

Then:

Condition A	Condition B
$T_1 = \frac{(1.25)(44.14)}{0.86}$	$T_1 = \frac{(1.25)(37.08)}{0.86}$
= 64.16 lb	= 53.90 lb
T_2 = 64.16 - 14.12	T_2 = 53.90 - 16.77
= 50.04 lb	= 37.13 lb
T_3 = 50.04 - 30.02	T_3 = 37.13 - 20.31
= 20.02 lb	= 16.82 lb

The allowable tension ratio (TR) for each loaded dN sheave can be determined from Figure 3. This value can then be compared to the calculated tension ratio at each sheave. If the calculated tension ratio is greater than the allowable tension ratio for any loaded driveN sheave, the tension in the belt must be increased for this condition according to Equation (5).

	Condition A		Condition B	
	Allowable TR	Calculated TR	Allowable TR	Calculated TR
Fan and water pump	3.05	1.23	3.05	1.45
Alternator	2.57	2.50	2.57	2.21

$$T_A = \frac{T_T - (R_A \times T_S)}{R_A - 1} \tag{5}$$

where

T_A = additional tension (lb)
T_T = tight span tension at sheave (lb)

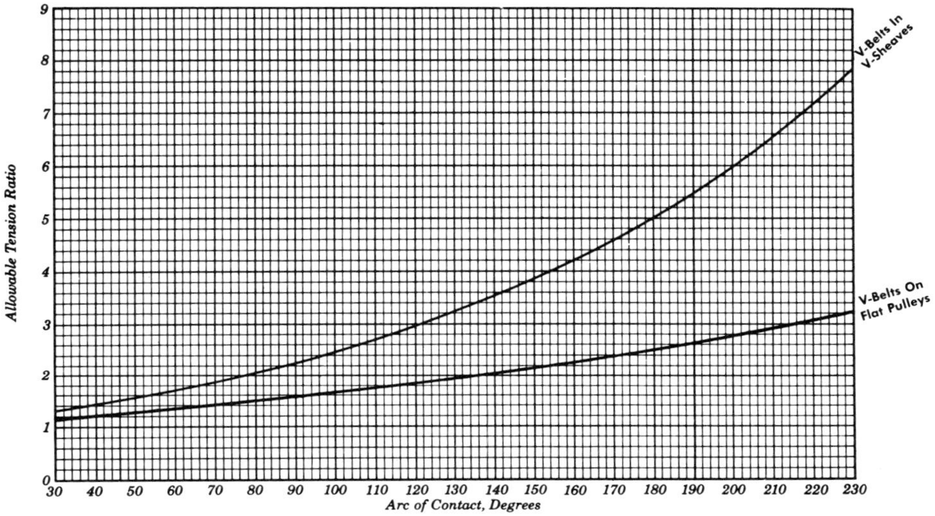

FIGURE 3 Allowable tension ratio.

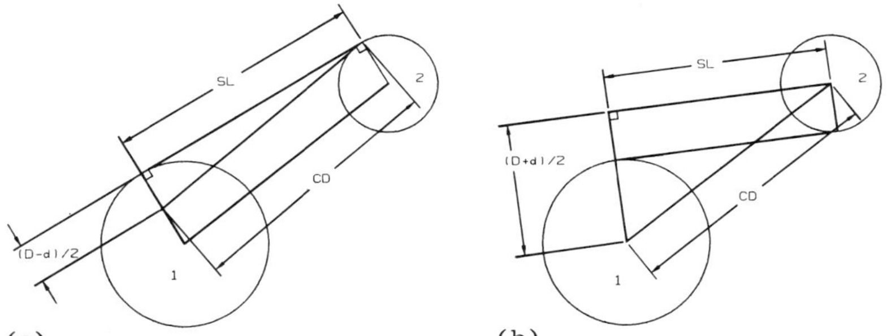

(a) (b)

FIGURE 4 Geometry for succesive sheaves (a) on the same side of the belt, (b) on opposite sides of the belt.

T_S = slack span tension at sheave (lb)
R_A = allowable tension ratio at sheave

In this example, use of Equation (5) is not necessary.

The highest required tension has now been identified to exist at condition 1. Since this is a statically tensioned drive, it must be designed for the peak tension requirement. The necessary static tension for the worst case condition can be calculated from Equation (6).

$$T_S = 0.9\left(\frac{(T_1)(t_1) + (T_2)(t_2) + (T_3)(t_3) + \text{etc.}}{t_1 + t_2 + t_3 + \text{etc.}} + \frac{mV^2}{10^6}\right) \tag{6}$$

where

T_1, T_2, etc. = operating tension per belt in span 1, 2, etc. (lb)
t_1, t_2, etc. = length of span 1, 2, etc. (in.)
m = centrifugal force constant, dependent on cross section
V = belt speed (ft/min)

For a SAE 0.380 belt from a particular manufacturer, the centrifugal force constant is 0.42. Then for this example,

$$T_S = 0.9\left(\frac{(64.6)(6.98) + (50.04)(8.76) + (20.02)(3.23)}{6.98 + 8.76 + 8.23} + \frac{(0.42)(1869)^2}{10^6}\right)$$

$$= 40.8 \text{ lb}$$

This tension must be imposed on the other conditions. This will increase the tensions present during operation at these other conditions. To determine the distribution of that increased tension, the static tension Equation (6) and effective pull Equation (2) must be solved backward. Note that the centrifugal tension term should not be part of the constant static tension. Its effect can be eliminated by use of Equation (7).

$$T_{LC} = T_S - \frac{0.9mV^2}{10^6} \tag{7}$$

where

T_{LC} = peak static tension less centrifugal effect (lb)
T_S = peak static tension (lb)

m = mass factor
V = belt speed at condition being considered (ft/min)

For this example,

$$T_{LC} = 40.8 - \left(\frac{(0.9)(0.42)(3738)^2}{10^6}\right)$$

$$= 35.52 \text{ lb}$$

Solving Equation (6), less the centrifugal tension term, for T_1 yields Equation (8). This equation can be used to find the new tight side tension for the condition

$$T_{1C} = \frac{T_{1c}}{0.9} - \left[\frac{(t_2 + t_3 + \ldots + t_n)EP_2 + (t_3 + t_4 + \ldots + t_n)EP_3 + \ldots + t_nEP_n}{(t_1 + t_2 + \ldots + t_n)}\right] \quad (8)$$

where

T_{LC}	= corrected tight side tension for the condition (lb)
T_{1C}	= peak static tension less centrifugal effect (lb)
t_1, t_2, etc.	= length of span 1, 2, etc. (in.)
EP_1, EP_2, etc.	= effective pull at span 1, 2, etc. (lb)
n	= number of sheaves in the system

So,

$$T_{1C} = \frac{35.52}{0.9} + \left(\frac{(8.76 + 8.23)16.77 + (8.23)(20.31)}{6.98 + 8.76 + 8.23}\right)$$

$$= 58.32 \text{ lb}$$

Knowing the required effective pull at each sheave, T_2 and T_3 for condition B can now be recalculated.

$$T_{2C} = 58.32 - 16.77$$

$$= 41.55 \text{ lb}$$

$$T_{3C} = 41.55 - 20.31$$

$$= 21.24 \text{ lb}$$

A fatigue rate can be calculated for each sheave at each drive condition. The fatigue rate at a sheave is determined by use of Equation (9). Note that the constants K_1, K_2, K_3, and K_4 are dependent on the belt construction and cross section. The values used here are for one supplier's SAE 0.380 belt. Manufacturers' design manuals should be consulted for fatigue constants peculiar to their belts.

$$\log(FR) = K_1\left[T_t + \frac{k_2}{d} + K_3\left(\frac{V}{100}\right)^2\right] + \log\left(\frac{V}{1000}\right) - K_4 \tag{9}$$

where

K_1, K_2, K_3, K_4 = constants, dependent on belt cross section
FR = fatigue rate
V = belt speed (ft/min)
T_t = tight span tension at sheave (lb)
d = pitch diameter of sheave (in.)

For a SAE 0.380,

$K_1 = 0.025495$

$K_2 = 443.08$

$K_3 = 0.43306$

$K_4 = 6.1427$

Condition A		Condition B	
	Fatigue rate (FR)		Fatigue rate (FR)
Crank shaft	0.10	Crank shaft	0.10
Fan and water pump	0.10	Fan and water pump	0.10
Alternator	0.72	Alternator	1.14
Total FR	0.92	Total FR	1.34

The total fatigue rates at each condition must be weighted according to the percentage of time the drive operates at that condition.

	% of total operating time	Weighted FR
Condition A	70	0.644
Condition B	30	0.402
	Sum of FR =	1.046

The final step is to calculate an average belt life for this application. The appropriate equation for this step (Equation 10) follows.

$$h = \frac{(100)(PL)}{\text{sum of FR}} \tag{10}$$

where

h = calculated average belt life (hr)
PL = belt pitch length (in.)

So,

$$h = \frac{(100)(41.17)}{1.046}$$

$$= 4000 \text{ hours (rounded to the nearest 100 hours)}$$

Note: This belt life corresponds to a statistical B-50. For other failure life relationships (i.e., B-10), belt manufacturers can be contacted for reliability data.

The repetition of tasks involved with this two-condition, three-sheave drive is obvious. It is easy to see that analyzing a drive with more sheaves and operating conditions could become an overwhelming task. Additionally, the designer's work may or may not be finished at this point. A final design is seldom achieved through a single iteration.

It might be desirable to examine the effect moving a sheave, increasing diameters, or changing the duty cycle will have on belt performance. Perhaps the drive will be automatically tensioned by use of a spring-loaded idler, thus allowing design at a higher tension ratio. If done by hand, any of these changes would require repeating the calculations outlined above with the new information. This example should illustrate the capacity a computer has to increase the efficiency and thoroughness of the life-in-hours drive design process.

C. Geometric Drive Design

Another part of the drive design process is determination of the correct belt length. This, too, is an aspect of design which can be handled more efficiently by a computer.

Generally, what is known about geometry are the shaft locations (in X-Y coordinates) and the sheave diameters. In the case of movable sheaves, minimum and maximum movement positions are often also known. With this information, calculation of belt length for a two-sheave drive is a simple matter. The pertinent equation for a two-sheave drive belt length is Equation (11).

$$L = 2C + 1.57\,(D + d) + \frac{(D - d)^2}{4C} \tag{11}$$

where

L = belt length (in.)
C = center distance (in.)
D = diameter of larger sheave (in.)
d = diameter of smaller sheave (in.)

For a crossed belt drive, the two-sheave length equation becomes

$$L = 2C + 1.57\,(D + d) + \frac{(D + d)^2}{4C} \tag{12}$$

It should be noted that using pitch diameters in these equations will result in pitch length. If effective outside diameters are used, effective length will be obtained.

Establishing belt length for drives with three or more sheaves can be a lengthy process. There are two methods available to the designer to perform this task. The first is to lay the drive out to scale and measure arcs of contact and span lengths. The total sum of the arc values must be 360° (if there are backside pulleys, the arcs for these should be subtracted from the total of the inside sheaves). The amount of belt on a sheave is determined by Equation (13).

$$AL = \frac{\pi}{360(d)(\theta)} \tag{13}$$

where

AL = arc length (in.)
d = diameter of sheave (in.)
θ = arc of contact (deg)

Belt length is found by summing arc and span lengths.

The second method is to calculate the belt length directly. The equations needed to do this will be developed below. These equations can be used in a computer program to solve drive geometry for belt arcs and spans.

The span length between any two sheaves depends on several factors. These factors are sheave diameters, distance between sheave centers, and whether the sheaves operate on the same or opposite sides of the belt. Figures 4a and 4b show the geometrical relationship between diameters, center distance, and span length for successive sheaves. Equations (14) and (15) are derived from these geometrical relationships to provide span lengths from known values.

$$SL = \left[CD^2 - \frac{(D - d)^2}{2}\right]^{1/2} \tag{14}$$

$$SL = \left[CD^2 + \frac{(D + d)^2}{2}\right]^{1/2} \tag{15}$$

where

SL = span length (in.)
CD = center distance (in.)
D = diameter of larger sheave (in.)
d = diameter of smaller sheave (in.)

Calculation of the arcs of contact is somewhat more involved. Three angles must be determined for each sheave in the system. The magnitudes of these angles are dependent on sheave diameters, center distances, span length, and whether successive sheaves operate on the same or opposite sides of the belt.

Figure 5 illustrates three sheaves that are part of a larger belt drive system. This example shows the three angles, α_{n-1}, α_{n+1}, and α_{CD} that are needed to calculate arc of contact on a sheave. It also shows the two situations that can exist between sheaves acting either on the same or on opposite sides of the belt. The equations developed for this example can be used for any sheave in any system.

The first angle to be determined will be α_{CD}. This is the angle between the line connecting the centers of sheaves n and n+1 and the one between the centers of sheaves n and n-1. The distances between centers will be needed and can be calculated by Equations (16), (17) and (18).

$$CD_{n,n-1} = [(X_n - X_{n-1})^2 + (Y_n - Y_{n-1})^2]^{1/2} \tag{16}$$

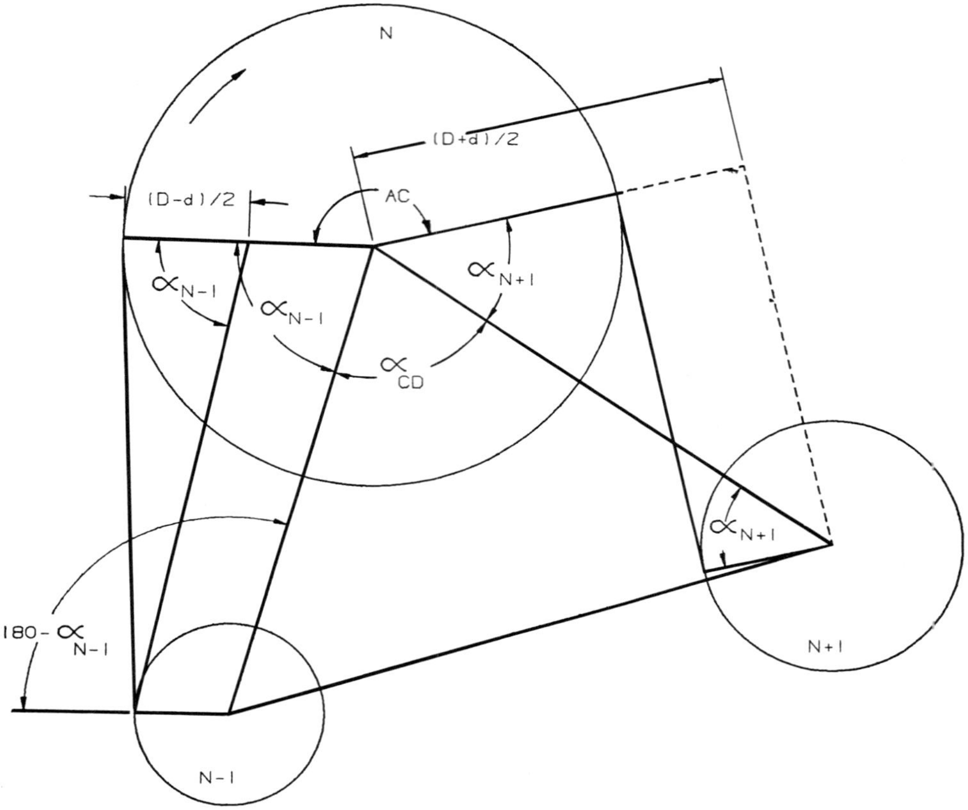

FIGURE 5 Diagram showing angles required to calculate arc of contact.

where

$CD_{n,n-1}$ = center distance between sheaves n and n-1 (in.)
X_n, X_{n-1} = X coordinates of sheaves n, n-1
Y_n, Y_{n-1} = Y coordinates of sheaves n, n-1

Similarly,

$$CD_{n,n+1} = [(X_n - X_{n+1})^2 + (Y_n - Y_{n+1})^2]^{1/2} \tag{17}$$

and

$$CD_{n-1,n+1} = [(X_{n-1} - X_{n+1})^2 + (Y_{n-1} - Y_{n+1})^2]^{1/2} \qquad (18)$$

Once the center distances are known, α_{cd} can be determined by the law of cosines (see Equation (19).

$$c^2 = a^2 + b^2 - 2ab\cos(\alpha) \qquad (19)$$

The resulting equation is

$$CD^2_{n-1,n+1} = CD^2_{n,n-1} + CD^2_{n,n+1} - 2(CD_{n,n-1})(CD_{n,n+1})\cos(\alpha_{CD}) \qquad (20)$$

Solving for α_{cd} yields

$$\alpha_{CD} = \cos^{-1}\left[\frac{CD^2_{n,n-1} + CD^2_{n,n+1} - CD^2_{n-1,n+1}}{2CD_{n,n-1}\,CD_{n,n+1}}\right] \qquad (21)$$

The other two required angles are those angles which exist between the perpendiculars to the belt spans through the tangent points and the lines connective sheave centers. Referring to Figure 5, it can be seen that:

$$\alpha = \tan^{-1}\left[\frac{2(SL)}{(D - d)}\right] \qquad (22)$$

where

SL = span length (in.)
D = diameter of larger sheave (in.)
d = diameter of smaller sheave (in.)

for sheaves acting on the same side of the belt, and

$$\alpha = \tan^{-1}\left[\frac{2(SL)}{(D - d)}\right] \qquad (23)$$

for sheaves acting on opposite sides of the belt.

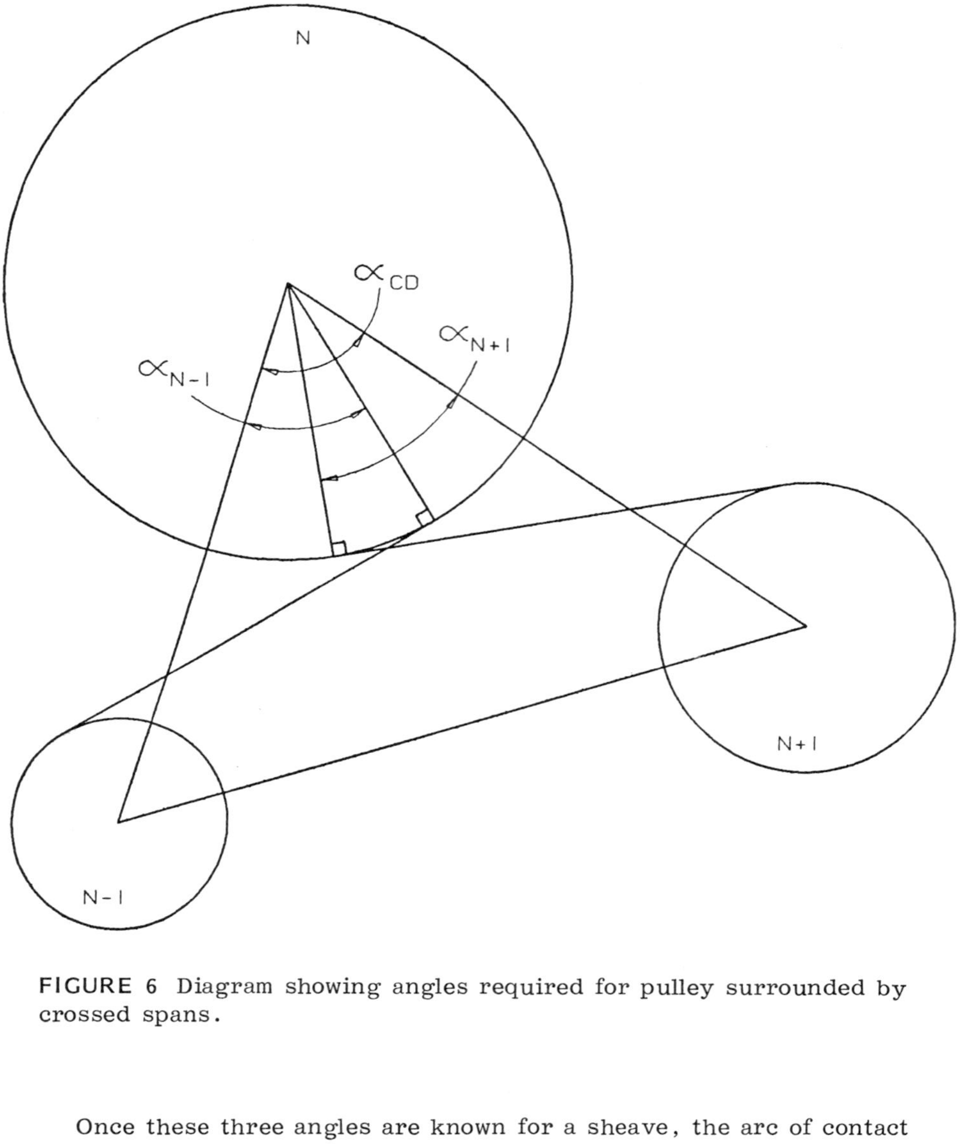

FIGURE 6 Diagram showing angles required for pulley surrounded by crossed spans.

Once these three angles are known for a sheave, the arc of contact can be calculated by Equation (24):

$$AC = 360 - (\alpha_{cd} + \alpha_{n-1} + \alpha_{n+1}) \tag{24}$$

If the sheave in question is surrounded by crossed spans as in Figure 6, the arc of contact is found by Equation (25)

$$AC = 360 - (\alpha_{n-1} + \alpha_{n+1} - \alpha_{cd}) \tag{25}$$

For the graphical method to be accurate, care must be taken when laying the drive out to scale. Calculating the belt length using the equations outlined above will take time and require repeated procedures. If a pulley location or diameter is changed even slightly, both the graphical and numerical methods will have to be repeated to get the new length. The alternative is to use the equations outlined above once to create a belt geometry program. This program can then be used to quickly check the installation and take-up allowance and the effect geometry changes have on belt length.

D. Capacity Design

The belt life value calculated by the life-in-hours procedure may or may not be acceptable. At the same time, it may not be possible to increase sheave diameters, decrease loading, or take other action that would reduce belt tensions. The capacity of a single belt to provide adequate service for that application has been exceeded. It would be desirable at this point to see if two or more belts would suffice. This capacity design would involve repeating belt life calculations with a greater number of belts.

For the earlier example, suppose the 4,000 hr life calculated for the drive was not acceptable. Perhaps the application requires at least 25,000 hr belt life. The next step would be to try two belts on the drive. To do this, the designer would have to return to the step where span tensions were first calculated. These span tensions would be divided by the number of belts being considered, in this case, two. Then, the following steps, including calculating peak static tension and imposing that tension on the other conditions, calculating belt fatigue and finally belt life, would have to be repeated.

Though these steps will not be shown here, the belt life for two SAE 0.380 belts is 18,500 hr. So the steps would have to be repeated again. This time for three belts and so on until the required belt life of 25,000 hr was reached. Once again, it is easy to see how use of the computer would make the process quicker and simpler.

III. INDUSTRIAL POWER RATING DESIGN METHOD

A. Use

Also known as the HP-rating method, this is the second basic type of drive design mentioned earlier in this chapter. This design method is used exclusively with two-sheave, heavy-duty industrial drives.

In order to calculate a HP rating, it is necessary to establish some value for belt life. This is expressed as 3–5 years continuous service because belt life cannot be predicted exactly. Variability in the loads, tension, and environmental conditions that a V-belt sees during its use make exact life prediction impossible.

The equations and design procedure pertinent to industrial power rating design are outlined in Chapter 3. When combined with specific information from belt manufacturers design catalogs, the procedure outlined in Chapter 3 can be used to create a design program.

B. Advantages of Computerization

There typically is more than one drive which will satisfy the requirements of a particular application. How do the designers insure that they have come up with the best one? They could sit down and, using a manufacturer's drive design manual, come up with every industrial V-belt drive that meets their needs. This task would involve many hours of work. When done, the information would still have to be put into a form which allowed easy comparison of the alternatives.

This points out the major advantage that computerization can bring to industrial power rating drive design. By creating databases which contain information on all stock belts and sheaves, a program can be written to put together all possible drive combinations. Each combination can then be checked against input data to see if an acceptable design has been achieved. The computer will rapidly select all acceptable belt drives for a particular application.

The information generated by an industrial power rating drive design program can be tabulated in a computer report similar to that shown in Figure 7. Using this report, an informed decision can be made as to which drive is best for the job. The computer program allows a designer the flexibility to choose a drive based on the criteria – space limitations, shaft pull, cost, and so forth – which are most important to his design.

As an example, consider the following application. Given:

1. An internal combustion engine driving a fan, continuous service
2. Service factor = 1.3
3. Engine speed = 2000 rpm
4. Engine rating = 75 HP, contimuous service
5. Desired fan speed = 1100 rpm
6. Desired center distance = 43 in.

An industrial power-rating computer run for this application results in the output shown in Figure 7. If there were no further restrictions, the natural choice would be the least costly

```
APPLICATION INFORMATION -

        ID            : Fan
        Motor HP      :      75.00
        dR RPM        :    2000
        SF            :    1.3
        Belt Types    : SUPER HC MN, SUPER HC, HI-POWER II & TRI-POWER V-BELTS

ACCEPTABLE BELT DRIVES -
```

dN RPM	SMALL DIA	LARGE DIA	--BELT-- SECT	LNG	N	- ACTUAL CD AND - -- ADJUSTMENT --		BELT DEF.	DEF.FORCE MIN	MAX	BELT PULL	FACE WIDTH	REL. COST
1143	8.00	14.00	3VX1180		8	41.62	+1.8/-0.80	0.65	4.8	7.1	1191	3.53	1.08
1111	10.60	19.00	3VX1320		6	42.56	+2.2/-0.80	0.66	5.1	7.6	907	2.72	1.03
1136	5.90	10.30	5VX1060		6	40.22	+1.5/-1.00	0.63	9.0	13.0	1622	4.44	1.26
1105	6.30	11.30	5VX1120		5	42.11	+1.8/-1.00	0.66	10.0	14.6	1520	3.75	1.17
1130	6.70	11.80	5VX1120		5	41.40	+1.8/-1.00	0.65	9.5	13.9	1429	3.75	1.18
1130	7.10	12.50	5VX1120		5	40.52	+1.8/-1.00	0.63	9.2	13.4	1350	3.75	1.24
1130	7.50	13.20	5VX1180		4	42.66	+1.8/-1.00	0.67	10.7	15.6	1277	3.06	1.10
1136	8.00	14.00	5VX1180		4	41.62	+1.8/-1.00	0.65	10.2	14.9	1199	3.06	1.20
1130	8.50	15.00	5VX1180		4	40.42	+1.8/-1.00	0.63	9.8	14.3	1131	3.06	1.24
1117	9.00	16.00	5VX1250		3	42.73	+1.8/-1.00	0.67	12.0	17.7	1068	2.38	1.03**B**
1099	10.30	18.70	5VX1320		3	43.03	+2.2/-1.00	0.67	11.1	16.3	937	2.38	1.13**C**
1064	11.30	21.20	5VX1320		3	40.18	+2.2/-1.00	0.62	10.7	15.7	862	2.38	1.26**D**
1111	11.80	21.20	5VX1320		3	39.82	+2.2/-1.00	0.62	10.5	15.4	823	2.38	1.27**E**
1093	6.00	11.00	B	108	10	41.48	+2.5/-1.25	0.65	5.5	8.0	1572	7.75	1.56
1099	6.80	12.40	B	112	8	41.73	+2.5/-1.25	0.65	6.1	8.9	1390	6.25	1.21
1117	8.60	15.40	B	120	6	41.92	+3.0/-1.25	0.65	6.8	9.9	1105	4.75	1.00**A**
1143	8.00	14.00	C	115	6	41.57	+3.0/-1.50	0.65	8.4	12.0	1184	6.38	1.68
1064	8.50	16.00	C	120	6	42.05	+3.0/-1.50	0.65	8.3	11.9	1121	6.38	1.76
1124	9.00	16.00	C	120	5	41.68	+3.0/-1.50	0.65	9.2	13.3	1057	5.38	1.47
1058	9.50	18.00	C	128	5	43.66	+3.0/-1.50	0.68	9.2	13.2	1006	5.38	1.58
1111	10.00	18.00	C	128	5	43.29	+3.0/-1.50	0.67	9.2	13.2	954	5.38	1.58
1099	11.00	20.00	C	128	4	40.87	+3.0/-1.50	0.63	10.5	15.2	873	4.38	1.43
1093	6.00	11.00	BX	108	10	41.48	+2.5/-1.25	0.65	5.5	7.9	1572	7.75	1.73
1099	6.80	12.40	BX	112	8	41.73	+2.5/-1.25	0.65	6.1	8.8	1390	6.25	1.35
1117	8.60	15.40	BX	120	6	41.92	+3.0/-1.25	0.65	6.7	9.8	1105	4.75	1.12
1075	7.00	13.00	CX	115	6	43.15	+3.0/-1.50	0.67	8.5	12.2	1352	6.38	1.77
1070	7.50	14.00	CX	115	5	41.95	+3.0/-1.50	0.65	9.4	13.7	1265	5.38	1.48
1143	8.00	14.00	CX	115	5	41.57	+3.0/-1.50	0.65	9.1	13.2	1184	5.38	1.54
1064	8.50	16.00	CX	120	5	42.05	+3.0/-1.50	0.65	9.0	13.0	1121	5.38	1.65
1124	9.00	16.00	CX	120	4	41.68	+3.0/-1.50	0.65	10.3	15.0	1057	4.38	1.46
1058	9.50	18.00	CX	128	4	43.66	+3.0/-1.50	0.68	10.2	14.8	1006	4.38	1.51
1111	10.00	18.00	CX	128	4	43.29	+3.0/-1.50	0.67	10.0	14.5	954	4.38	1.53
1099	11.00	20.00	CX	128	4	40.87	+3.0/-1.50	0.63	9.9	14.4	873	4.38	1.58

FIGURE 7 Sample report of an industrial power rating drive design program.

drive, indicated by A. Suppose, however, that there were two further restrictions:

1. Belt shaft pull cannot exceed 1200 (lb).
2. Pulley face width must be less than 2.5 in.

This narrows the choice even further to those drives labeled B, C, D, and E. Once again, it would be natural to choose the least expensive of these alternatives, B, all else being equal.

Perhaps, however, the designer already stocks 5VX 1320 belts and 10.30 in., 3-groove 5V sheaves for another application. In this case, though drive C may cost more initially, its selection could save money in the long run by cutting down on MRO stores. Allowing the user to select drives based on a commonality of parts is another advantage.

A program of this type can get complex and lengthy. Some belt manufacturers offer use of such a program to customers as a free design service.

This example illustrates the power of computerized industrial power-rating design. It puts the information needed to make the proper drive selection in the hands of the designer, and it does it quickly. The result is efficient, accurate V-belt drive design.

A flow chart of a simplified version of this program is in the appendix. This program flow chart was created for programmable hand-held calculators. It can easily be modified for use with computers where storage and input/output capability are greatly enhanced.

APPENDIX: COMPUTERIZED V-BELT DRIVE DESIGN FLOW CHART

A. Abbreviated Terms

CS	Belt Cross Section
D	Pitch Diameter of Large Sheave
d	Pitch Diameter of Small Sheave
ND*	Nominal Diameter of Large Sheave
Nd*	Nominal Diameter of Small Sheave
ND TO PD	The value obtained by subtracting the pitch diameters from the nominal diameter
R*	RPM of the Small Diameter
r	RPM of the Small Diameter divided by 1000
SR	Speed Ratio
RS	Rim Speed
MRS*	Maximum Rim Speed
HP	HP being transmitted by the belts
SF	Service Factor
HPR*	HP Rating per belt
BL	Belt Length
C	Center Distance
LCF	Length Correction Factor

*Further explanation below.

G*	Arc of Contact Correction Factor
N	Number of Belts
V	Belt Velocity
TT	Tight Side Tension
ST	Slack Side Tension
UST	Upper Static Tension
LST	Lower Static Tension
UFD*	Upper Force Deflection
LFD*	Lower Force Deflection
t*	Span Deflection
TBP*	Total Belt Pull

Belt Constants	K1 - K5, M, Y, L, A, ND to PD, MRS
Dummy Variables	W, X, Z, Q, S1-S10, SRG1 - SRG9
ND, Nd	Nominal diameters are the diameters given in the Heavy Duty Catalog. For Super HC sheaves the nominal diameter is equal to the outside diameter, for classical sheaves the ND is equal to the pitch diameter.
R	All drive design is based on the speed of the small sheave. For speed down drives this will be the driveR sheave, for speed up drives this will be the driveN sheave.
MRS	There is a speed limitation on sheaves. For Super HC sheaves this value is 6500 feet per minute, for classical sheave it is 6000 feet per minute.
HPR	This is the HP one belt can transmit. It has not been corrected for arc of contact or length.
G	All drives are designed for 180° of contact. This is the value which accounts for drives which the driveR does not have 180° arc of contact
TT, ST	These are dynamic tensions.
UFD, LFT, t	These are values used to tension the drive by the force deflection methods of tensioning.
TBP	This is the dynamic loading on either shaft caused by the belts when the drive is properly tensioned. (It does not include the weight of the sheaves.)
**	There are 11 constants which are used to calculate the transmission capability, speed limitations and belt tension for each belt cross section. These constants must be entered into storage areas at the beginning of the program. However, there is one constant (K_5) which is a function of the speed ratio and must be determined after the speed ratio is calculated.

B. Symbols

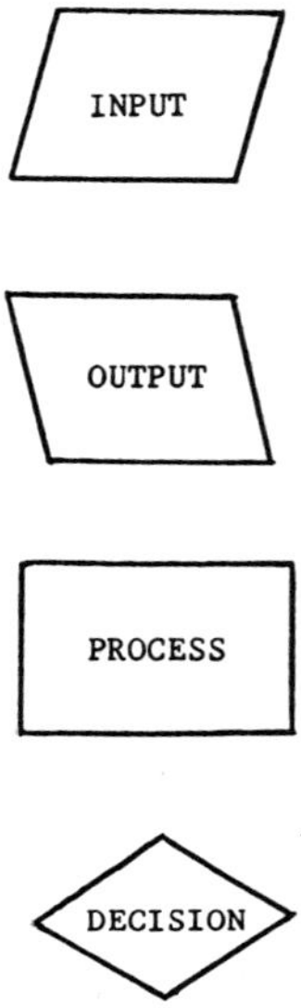

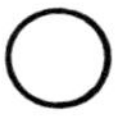

CONNECTOR - This is used to move from one part of the flow chart to another.

C. Flow Chart

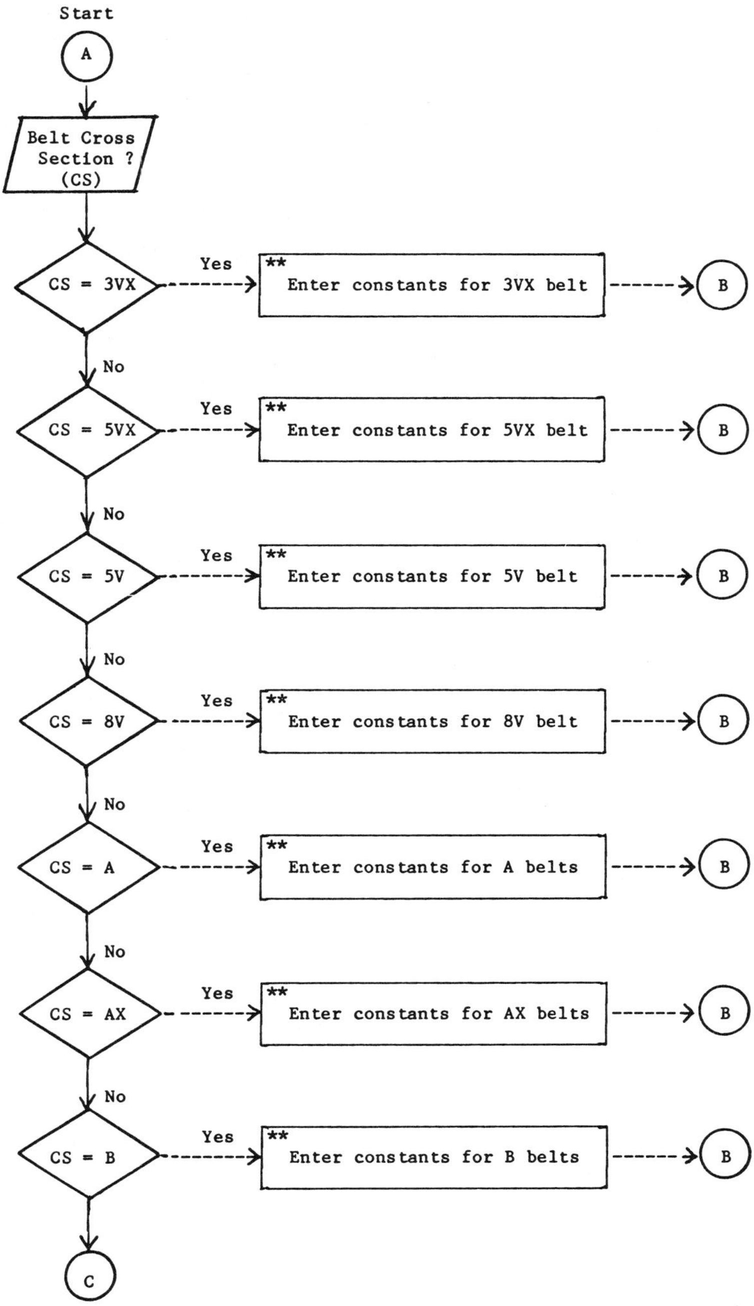
Start
A
Belt Cross Section ? (CS)
CS = 3VX
Yes
** Enter constants for 3VX belt
B
No
CS = 5VX
Yes
** Enter constants for 5VX belt
B
No
CS = 5V
Yes
** Enter constants for 5V belt
B
No
CS = 8V
Yes
** Enter constants for 8V belt
B
No
CS = A
Yes
** Enter constants for A belts
B
No
CS = AX
Yes
** Enter constants for AX belts
B
No
CS = B
Yes
** Enter constants for B belts
B
C

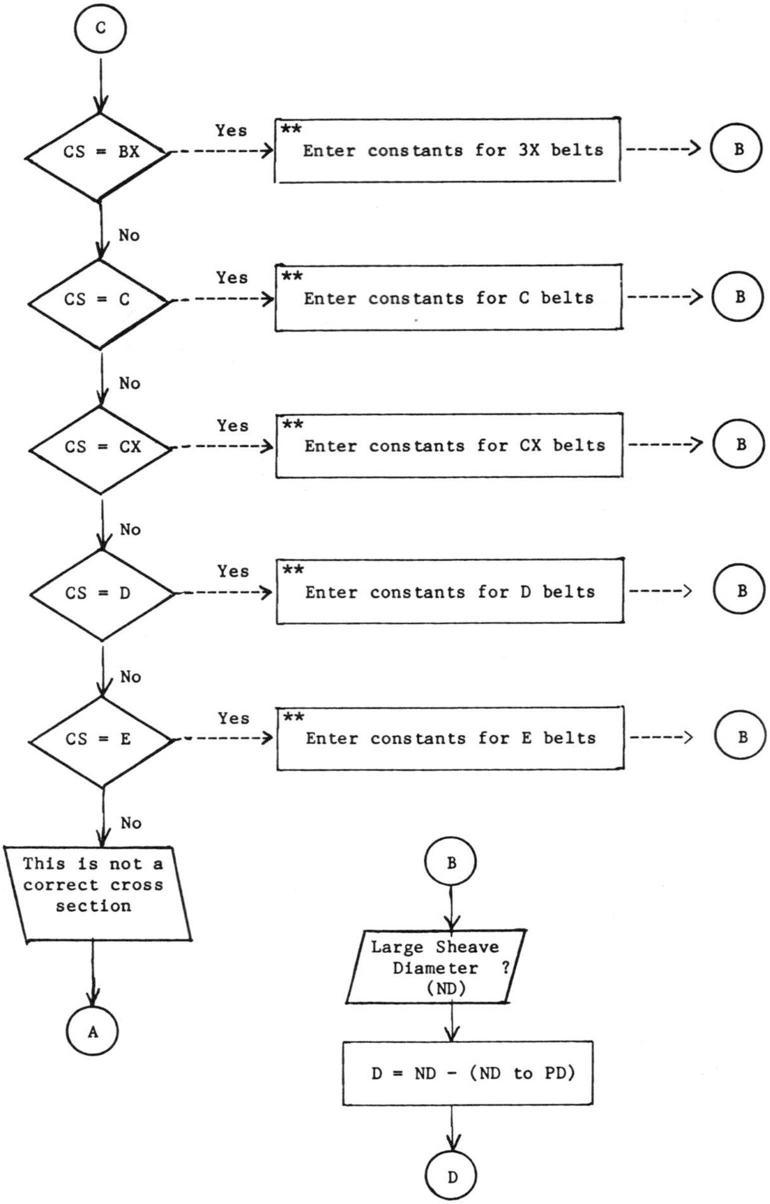
C
CS = BX
Yes
** Enter constants for 3X belts
B
No
CS = C
Yes
** Enter constants for C belts
B
No
CS = CX
Yes
** Enter constants for CX belts
B
No
CS = D
Yes
** Enter constants for D belts
B
No
CS = E
Yes
** Enter constants for E belts
B
No
This is not a correct cross section
A
B
Large Sheave Diameter ? (ND)
D = ND - (ND to PD)
D

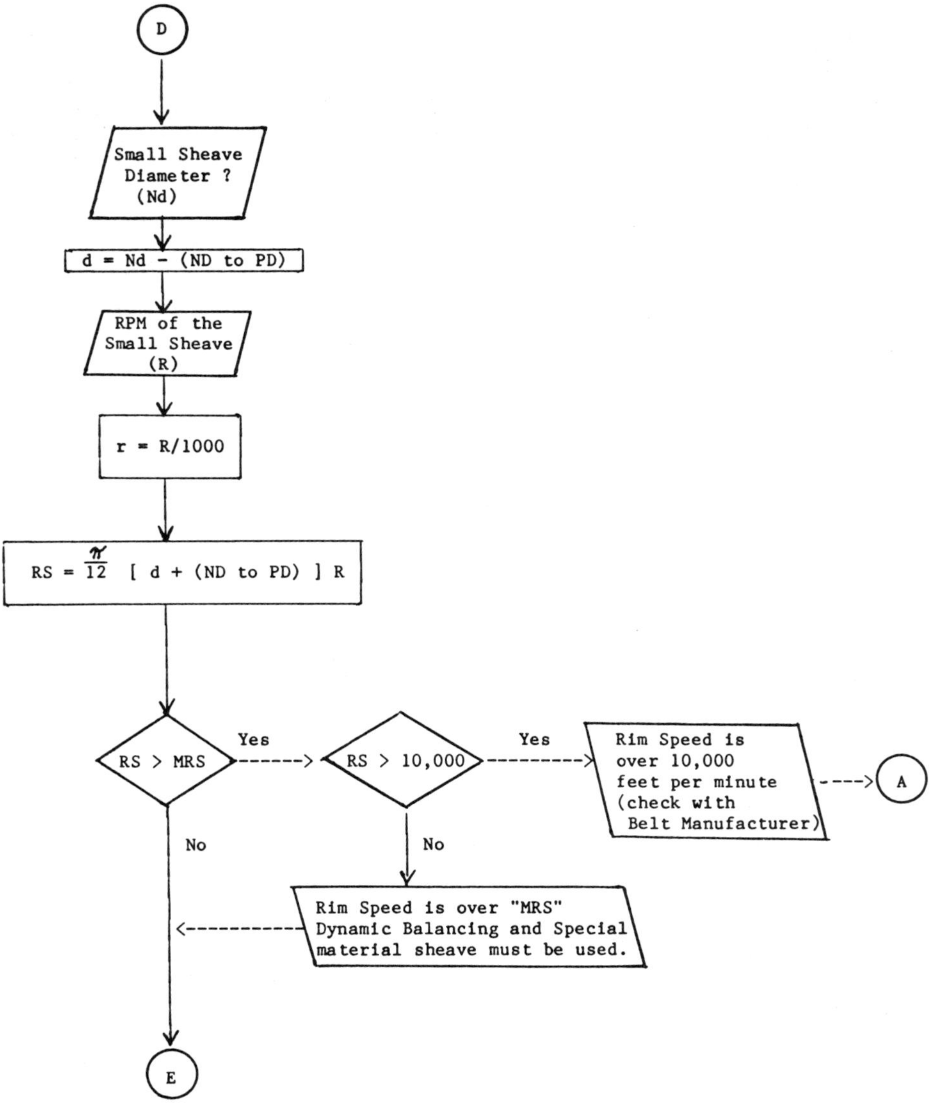
D
Small Sheave Diameter ? (Nd)
d = Nd - (ND to PD)
RPM of the Small Sheave (R)
r = R/1000
RS = π/12 [d + (ND to PD)] R
RS > MRS
Yes
No
RS > 10,000
Yes
No
Rim Speed is over 10,000 feet per minute (check with Belt Manufacturer)
A
Rim Speed is over "MRS" Dynamic Balancing and Special material sheave must be used.
E

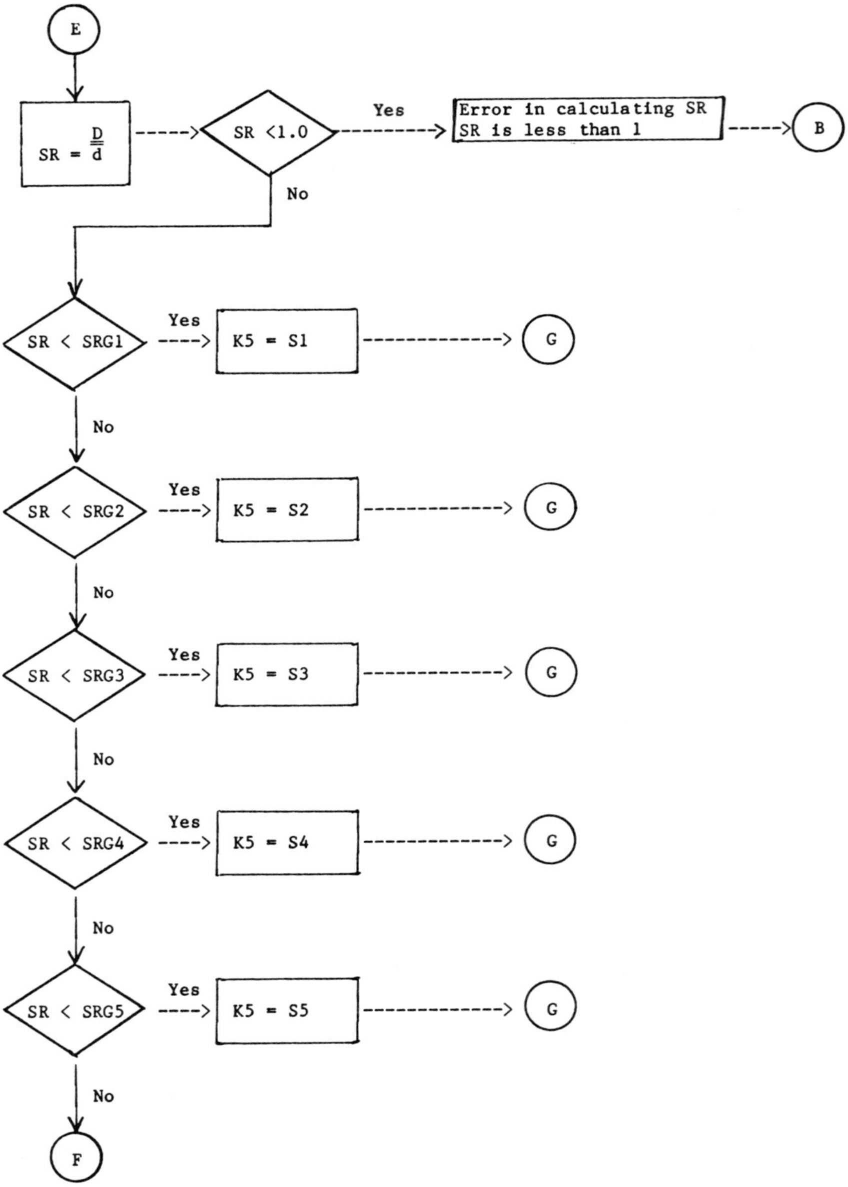
E
SR = D/d
SR <1.0
Yes
Error in calculating SR
SR is less than 1
B
No
SR < SRG1
Yes
K5 = S1
G
No
SR < SRG2
Yes
K5 = S2
G
No
SR < SRG3
Yes
K5 = S3
G
No
SR < SRG4
Yes
K5 = S4
G
No
SR < SRG5
Yes
K5 = S5
G
No
F

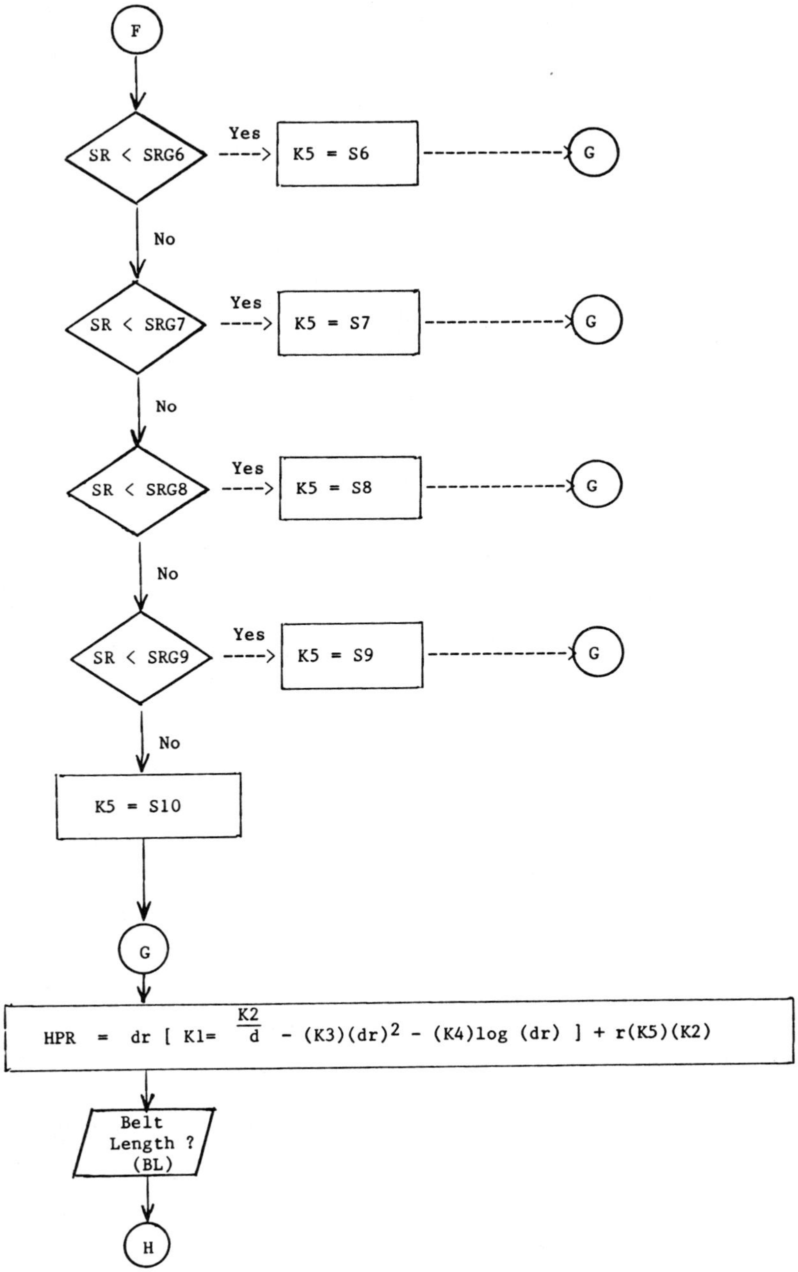
F
SR < SRG6
Yes
K5 = S6
G
No
SR < SRG7
Yes
K5 = S7
G
No
SR < SRG8
Yes
K5 = S8
G
No
SR < SRG9
Yes
K5 = S9
G
No
K5 = S10
G
HPR = dr [K1= K2/d - (K3)(dr)^2 - (K4)log (dr)] + r(K5)(K2)
Belt
Length ?
(BL)
H

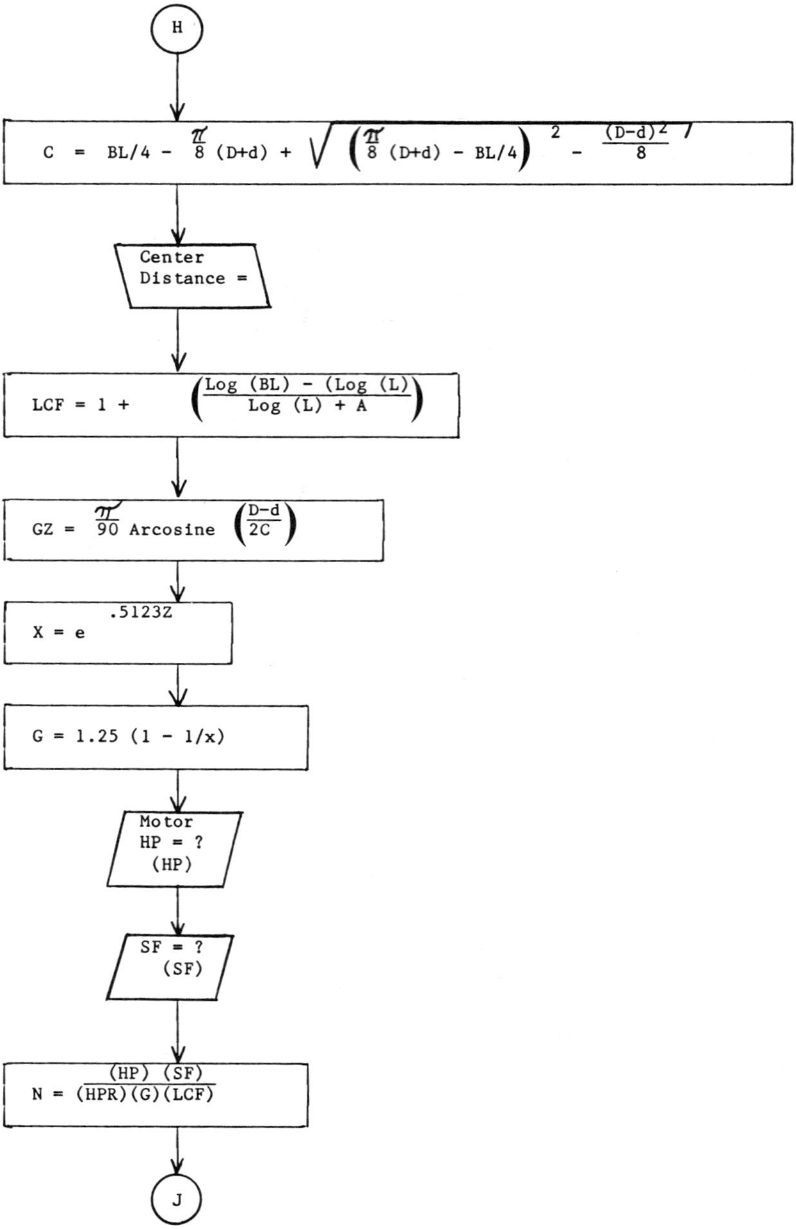
H
C = BL/4 - π/8 (D+d) + √((π/8 (D+d) - BL/4)² - (D-d)²/8)
Center
Distance =
LCF = 1 + ((Log (BL) - (Log (L))/(Log (L) + A))
GZ = π/90 Arcosine ((D-d)/2C)
X = e^.5123Z
G = 1.25 (1 - 1/x)
Motor
HP = ?
(HP)
SF = ?
(SF)
N = (HP) (SF)/(HPR)(G)(LCF)
J

The computer calculates the number of belts needed. Then asks for the number of belts to be used. (Usually the number calculated rounded up to the next higher whole number).

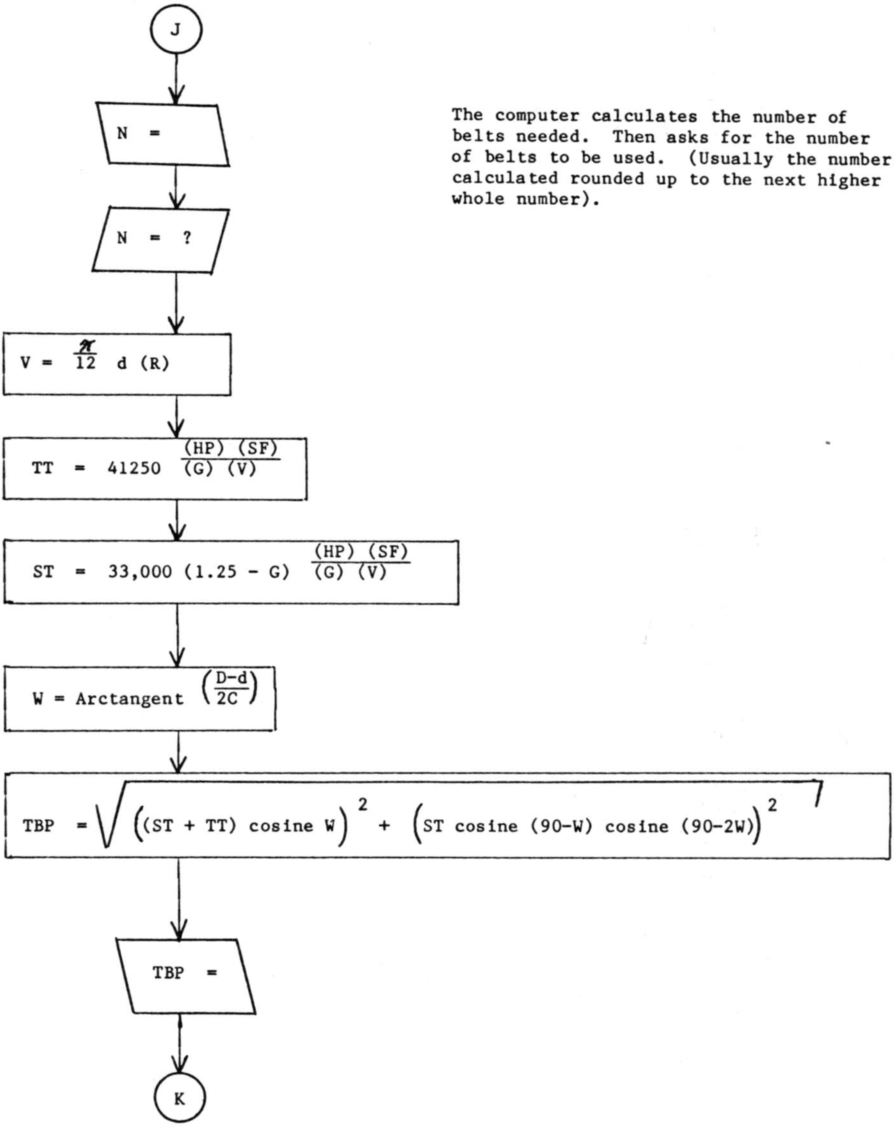

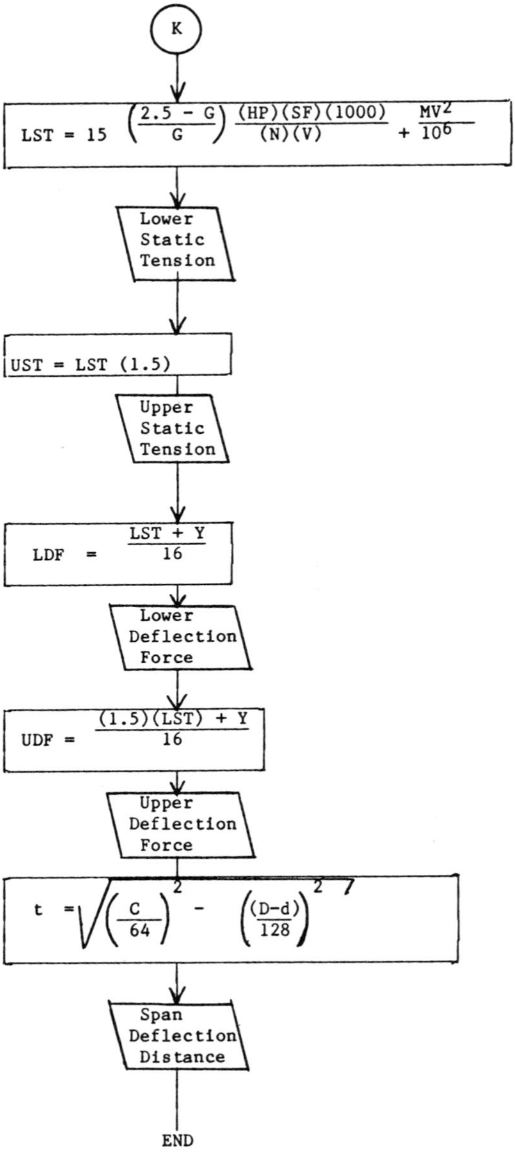
K
LST = 15 ((2.5 - G)/G) ((HP)(SF)(1000))/((N)(V)) + MV2/106
Lower Static Tension
UST = LST (1.5)
Upper Static Tension
LDF = (LST + Y)/16
Lower Deflection Force
UDF = ((1.5)(LST) + Y)/16
Upper Deflection Force
t = √[(C/64)2 - ((D-d)/128)2]
Span Deflection Distance
END

16

Metrication of Power Transmission Components

WILLIAM L. WESTHOFF *Automotive Products Technical and Development Division, The Gates Rubber Company, Denver, Colorado*

I. BASIC PHILOSOPHY ON METRICATION

A. As Applied to Power Transmission Items as Component Parts

Power transmission components not only concern the belting but the necessary pulleys and pulley mounting systems as well. The belting must be able to function properly in the specified grooves as in the English system and the pulleys must be mounted to specified metric shafting with keyways. The use of components in the English and metric system is also a prime consideration. This allows metric belting to function in English pulleys and vice versa. This allows interchangeability of new and old components in a usable system.

B. Industry Standardization Philosophy on Belting

A standard nomenclature has been established for metric belting. This nomenclature allows a distinct description of the belting that distinguishes the belt as a metric component according to the described procedure. The basic system allows for the belt top width to be described in mm, a letter or letters to describe the belt type, and the belt nomenclature length in mm.

A conversion process is used to obtain the metric belt length from the present English length. The basic process is to convert the specified English in. length to mm and round to a nearest even 5, 10, or 25 mm dependent on the length of the belt. This is a soft conversion process necessary to assure English and metric part interchangeability.

This procedure of soft conversion makes the lengths basically equivalent to the English counterparts but does not address the International Standards Organization procedures. The ISO procedure addresses a system of preferred lengths according to a geometric progression series. This geometric progression allows a system of belt lengths that are evenly spaced in length increments as a percentage of length. The chosen nomenclature of the belt section and type distinguish the English converted belts from those of the ISO system. The interchangeability of the belts is sometimes possible with the assistance of a knowledgeable belt application engineer.

II. SPECIFIC METRICATION ACTIVITY

A. Publication of Standards in Dual Units

The present converted industry standards contain two distinct sections. The first section of the standard is the customary English units and the second section is the metric portion. There is no mixing of the two systems and each is a self-contained standard covering the required aspects of all components.

At present, the English portion of the standard is the front portion of the standard since this is the prevalent system in today's marketplace. The sections will be reversed once the majority of the users are on the metric system. The English portion can be completely eliminated once the users recognize only the metric system.

B. Specific Examples of Metrication that Is Other than Soft Metric Conversion

An example of the conversion process that entails conversion that is not of the soft type is the classical multiple belt standard. Here the conversion to the effective system from the pitch system described in Chapter 1 has taken place. Since this process involves the changing of the reference point for the sheave groove, a logical step was taken to provide an integer metric reference for the dimensions. An example is the groove top width for the A-section which is as follows:

Groove angle	English units (pitch system, in.)	Metric units (effective system, mm)
34	0.494	13
38	0.504	13

Within the same standard, the conversion to belt lengths was accomplished by other than soft conversion. The classical belt nomenclature lengths had been established based on a pseudo inside length measurement when they were established over 40 years ago. This led to a misnomer as to the belt's true length as the years proceeded and changes were established. Again, during metrication, the recommended lengths were established that would indicate true metric effective length. The new lengths established were chosen so as to follow the preferred number series in order to geometrically evenly space the recommended belt lengths by percent length. During this process, the existing English interchangeability was not ignored and was kept through the use of the allowable length tolerance from ideal dimension. The newly chosen metric belt lengths would fall within the normal allowable length tolerance for their English counterparts, hence maintaining interchangeability of parts.

The American Society of Agricultural Engineers (ASAE) standard is also an example of metric conversion without a direct soft English equivalent. Here the groove dimensions were established based on logical integer metric values. The effective system was in use in this standard but since there are no basic preferred standard belt lengths, the change from English to metric would not affect proper calculation using the appropriate sheave grooves.

The tension at which the effective length of the belt is determined, referred to as the measuring tension, was also a hard metric conversion. In place of a direct soft conversion, a tension in Newtons was chosen to better indicate the belt length as it is to be applied to the user.

III. LISTING OF ALL STANDARDS THAT HAVE BEEN CONVERTED AS OF JANUARY 1, 1985

The work on converting the industry standards has proceeded according to user demand and continues as this is published.

The list of the converted standards to date is as follows:

IP-20	Classical Multiple V-Belt Standard
IP-22	Narrow Multiple V-Belt Standard
IP-25	Variable Speed V-Belt Standard
IP-24	Synchronous Belt Standard
IP-26	V-Ribbed Belt Standard
API Spec 1B	API Oilfield Belting Standard
J1313 & J1278	SAE Synchronous Belt Standards

Standards that have not been converted to date are as follows:

Single V-Belt Standard
Double V-Belt Standard

IV. DUAL LABELING OF POWER TRANSMISSION BELTS

The conversion process in the industry is a dual-stage process of producing metric parts and getting the users to recognize the new designations. In an effort to provide some effort toward user recognition, some belts are being dual labeled with both the customary English nomenclature and the established metric nomenclature on the same part. This allows the user to recognize the metric nomenclature and become familiar with it and also to specify the part in the metric system when establishing part numbers or ordering replacements.

17

Industry Standardization of Power Transmission Belts

JAMES D. SHEPHERD *V-Belt Product Application Department, The Gates Rubber Company, Denver, Colorado*

I. PURPOSE OF STANDARDIZATION

Standardization has been in existence since the turn of the century. For example, the American Society of Agricultural Engineers (ASAE) was formed in 1910 and the predecessor of the American National Standards Institute (ANSI) was established in 1918. Even in these early years, it was recognized that there was a need for standardization in the industry in order to increase parts interchangeability, improve manufacturing processes, and generally serve the using public.

The United States has participated in development of both national and international standards in order to promote interchangeability of power transmission belts in various applications and to provide availability of replacement parts on an international basis. With the increase of multinational companies, the international aspect of standardization has become increasingly important in the last decade. International standards activity for industrial belts is organized by the International Organization for Standardization (ISO). The activities of this group will be described later in this chapter.

The United States is rather unique in that its standards-generating organizations are not related to government agencies. This is in contrast to many foreign countries where standards are often written solely by the government. Several congressional bills have been drafted to alter the relationship between independent standard groups and government agencies. It has been proven that each of the individual groups is capable of writing standards consistent with good user practice and has, at the same time, dealt effectively with the issues of health and safety. This has resulted in the complete separation and autonomy of the nongovernment drafting groups.

II. MAIN STANDARDIZATION BODIES IN THE UNITED STATES

A. American National Standards Institute

Virtually all domestic standards-writing organizations are governed by the American National Standards Institute (ANSI). ANSI is a voluntary, nonprofit federation of scientific, technical, trade, professional, consumer, and labor organizations. These groups result in roughly 90% of the standards-developing capability in the private sector based on an analysis of the NBS directory of organizations. ANSI's function is to manage and coordinate the voluntary development of national standards in the United States. More than 11,000 American national standards now exist to serve the needs of industry, government, consumers, labor, and the general public.

While ANSI does not generate standards, it assures that standards written by various associations are compatible with the industry needs and that they do not conflict with individual manufacturers or user groups. These standards are commercial and are not enforceable by law.

ANSI also serves as the U.S. representative to the International Organization of Standardization (ISO). All belt-related member associations of ANSI participate in the drafting and reviewing of ISO standards. The delegation attending the ISO meetings generally consists of members of the Rubber Manufacturers Association (RMA) Power Transmission Belt Technical committee.

In the case of power transmission belts, any national standard written by an association is submitted to a canvas method before it is approved as an industry (ANSI) standard. The canvas method consists of submitting the draft proposal to all interested manufacturers and users group — a list that has been agreed upon by the ANSI officials and the standards-writing organization. Once the canvas procedure has been completed, the results are forwarded to the standards-writing technical committee for review and resolution of all comments and objections. After this review has been completed, the standard is submitted for certification by ANSI.

The following groups are the actual writers of belt and sheave power transmission standards in the United States:

1. Rubber Manufacturers Association (RMA)
2. Mechanical Power Transmission Association (MPTA)
3. Society of Automotive Engineers (SAE)
4. American Society of Agricultural Engineers (ASAE)

Each of these groups has, as their main participants, the belt manufacturers. However, the users are invited to participate and in some cases, are quite active in the drafting of power transmission standards.

B. Rubber Manufacturers Association

The major writer of belt power transmission standards for the United States is the Rubber Manufacturers Association (RMA). The drafting group for RMA is the Power Transmission Belt Technical Committee (PTBTC) and has as its main membership each of the domestic manufacturers of power transmission belts. Each of the standards is written such that all manufacturers contribute to and agree upon the contents of each standard.

In many cases, the PTBTC of RMA contributes to the drafting of standards written by other organizations listed above. In addition, they also review each of the standards written by the other organizations to assure compatibility with RMA standards.

PTBTC also works very closely with the Mechanical Power Transmission Association (MPTA) in drafting their standards since this is the association dealing with the V-belt sheaves and synchronous belt sprockets to which the power transmission belts are applied. Further, the RMA groups work closely with the Rubber Association of Canada (RAC) in drafting standards that are compatible with Canadian procedures and manufacturing processes.

A description of the RMA published standards is shown in Table 1. These standards are reviewed on a 5-year basis for consideration of revision or total rewrite. RMA attempts to keep the standards as current as possible.

In addition to the published standards, RMA also publishes several bulletins pertaining to various functions and design methods of power transmission belts. These bulletins are listed in Table 2.

TABLE 1 RMA Standards

Publication number	Description
IP-20	Specifications: CLASSICAL MULTIPLE V-BELTS (A, B, C, D, and E Cross Sections)
IP-21	Specifications: Double-V (Hexagonal) Belts (AA, BB, CC, DD Cross Sections)
IP-22	Specifications: NARROW MULTIPLE V-BELTS (3V, 5V, and 8V Cross Sections) (1983)
IP-23	Specifications: SINGLE V-BELTS (2L, 3L, 4L, and 5L Cross Sections)
IP-24	Specifications: SYNCHRONOUS BELTS (MXL, XL, L, M, XH, and XXH Belt Sections)
IP-25	Specifications: VARIABLE SPEED V-BELTS
IP-26	Specifications: V-RIBBED BELTS (H, J, K, L, and M Cross Sections)

TABLE 2 RMA Bulletins

Publication Number	Description
IP-3-1	V-Belt Heat Resistance
IP-3-2	V-Belt Oil Resistance
IP-3-3	Static Conductive V-Belts
IP-3-4	Storage of V-Belts
IP-3-6	Effect of Idlers on V-Belt Performance
IP-3-7	V-Flat Drives
IP-3-8	High Modulus Belts
IP-3-9	Joined V-Belts
IP-3-10	V-Belt Drives with Twist and Non-Alignment Including Quarter Turn
IP-3-13	Mechanical Efficiency of Power Transmission Belt Drives
IP-3-14	A Drive Design Procedure for Variable Pitch V-Belt Drives

C. Mechanical Power Transmission Association

The Mechanical Power Transmission Association (MPTA) is the standards organization for manufacturers and users of V-belt sheaves and synchronous belt sprockets. A working relationship exists between MPTA and RMA in that each reviews the others standards for compatibility. In general, at least one joint meeting is held each year to review the standards of each organization.

D. Society of Automotive Engineers

The Society of Automotive Engineers (SAE) is an engineering society dedicated to advancing mobility and includes automotive applications as well as related fields including aerospace, trucks and buses, and farm machinery. Under their standardizations organization, a V-belt

TABLE 3 SAE Standards

Number	Name
J636C	V-Belts & Pulleys
J637B	Automotive V-Belt Drives — Recommended Practice
J1278	SI (metric) Synchronous Belts & Pulleys
J1313	Automotive Synchronous Belt Drives
J1459	V-Ribbed Automotive Belts

committee exists to guide the development of standards relating to application and testing of power transmission belts used in automotive applications. Each of the individual standards is written by a subcommittee consisting of manufacturer and user members. The standards written by SAE are listed in Table 3.

E. American Society of Agricultural Engineers

The American Society of Agricultural Engineers (ASAE) is an engineering society created to advance the theory and practice of engineering in agricultural and the allied sciences and arts. As with previous societies, the various committees are manned by both manufacturers and users. For the most part, the belt standards are drafted by RMA and submitted to ASAE for final revision and modification. Table 4 lists the two standards supported by this organization.

TABLE 4 ASAE Standards

Number	Name
S 211.3	V-Belt Drives for Agricultural Machines
S 212.1	Laboratory Procedure for Testing V-Belts

III. INTERNATIONAL STANDARDIZATION ACTIVITY

A. The International Organization for Standardization

While many nations have their own standardization groups, the International Organization for Standardization (ISO) is the only international standards organization which is responsible for and supported by most manufacturing nations in the world.

International standards benefit both the producer and the consumer in that a common technical language and, therefore, the basis of a worldwide market, is provided. Indeed, one of ISO's objectives is to: "...promote the development of standards in the world with the view to facilitate the international exchange of goods and services and to develop a mutual cooperation in the sphere of intellectual, technical and economic activity."*

ISO is the largest of many international voluntary groups for industrial and technical cooperation. ISO is a nongovernmental organization, and its standards have no legal force. However, more than half of its members are governmental agencies or bodies incorporated by public law. As of 1980, ISO had over 160 technical committees and more than 1900 subcommittees and working groups. It has established approximately 4000 international standards, nearly half of them published or revised since 1976.

Belt power transmission activity is the function of ISO Technical Committee 41. Under this technical committee are two subcommittees dealing with power transmission belts:

Subcommittee 1: V-belt standards
Subcommittee 4: Synchronous belt standards

B. ISO Belt Standards

The belt standards presently published by ISO are listed in Table 5. Each of these standards is formally reviewed on a 5-year cycle in order to keep it current with present technology and industry procedure.

While none of the ISO standards is officially used in the United States, the U.S. representatives have been diligent in efforts to

---

*ISO Organizational Catalogue, 1985, International Standards Organization of the Rubber Manufacturers Association.

TABLE 5 International Organizations for Standardization (ISO) Standards

Number and date	Name
ISO 255-1981	Pulleys for classical and narrow V-belts – Geometrical inspection of grooves
ISO 1081-1980	Drives using V-belts and grooved pulleys – Terminology
ISO 2790-1982	Narrow V-belt drives for the automotive industry – Dimensions
ISO 3410-1978	Agricultural machinery – Endless variable-speed V-belts and groove sections of corresponding pulleys
ISO 4183-1980	Grooved pulleys for classical and narrow V-belts
ISO 4184-1980	Classical and narrow V-belts – Lengths
ISO 5287-1978	Narrow V-belt drives for the automotive industry – Fatigue test
ISO 5289-1978	Endless hexagonal belts for agricultural machinery, and groove sections of corresponding pulleys
ISO 5290-1978	Grooved pulleys for joined narrow V-belts – Groove sections 9J, 15J, 20J and 25J
ISO 5294-1979	Synchronous belt drives – Pulleys
ISO 5296-1978	Synchronous belt drives – Belts
ISO 5288-1982	Synchronous belt drives – Vocabulary

guide the standards into practices consistent with U.S. technology and procedures. In general, none of the standards conflict with U.S. procedure. However, several cross sections are standardized by ISO which are not consistent with cross sections used in the United States. This has not presented a major problem with the exception of retrofitting equipment imported from outside the United States.

C. Worldwide Use of ISO Standards

Because of international standards, the growing tendency toward global products and vast amounts of import and export items, belt power transmission design and belt replacement have not been major problems. The equipment intended for international manufacture and use is generally designed with belts consistent with international application and thus the ISO standards prevail worldwide.

Any company intending to manufacture equipment to be used outside of the United States should carefully review the standards indicated in this chapter to ensure that their product will have belt parts that are interchangeable anywhere in the world. As an alternative, a domestic manufacturer could consult with any of the belt manufacturers in order to assure that his product will have universal application.

18

Belt Drive Efficiency

DAVID E. ROOS *Synchronous Application Engineering, The Gates Rubber Company, Denver, Colorado*

I. BELT EFFICIENCY

A. Definition

There are several methods of power transmission available for today's machinery – gears, chains, and belts – but when the speed of the driveN shaft is different from the speed of the driveR, or when the shafts must be separated by some distance, belts have proven to be the preferred method of power transmission.

Drive designers are working with higher speed prime movers and the necessity of a speed ratio to achieve a slower, useful driveN speed is also causing a significant increase in the use of belts.

Belt drives that are well-designed and maintained are very good transmitters of power, ranging from 95 to 98% efficiency. According to industry sources, the efficiency of chain is in the 95 to 99% range and efficiency of worm and spur gears is in the 92 to 96% range.

1. Basic Advantages

Besides efficiency, belt drives have a number of advantages over other power transmission approaches including:

Low initial cost
Cleanliness
Quietness
No lubrication requirement
Low maintenance costs
Easy installation
Transmission of a wide range of power
Smoothing out of shock loads
Easy adaptation to special purposes such as clutching and speed variations (V-belts)

2. Formulas

Before examining various aspects of belt efficiency, a definition of belt efficiency is beneficial. Simply stated, belt efficiency is:

$$E = \frac{\text{HP out}}{\text{HP in}} \times 100 \quad \text{or} \quad E = \frac{\text{T out} \times \text{rpm out}}{\text{T in} \times \text{rpm in}} \times 100 \qquad (1)$$

where

E = efficiency (%)
HP = horsepower (output or input)
T = torque (output or input)
rpm = revolutions per minute

The first formula is most frequently used, but to study the details of belt efficiency, the second equation is more helpful. When discussing the source of energy lost in the form of heat, it is important to examine these losses in the terms of torque and speed.

Torque losses on belt drives are due to hysteresis losses resulting from bending stresses and friction at the belt/sheave interface. Also, there can be losses due to windage – caused by the belt and sheaves moving through the air.

Speed losses are the result of slip and V-belt creep. Creep results from the difference in tension from tight to slack side of a sheave or pulley. Since a belt elongates slightly under tension, it is slightly longer and narrower as it exits the driveN pulley compared to when it entered the pulley. This makes the driveN shaft rotate at slightly less rpm than it would if there were no creep. The amount of creep varies with load, speed, sheave diameter, and belt construction, and is generally in the range of 0.5 to 1.0% for most well-designed V-belt drives. Synchronous belts do not exhibit these speed losses.

3. Laboratory Measurements

Figure 1 shows a laboratory method for accurately measuring belt efficiency using a motor and generator (both mounted in torque-reactive cradles) and two speed transducers. Because of the relatively high efficiencies involved, all of the equipment must be set up carefully and calibrated accurately to produce meaningful data.

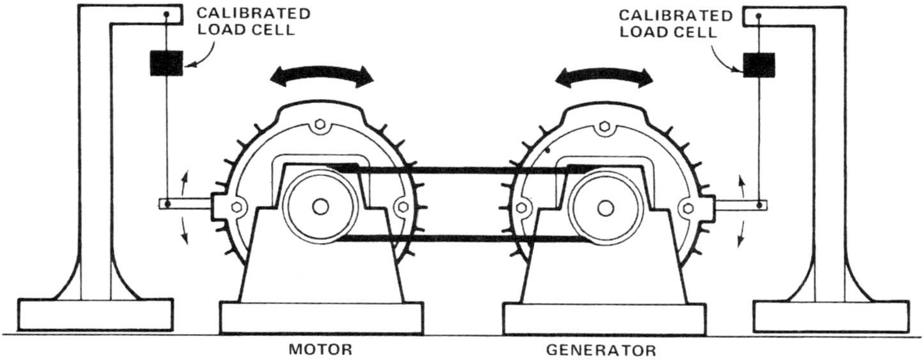

FIGURE 1 Laboratory equipment for measuring belt efficiency.

Today's V-belts are designed to provide maximum efficiency within their design range. This is achieved through optimum compounding and component design which assures minimum bending stresses while offering maximum support for the load carrying tensile members. Maximizing tensile member support while decreasing bending stresses minimizes hysteresis losses. This optimization of belt construction, together with modern tensile cord materials, also helps reduce creep or speed losses.

Even though V-belts are very efficient, there is still room for improvement. This chapter is devoted to explanations of how added efficiencies can be obtained – either through redesign of the drives or improvements that can be made on existing drives.

II. BASIC DRIVE DESIGN FACTORS

Industrial V-belts operate at peak efficiency when designed according to recommended practices. The belt efficiency graph in Figure 2 illustrates the power loss when belts are underloaded and overloaded. For instance, efficiency losses become significant when a belt is run at 50% or less of its rated load capacity. V-belts operate most efficiently at their rated load capacity. Older drives should be examined to take advantage of uprated belts with an eye to improving efficiency. Belt manufacturers can be especially helpful in this area.

A. Sheave Diameters

One of the most important drive design considerations is the use of proper sheave diameters. Sheaves should be selected based on proper drive design methods with consideration given to belt type and cross section, National Electrical Manufacturers Association (NEMA) recommendations, space limitations, and efficiency. In general, the larger the sheave, the greater the drive efficiency. This is because there is less hysteresis (torque loss). There is also less creep and slip due to the diameter effect and lower operating tensions. If the drive requires a small diameter sheave, then the designer will sacrifice drive and belt efficiency and place added stress on the motor and driven equipment. NEMA specifies the minimum recommended diameters for sheaves used on electric motors in NEMA Standard MG 1-14.42. These recommendations are designed to protect the motors by limiting bearing loads and shaft stresses.

The Rubber Manufacturers' Association (RMA) also publishes information on recommended sheave diameters in various standards cited later in this article. The purpose of these guidelines is to promote acceptable belt performance. For example, smaller sheave diameters can

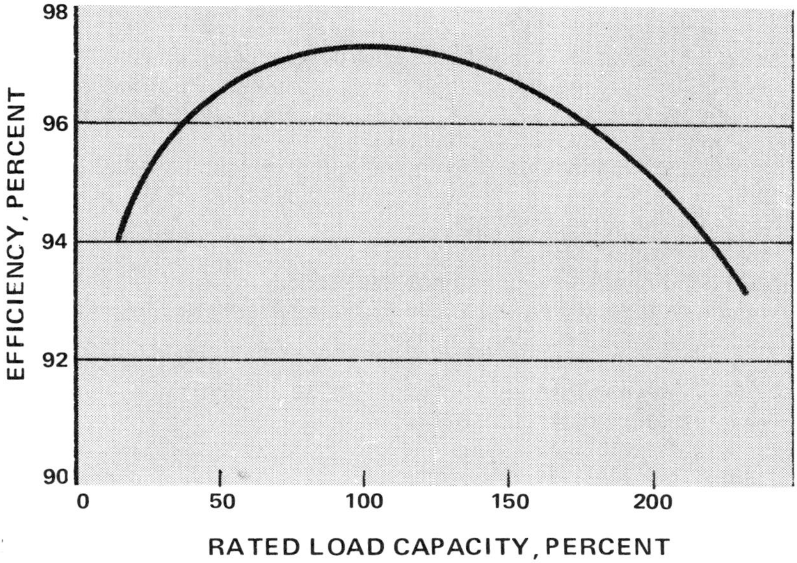

FIGURE 2 Graph showing efficiency-rated load curve.

lead to less efficient drives due to excessive bending stresses in the undercord and higher belt tensions. This, in turn, results in lower efficiency which is often manifested in higher internal belt temperatures. In addition to efficiency, this can also lead to reduced belt life.

B. Motor Sizing

Motors should be properly sized. While optimization of the belt drive is important, it is only one component of the drive package. Applications should be examined carefully to make sure the motor has sufficient capacity to meet the HP requirements, yet is not overdesigned for the application -- a condition that can lead to electrical inefficiencies.

Consider the use of high-efficiency motors. For the same HP, motors with higher speeds have higher efficiency at rated load than motors with lower speeds. Motor suppliers should be consulted when motor replacement or new motor applications are being examined.

C. Maintenance

Proper drive maintenance procedures can dramatically increase efficiency. For example, inadequate belt tension on a drive allows V-belt slip which can result in up to 10% efficiency loss. Details of drive maintenance are also covered in Chapter 12.

III. EFFECTIVE DRIVE REDESIGN

When there is the option of redesigning machinery belt drives, there are four different belt types to choose from (as discussed in previous chapters):

Classical (molded notch and conventional banded V-belts)
Narrow cross section industrial V-belts
Synchronous belts
V-ribbed belts

A. Classical V-Belts

As discussed in earlier chapters, classical V-belts are available in both banded and bandless molded-notch constructions from many belt manufacturers (See Figures 3 and 4).

When a belt drive is being redesigned, it may be possible to improve overall drive efficiency with the molded-notch versions of the classical sections. This would be particularly true if the spatial restrictions and drive ratios dictate the use of very small or subminimum diameter sheaves. The molded-notch V-belts show improved efficiency on subminimal diameter sheaves as shown in Figure 5. However, a molded-notch belt loses its efficiency advantage over banded belts as diameter increases. Its efficiency advantage is less than 1% above minimum recommended diameters. Justifiable energy savings should be compared against the premium price paid for molded-notch belts.

B. Narrow V-Belts

As with the classical sections, narrow V-belts (see Figure 6) are also available in both banded and bandless molded-notch constructions. Providing high power capacity in a smaller, narrower drive, they can be lighter, less costly, and yield improved drive efficiency. The lower inertia of these drives may permit the use of smaller motors and can reduce energy costs, particularly on machines which start and stop frequently.

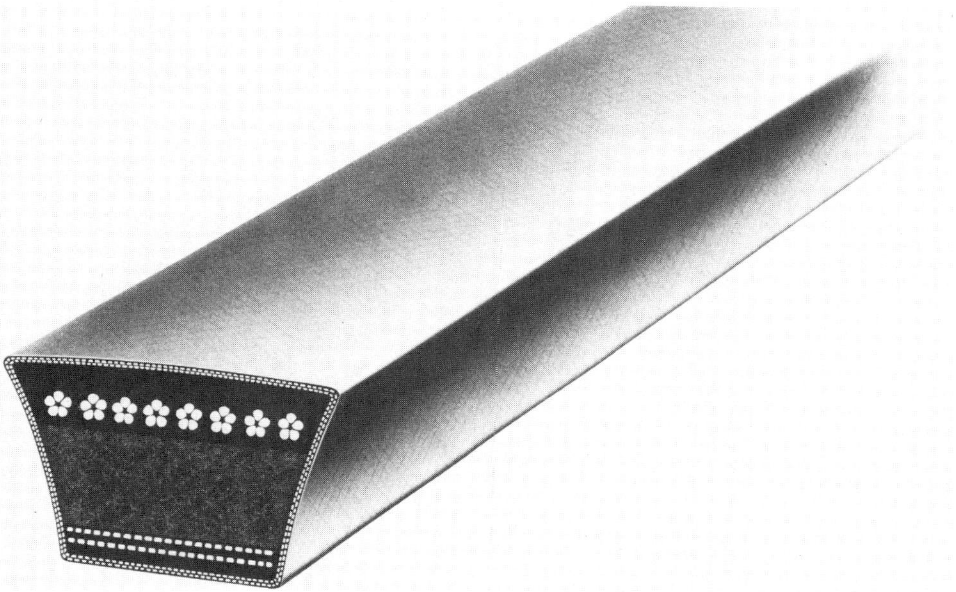

FIGURE 3 Classical V-belt cross section.

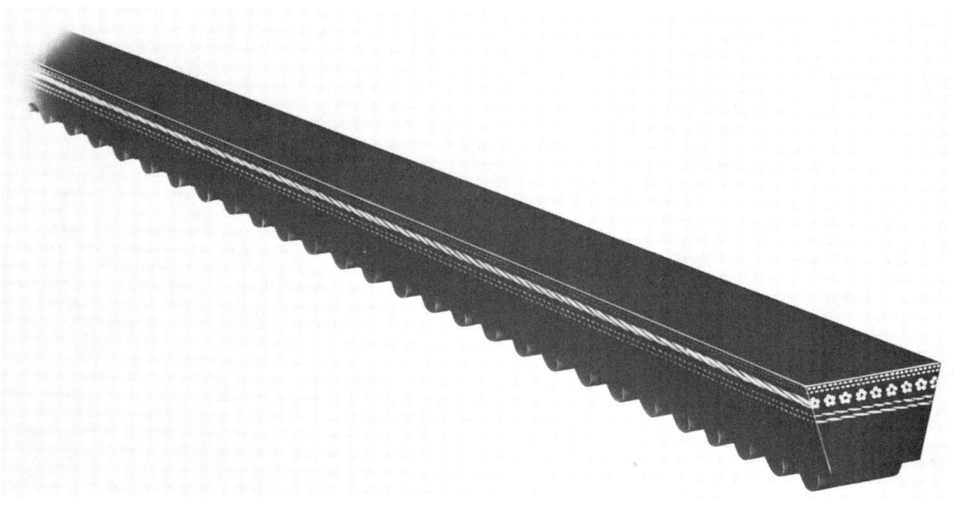

FIGURE 4 Classical molded-notch V-belt cross section.

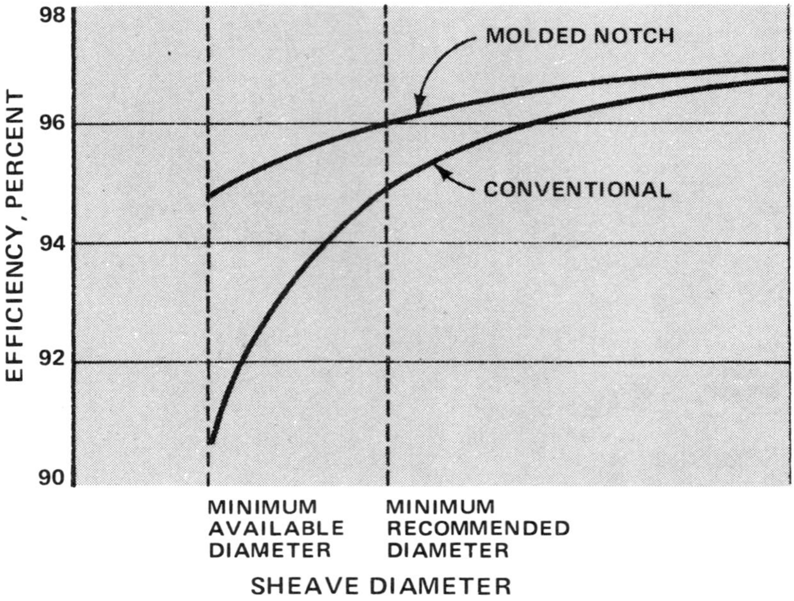

FIGURE 5 Graph comparing efficiency of conventional and molded-notch V-belts.

C. Synchronous Belts

Synchronous belts as shown in Figure 7 provide positive power transmission, can operate on relatively small diameters, and offer greater efficiency as compared to standard belts. This is due primarily to reduced bending stresses and lack of slip. Careful consideration should be given to the higher initial costs which may negate savings derived from the higher efficiency and resultant energy savings. Synchronous belts are also more sensitive than V-belts to misalignment. Consult your belt supplier for help on an energy savings analysis.

D. V-Ribbed Belts

Relatively thin cross sections of V-ribbed belts (see Figure 8) can provide a slight efficiency advantage over both standard banded and bandless molded-notch belts. They are able to operate over very

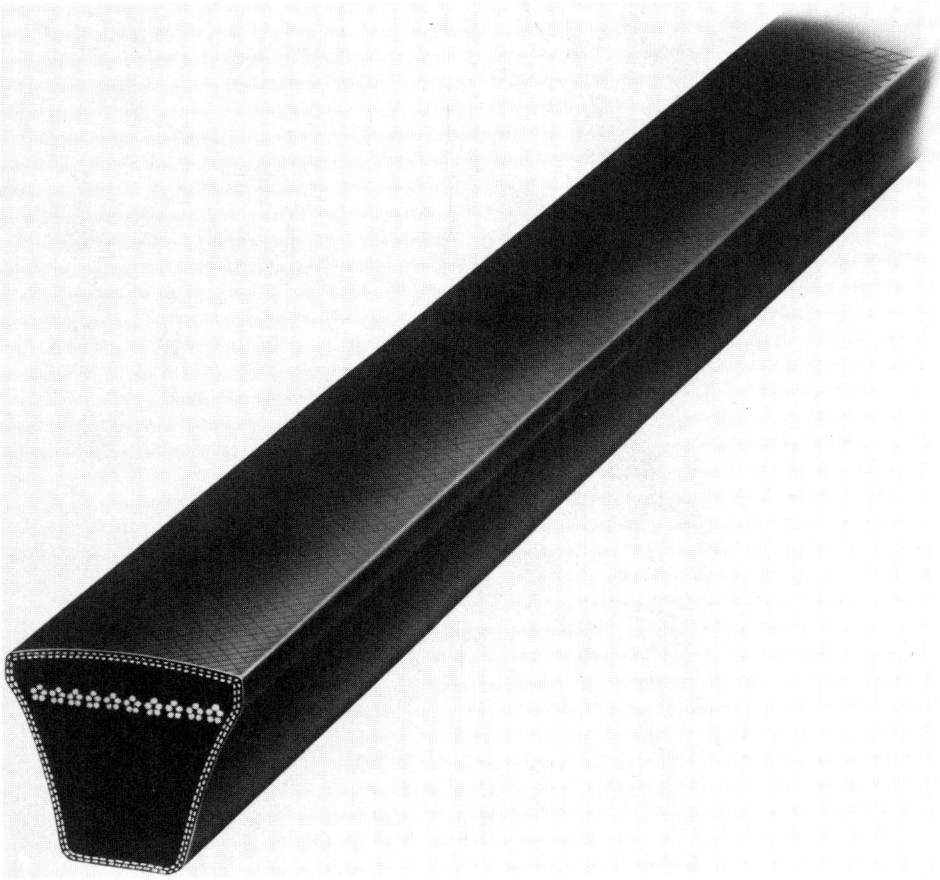

FIGURE 6 Narrow V-belt cross section.

small diameters with reduced bending stresses. Again, design considerations should be carefully reviewed against potential energy savings. V-ribbed belts require up to 20% higher tension as compared to conventional V-belts. This will result in higher operating tensions, imposing higher stresses on motor bearings, shafts, and other drive hardware.

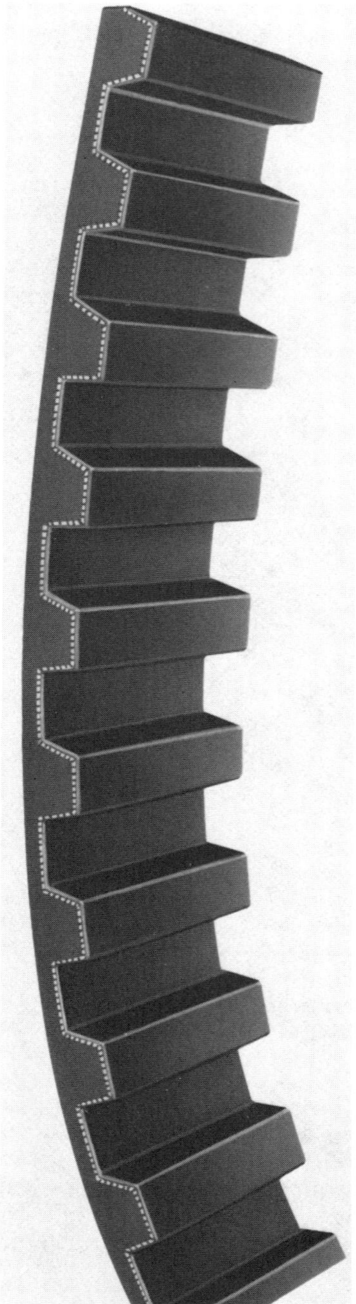

FIGURE 7 Timing belt cross section.

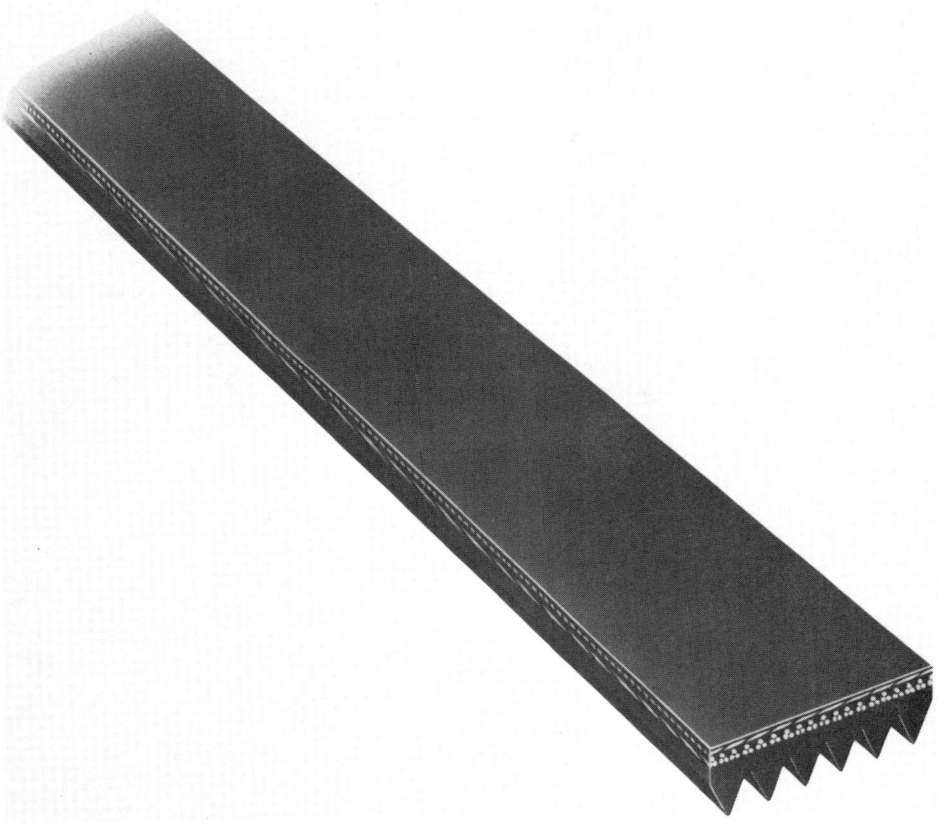

FIGURE 8 V-ribbed belt cross section.

IV. EVALUATING EXISTING DRIVES

If the drive cannot be redesigned and the existing installation must be used, then there are several options which can be evaluated.

A. Different Belt Types

As discussed earlier, the use of molded-notch V-belts can be considered if the drive has sheave diameters smaller than RMA minimum recommendations. These belts use the same sheaves as standard V-belts. A cost analysis should be made to determine if the expected energy savings justifies the added cost of the belts.

B. Overbelting Problems

The efficiency loss that accompanies overbelting presents an opportunity for the drive designer. A drive designed years ago to comply with ratings in existence at that time should be examined.

For example, consider a 30-HP belt drive operating at 1160 rpm. The drive has two 9-in. pitch diameter sheaves and C114 belts. Based on a 1964 RMA rating of 9.31-HP per belt, this drive required four belts. The capacity of belts has improved since 1964 and in 1977, RMA published new ratings. The rating for this belt has increased to 11.34-HP per belt. Now, only three belts are required to do the job. Overbelting is eliminated and a slight efficiency improvement may be achieved. This can also reduce bearing and shaft loads. *Note*: Using less than the full number of belts on an existing drive will result in uneven groove wear and may create problems when the next set of belts is installed.

C. Maintenance Procedures

Proper maintenance can be very effective in increasing V-belt drive efficiency. Improper belt tensioning, for instance, can result in efficiency losses of up to 10%. A number of factors beside tensioning can affect efficiency.

1. Replacing Belts

Belts become stretched and worn from use. If old and new belts are mixed on the same drive, the new belts will be tighter and will carry more than their share of the load. Belt life will likely be reduced. When replacing belts on an existing drive, always replace all of the belts. Also, use belts from only one manufacturer. This will result in maximum belt life and efficiency.

2. Tensioning

Check tensions frequently. The best tension for a V-belt is the lowest tension at which the belt will not slip under full load. Too little tension causes slippage and too much shortens belt life. Both conditions will reduce efficiency. Proper installation, pretensioning, run-in and retensioning, along with a tension maintenance program are ideal.

3. Worn Sheaves or Pulleys

Check sheaves and pulleys for wear and damage. Any significant wear will lead to improper support of the belt and unequal tensile cord loading. It may also be necessary to overtension the belts to effectively transmit the load and this is inefficient.

4. Belt Misalignment

Align the belts as accurately as possible, particularly synchronous and V-ribbed belts which require tighter alignment tolerances. Belt misalignment results in unequal cord loading. This leads to inefficiency.

5. Drive Guards

Guards are a safety precaution and are required by the Occupational Safety and Health Administration (OSHA) and several other regulatory agencies. Improper use of guards can lead to inefficient drives. Guards should allow for proper drive ventilation. An improperly ventilated drive can build up excess heat, contributing to loss of efficiency and reduced belt life. The guard will also protect the drive from environmental conditions, debris and tampering, all of which can reduce overall drive efficiency.

6. Trouble Signals

Listen for trouble signals. Improperly placed guards, loose belts, or excessive vibration can result in the belts rubbing against the guard. V-belt squeal can be caused by inadequately tensioned belts or foreign material such as grease, oil, or paint in the sheave grooves. These reduce drive efficiency. Look for problems. Watch the drive running, remove the guard, if necessary. If one or more belts appear to be abnormally loose or tight, check for improper tension, a damaged belt, excessive wear, or misalignment.

V. ONSITE EVALUATION OF BELT EFFICIENCY

As stated earlier, well-designed and maintained belt drives operate with efficiencies in the 95 to 98% range. Obviously, the belt drive is only one component of the equipment. It is virtually impossible to measure absolute equipment efficiency, let alone isolating the absolute belt drive efficiency. While energy input is relatively easy to measure, the work delivered by the driveN unit normally is not measurable to the accuracy required.

FIGURE 9 Power survey demand recorder used to determine drive efficiency.

A. Efficiency Measuring Methods

However, comparisons can be made between similar machines and the effects of drive changes can be evaluated. In order to properly make these comparisons, laboratory-quality measuring equipment is required. While an ammeter will indicate electrical current changes, this is only one of the AC power parameters. Instead, an electrical analyzer, such as the Esterline Argus Power Survey Demand Recorder could be used as shown in Figure 9. In addition to the analyzer, an accurate tachometer and temperature measuring device is desirable, as will be discussed later.

The electrical analyzer will quantify the change in power requirement for the equipment. However, extreme care must be exercised to determine the source of the power change. In order to isolate the drive component, the equipment variables must be minimized. For example, motor temperature can affect amperage readings. The system must be allowed to stabilize.

By far the most critical variable to consider is the driveN HP load. Many drives do not have a constant load: the load varies with time or operator methods, or both. Even with equipment that

FIGURE 10 Infrared temperature gun used to measure excess heat.

is considered to have a relatively constant load, slight changes affect the data. For example, fan drives have predictable loads, but ambient conditions such as air density or wind velocity can change the loads and thus the electrical readings. In fact, just changing belts on a fan drive (from one construction to another or even substituting a set of belts from a different manufacturer) can change the driveN speed slightly, and this can change the load significantly, remembering that fan HP requirements are a cubic function of the fan speed. An accurate tachometer should be used to measure driveN rpm.

B. Temperature Considerations

If it is determined that potential energy savings justify the cost of a drive efficiency study, there is a simple method to select a drive to be studied in detail. Remember that energy losses are manifested in the form of heat. Thus the drive that exhibits the highest temperature differential over the ambient temperature would be the most likely candidate to study.

Along the same lines, any well-designed drive that is operating efficiently will be only slightly warm to the touch immediately after being shut down, assuming ambient conditions are less than 100°F. If the drive belts are uncomfortable to hold in your hand (approximately 140°F or greater), then the drive design and maintenance should be reviewed. Temperature measuring equipment, such as an infrared gun, shown in Figure 10 can be useful.

Index